U0897292

CONTEMPORARY ARCHITECTS 3

当代建筑师 3

李壮 编

華中科技大学出版社
http://www.hustpas.com

图书在版编目（CIP）数据

当代建筑师 3 / 李 壮 编.
—武汉 ：华中科技大学出版社，2011.4
ISBN 978-7-5609-6804-9

Ⅰ. ①当… Ⅱ. ①李… Ⅲ. ①建筑设计－作品集－世界－现代 Ⅳ. ①TU206

中国版本图书馆CIP数据核字(2010)第236722号

当代建筑师 3

李 壮 编

出版发行：华中科技大学出版社（中国·武汉）
地　　址：武汉市珞喻路1037号（邮编：430074）
出 版 人：阮海洪
责任编辑：赵　萌
策　　划：吉典文化
装帧设计：陈　利
版式设计：周　宇　杜　璟
责任监印：张贵君
组　　稿：胡亚凤
编　　委：董　君　吕九芳　王　建　夏　屹　蔡春艳　林蜜蜜
李　秀　夏秀田　韦成刚　刘　云　姜　野　丛玲玲
仲　欣　柳　燕　付　军　于　辉　农丽媚
印　　刷：北京佳信达欣艺术印刷有限公司
开　　本：965 mm×1270 mm　1/16
印　　张：20
字　　数：160千字
书　　号：ISBN 978-7-5609-6804-9/TU・1002
版　　次：2011年4月第1版 第1次印刷
定　　价：268.00元（USD53.00）
销售电话：400-6679-118

（本书若有印刷质量问题，请向出版社发行部调换）

CONTENTS

ARATA ISOZAKI & ASSOCIATES

矶崎新工作室

公司合伙人及主要设计师

- 矶崎新　Arata Isozaki

1931年生于大分县
1961年东京大学数物系大学院建筑学博士
1963年设立矶崎新工作室
1963年出任东京大学、洛杉矶加州大学、哈佛大学、哥伦比亚大学等国内外的客座教授，并且在多个国际竞赛中担任审查员。不仅在世界各地举办过讲演及研讨会，而且在举办建筑展、美术展、个人展及等方面也开展着丰富多彩的活动

- 胡倩　Hu Qian

胡倩，1968年生于上海。早稻田大学建筑学硕士。日本一级注册建筑师。1998年于日本矶崎新工作室就职，2005年任矶崎新中国工作室负责人。期间作为项目建筑师或项目经理，参与了矶崎新工作室所有的中国项目的设计工作，主要有深圳文化中心、国家大剧院竞赛、中央美院美术馆、九间堂别墅、证大喜马拉雅艺术中心等。

矶崎新工作室（AI&A）

1963年，矶崎新于东京成立了矶崎新工作室（AI&A），在这里，矶崎新同时从事着建筑师和理论家的工作。长期以来，矶崎新的项目并不局限在日本，为此他建立了数个海外的工作室，包括巴塞罗那、米兰、上海、纽约（曾经）。矶崎新工作室是一个国际化的设计团队，以东京工作室为核心，共同提供建筑设计服务，包括城市规划、建筑设计以及设计监理。开辟海外事务所有助于保证建筑品质和细节设计，及更好地与客户和合作伙伴沟通。矶崎新工作室擅长于图书馆、博物馆、音乐厅、影剧院和教育设施等公共建筑的设计，其中也包括精品酒店、别墅等商业建筑。该团队包括建筑设计师、项目经理和设计监理工程师。从方案设计到施工配合阶段，每一环节都会得到矶崎新先生的直接参与和决策。在条件允许的情况下，矶崎新工作室会邀请各方专家和合作单位配合设计来满足项目的特殊要求。五十余年来，通过建成的作品或是未建成的作品，矶崎新作为AI&A的创立者和领导者，向全世界35个国家传达了先进的建筑和规划理念。

矶崎新上海工作室（AI&A Shanghai）

矶崎新上海工作室由矶崎新和工作室资深级员工胡倩创办，成立于2004年，规模为20人左右。矶崎新上海工作室目前同时负责多个在建和设计项目，其中包括文化中心、博物馆、音乐厅、校园建筑和区域规划。工作范围包括：原创的建筑和城市设计，配合东京工作室与客户、顾问、合作设计师之间进行沟通，及时收集施工现场的即时信息，基于当地的建筑设计规范进行设计监理。

目前在建或已建成的项目有深圳文化中心、南京佛手湖国际会议中心、上海喜玛拉雅中心、上海九间堂别墅、北京中央美院新美术馆、上海交响乐团音乐厅等。作为一个享有全球声誉、国际化的建筑设计事务所，矶崎新上海工作室寄希望于更好的团队合作和与各设计单位之间的配合，以传达最完美的设计理念。

Arata Isozaki & Associates(AI&A)

In 1963, Arata Isozaki set up the AI&A in Tokyo, where Arata Isozaki was engaged in the work of an architect and a theorist. For a long time, AI&A' s projects are not limited to Japan, so he has set up a number of overseas studios, including Barcelona, Milan, Shanghai, New York (has been removed). AI&A is an international design team, with its core studio in Tokyo. They undertake the architectural design services, including urban planning, architectural design and design supervision. The establishment of overseas offices can help to guarantee the construction quality and design details as well as better exchange between customers and partners. AI&A excel in public building designs including libraries, museums, concert halls, theaters and educational facilities as well as boutique hotels, villas and other cultural and commercial buildings. The team includes architects, engineers, project managers and design supervisors. From the design to construction, each link gets participation and decision making of Mr. Arata Isozaki. In proper conditions AI&A will invite experts and co-operation companies in the projects to meet special requirements. Over the past more than 50 years, through completed or uncompleted works, Arata Isozaki, as the founder and the leader, has been conveying advanced construction and planning concepts to 35 countries around the world.

Arata Isozaki & Associates Shanghai (AI&A Shanghai)

AI&A Shanghai was co-founded by Arata Isozaki and the senior-level staff Hu Qian in 2004,with the size of around 20 persons. AI&A Shanghai now is responsible for many projects under construction or design, including cultural centers, museums, concert halls, campus buildings and regional planning. Its scope of work includes: original architecture and urban design, to help the Tokyo studio to communicate with clients, consultants and the co-designers, to collect real-time information on the construction site as well as the design and supervision of the local building regulation.

The currently projects completed or under construction are cultural center of Shenzhen, Nanjing Foshou Lake International Convention Center, Shanghai Himalayan Center, the Villa in Shanghai Madarin Palace, Beijing Central Academy of Fine Arts Center, Shanghai Symphony Orchestra concert hall,etc.

As a global pretigeous international architectural design firm, AI&A Shanghai hopes for better teamwork and coordination with other deisgn companies, to deliver perfect design concepts.

中央美术学院美术馆

MUSEUM OF CENTRAL ACADEMY OF FINE ARTS

项目地点：中国 · 北京　用地面积：8641 m²　建筑面积：14 777 m²
建筑设计：矶崎新工作室
建筑师：矶崎新，青木宏，东福大辅，镰野良亮，胡倩，高桥邦明，川久保智康，原田真宏

LOCATION: Beijing, China　SITE AREA: 8,641m²　BUILDING AREA: 14,777 m²
DESIGN CORPORATION: AI&A
ARCHITECTS: Arata Isozaki, Hiroshi Aoki, Daisuke Tofuku, Ryosuke Kamano, Hu Qian, Kuniaki Takahashi, Tomoyasu Kawakubo, Masahiro Harada

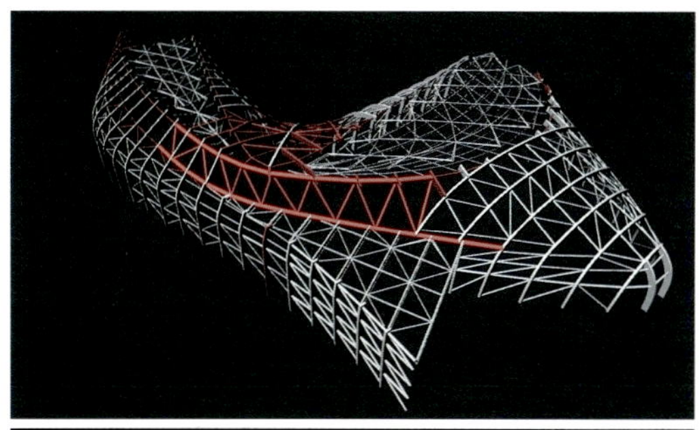

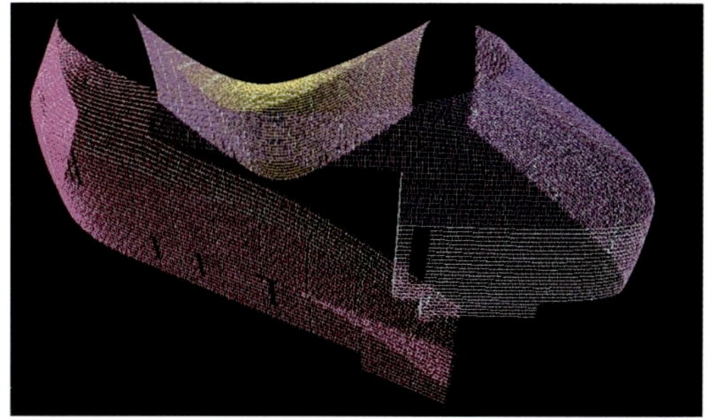

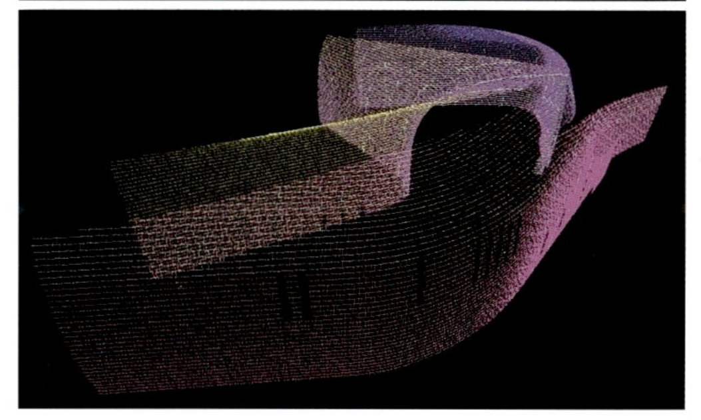

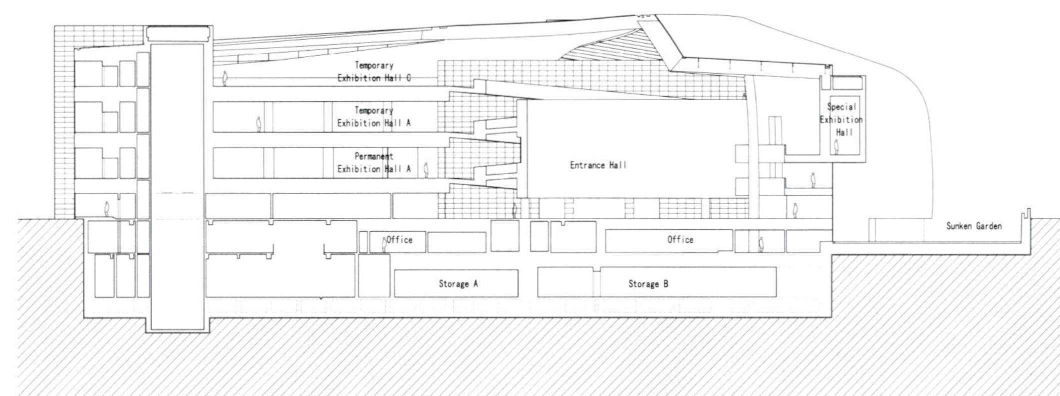

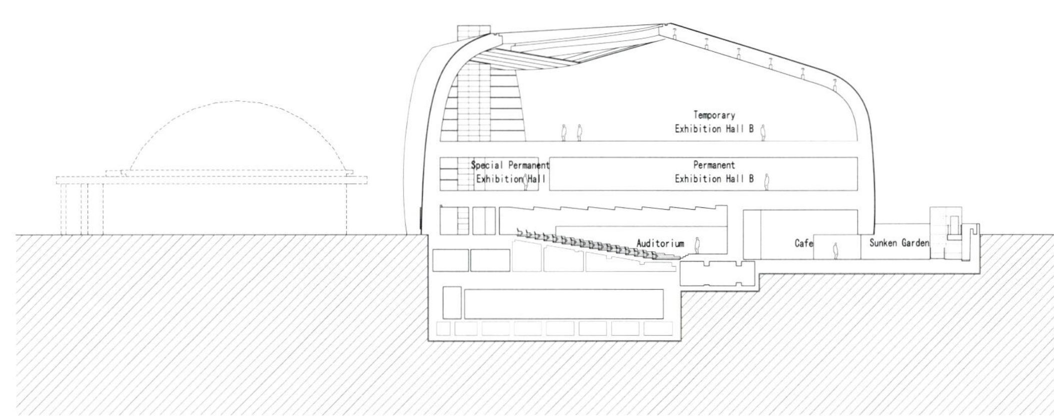

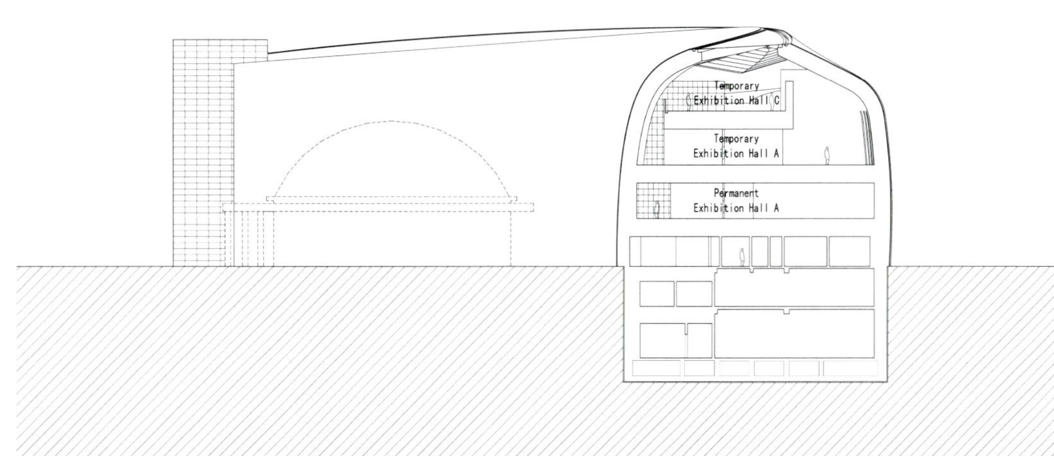

中央美术学院（Central Academy of Fine Arts，简称 CAFA）目前除了传统美术之外，还有美术印刷设计、产品设计、服装设计及建筑设计等学科，是包罗多种设计学科的综合性美术大学。学院自 2001 年开始分阶段地把校园从过去狭小的旧校址迁移到郊区，以校园为中心并联合周围 798 艺术区和酿酒厂旧址，形成了在东亚范围内屈指可数的大规模艺术区域。作为学院迁址的最后阶段，由矶崎新工作室设计并实行工程现场监督的中央美术学院附属美术馆，不仅可以展示该学院收藏的艺术珍品，还将承办来自国内外各界艺术家的展览。美术馆不仅是学院的象征性建筑，并期盼其带动周围艺术区，起到核心设施的作用。

798 艺术区具有代表性的艺术画廊是由多幢 20 世纪 50 年代建造的厂房和仓库加以改造利用的，其中有不少建筑有很高的层高并拥有高侧窗的自然采光。目前这些建筑对于商业艺术来说不失为理想的内部空间，但都没有超越 20 世纪的白色立方体型美术馆的空间范畴。然而作为引领中国艺术舞台的建筑，仅靠数幢雷同的白色立方体组合是不够的，需要更进一步深入地探索创意空间。

由于北京的街道基本上呈棋盘网格状，北京的现代建筑只能在四四方方的舞台上竞相展现它的形态。然而伴随着北京市区范围的急剧扩大，打破常规的城市规划相继出台。本地块呈现出的就是圆弧曲线勾勒出的“L”形。根据这种地形，设计首先考虑将临街部分的建筑设计为曲面墙体，然后采用既独立又相关的三个自由曲面来组合空间，最终形成目前的建筑体量。

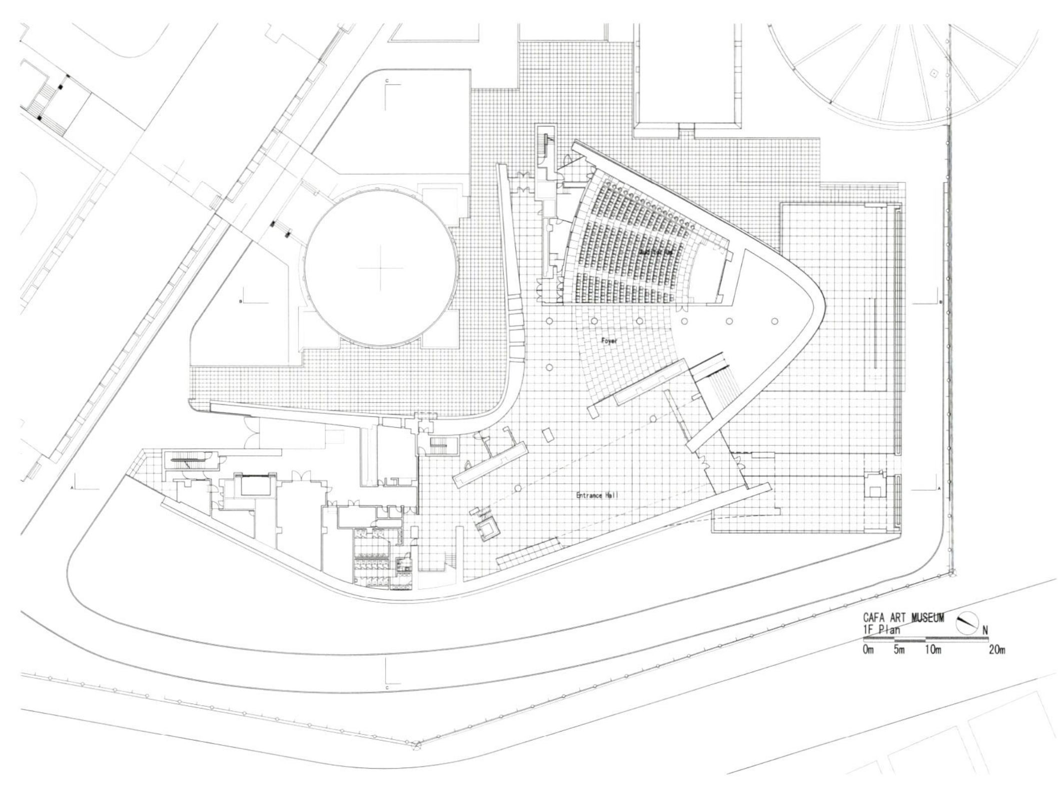

三个自由曲面的端部在平面布局上分别为主入口、连接报告厅的入口以及货物出入口，在垂直空间上则提供了可以引入自然光线的采光天窗。设备管道井、电梯和楼梯间等纵向交通空间则隐藏在几个矩形建筑体量内部，其作为独立的形体穿插于曲面墙体的内外。针对内部空间则利用水平的楼板进行垂直方向的分隔，或将夹层直接凌驾于空中来构成立体的展示空间，错落有致的楼层之间采用坡道平滑连接。

从采光天窗照射进来的阳光首先透过作为吊顶材料的玻璃纤维薄膜漫射开来，从而形成匀质的光线后进入展厅，光线衍射之处跟随墙体的自由曲面演绎出不同的表情。自然光所产生的这种效果，在展厅的不同部位衍生出微妙的变化。

通过架设在下沉式广场上方的天桥可以直接进入首层的入口大厅，该大厅是建筑内最高的空间。这里将用来展示装置作品和大型雕塑，且考虑到可以从各个楼层欣赏艺术品的可能性。另外，从大厅内电梯竖井处悬挑出来的讲台，可以在开幕式中使用。

二层固定展厅的主要功能是展示学院收藏的艺术品。因主要展品为国画及历史名作，所以内装修材料采用了木质板材、石材、织物以及清水混凝土等具有天然气息的材料。

三层临时展厅是能够适应现代艺术展需求、拥有馆内最大面积的展厅。展厅周围无接缝的曲面墙体可以作为装置作品的穹窿布景般的背景，也可将展品直接固定在墙体上，此外还有一个小型展厅作为夹层设置在上方。

从首层通向地下一层下沉广场的大台阶一部分用玻璃隔离开来，作为报告厅。

在施工图设计和施工过程中，除了一部分美术馆重要的设备之外，其余均采用中国当地的建筑材料。尽管中国处于建设高峰期，建筑成本快速上涨，但相比之下建筑成本仍然没有发达国家高，其中人工费价格低廉，即使在施工中发生大量手工作业的情况，也不会像日本那样造成建筑成本急剧上涨的现象。反而言之，中国建筑领域存在着过多依赖人工的问题，对国外常用的单元化生产则显得力不从心。

外装修采用干挂叠压式中国国产板岩，同样厚度的石材干挂项目还有日本奈良的百年会馆（Nara Centennial Hall, Nara, Japan）、静冈县国际会议艺术中心（豪华巨轮）（Shizuoka Convention Arts Center, Shizuoka, Japan）以及西班牙拉克鲁尼亚人类科学馆（Interactive Museum about Humans, La Coruna, Spain），这些建筑外形大致上都为圆形或椭圆形，外装修材料易于单元化生产。但由于这次项目的自由曲面难于进行单元化生产，通过对中国施工工艺的研究，决定采用不同宽度的板岩结合现场情况随机张贴的办法。在这种情况下，由于要靠施工人员悬吊在高空中进行手工操作，很难确保稳定的质量，因此事先制作了 1 ： 1 实体模型，来探讨施工方法和节点处理，在此基础上再进行施工。

在绘制施工图方面，三维模型的应用伴随始终，不仅数量巨大的石材分割、钢结构以及混凝土的形状需要利用三维模型进行推敲，甚至连风管、水喷淋的设置也都进行了三维定位。板岩作为制作砚台的一种常用材料来概括以中国传统美术教育为基础的美术学院，可以说是非常贴切的，具有现代风格且自由奔放的曲面形体，则是当代先进的美术教育机构的象征。

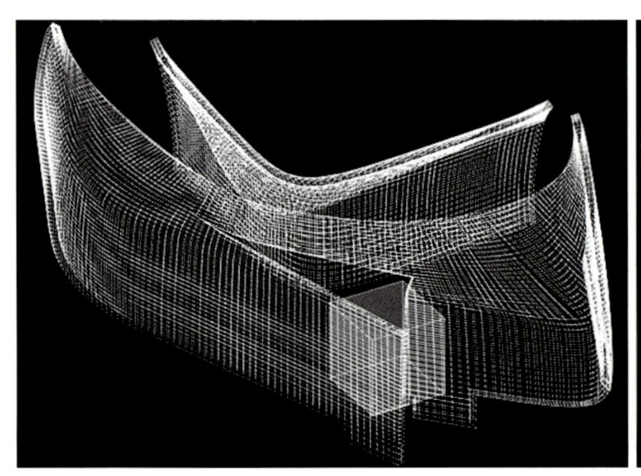

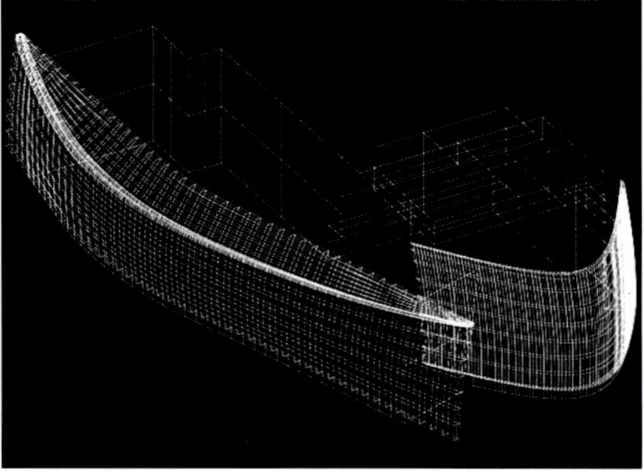

Central Academy of Fine Arts (referred to as CAFA) was originally a small professional art school. Since the enhancement of the reform and opening-up policy in 1990, Chinese universities has been self-managed, resulting in expansion of subjects, and the number of students doubled. Now in addition to traditional art, there are graphic design, product design, costume design and other construction disciplines. It has become a comprehensive Art University with a wide variety of design disciplines. Since 2001, the campus has been moved to the suburban areas by different stages from the old small site. It's centered by the 798 Art District and the old brewery site and they formed one of the few large-scale art areas in East Asia. As the final stage of the relocation, the subsidiary Museum of CAFA designed and supervised by AI&A, not only demonstrates the College collection of art treasures,but will also contract exhibitions of works of international and domestic artists. The Art Museum is not only the symbolic building of the college,but will expectedly be the locomotive of the surrounding art districts.

798 Art District, a representative of the Art Gallery,was transformed by the factories and warehouses in 1950s, many of which are high-storey buildings with windows of natural light. So far the buildings are exactly ideal internal spaces for commercial arts, but without surpassing the white cube space of 20th-century art museums. However, as the leading construction of Chinese art scene, it's inadequate to build some blocks of identical white cubes and requires further explorations in creative spaces.

As the streets of Beijing basically resemble chessboard grids, Beijing's modern constructions can only compete in the boxy stage to display their shapes. However, with the rapid expansion of urban area of Beijing, the plans to break the routine successively come out. This plot is just the L-shape outline by curve lines. According to this topography, the design should first consider the street part of the building as curved wall surface, then use three semi-related free surfaces to build the space, and eventually form the current building volume.

In terms of the plane layout, at the ends of the three free surfaces are respectively the main entrance, the lecture hall connecting the entrance , and the passage way for goods transportation. The vertical space provides skylights introducing natural light. The tube wells, elevators and staircases and other vertical transportation spaces are hidden in the interior of several rectangular body masses, and were inserted in the inner and outer areas of the wall surfaces as an independent body. While the interior space is divided by level floorslabs in the vertical direction, or directly frame the interplayer in the air to form three-dimensional display space, the well proportioned foloors are connected by smooth ramps.The sunlight shining through the skylights first diffuse widely through the glass fiber film of the ceiling materials to form a homogeneous light entering the hall. The light diffraction together with the free surfaces display different expressions. This natural light effect derives subtle changes in different parts of the exhibition hall.

Through the footbridge erected above the basement square one can directly enter the entrance hall in the first floor, the highest space in the buiding, which will be used to display installation works and large-scale sculpture, also taking into account the possibility of appreciating the works of art from of each floor. In addition, the cantilevered platform stretching out from the lobby elevator can work at the opening ceremony.

The main function of the fixed exhibition hall of the second floor is to showcase art collections of the college. Because most of the exhibits are landscape paintings and historical masterpieces, the interior maily uses wooden plates, stones, fabric,concrete and other materials with natural flavor.

The third floor exhibition hall is capable of meeting the needs of the contemporary art exhibitions with the largest size of the museum. The

中央美術學院
Instinctive

seamless curved wall surfaces enclosing the hall can be set as the cove scene background of the installation works, or they can be directly fixed to the wall. In addition, a small hall was set at the top as an interlayer.

The part separated by glass at the large stair from the first floor to the basement square is used as the lecture hall.

In the processes of the drawing design and the construction, all building materials are local in China except some of the important museum facilities. Although China is at its building peak when the construction costs rapidly rise, it's still far less than that of developed countries. Because China has its advantage of cheap labor costs, even if there's a large amount of manual operation,the construction costs won't soar as it did in Japan. Otherwise there are too many handwork staff in Chinese construction, which is inadaptable to the foreign style unitized production.

The external decoration uses hanging marbles and laminated Chinese-made slate. More hanging items of slates with the same thickness include Nara Centennial Hall, Nara, Japan, Shizuoka Convention Arts Center, Shizuoka, Japan as well as the Interactive Museum about Humans, La Coruna, Spain, whose physical appearance are generally round or oval-shaped and the exterior decoration materials are easy to be mass-produced. However because the free surfaces of this project are inadaptable to unitized production, after the research to the Chinese construction techniques, we decided to post the slates of different widths timely in different situations. In this case, as workers are demanded of manual operation hanging high above the ground, it's hard to ensure stable quality, therefore we made the 1:1 model to explore the construction approach and the nodes' treatment, and then began construction.

The three-dimensional model was applied throughout the working drawings. Not only the scrutiny of spliting the large stones, building the steel and concrete structures are dependent to the use of the three-dimensional model, even the wind pipes and the water sprays were positioned by three-dimensional method.

It's quite appropriate to use slate to define the Chinese traditional art education-based Academy of Fine Arts, because it's the basic ink-stone material. While the modern and free style of the surface shape symbolizes the contemporary advanced art educational institutions.

证大喜马拉雅艺术中心

ZENDAI HIMALAYAN ART CENTER

项目地点：中国 · 上海　用地面积：28 893 m^2　建筑面积：154 900 m^2

建筑设计：矶崎新工作室

建筑师：矶崎新，青木宏，胡倩，高桥邦明，饭岛刚宗，李卓

LOCATION: Shanghai, China　SITE AREA: 28,893 m^2　BUILDING AREA: 154,900 m^2

DESIGN CORPORATION: AI&A

ARCHITECTS: Arata Isozaki, Hiroshi Aoki, Hu Qian, Kuniaki Takahashi, Yoshitoki Iijima, Li Zhuo

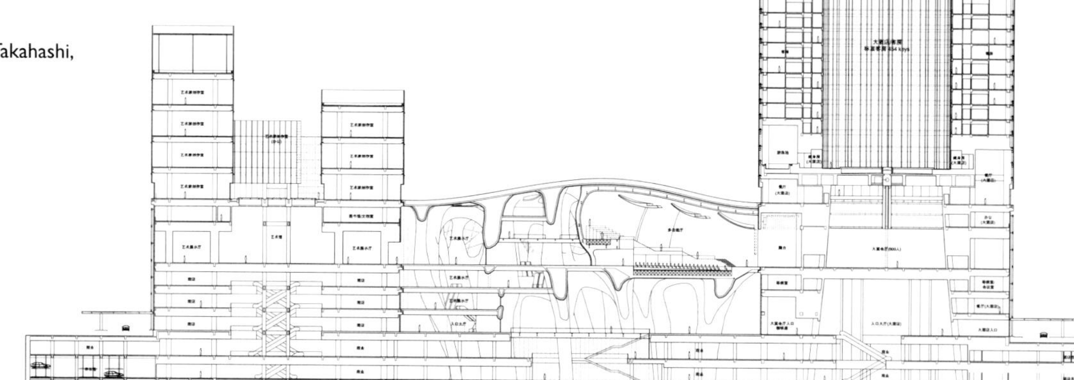

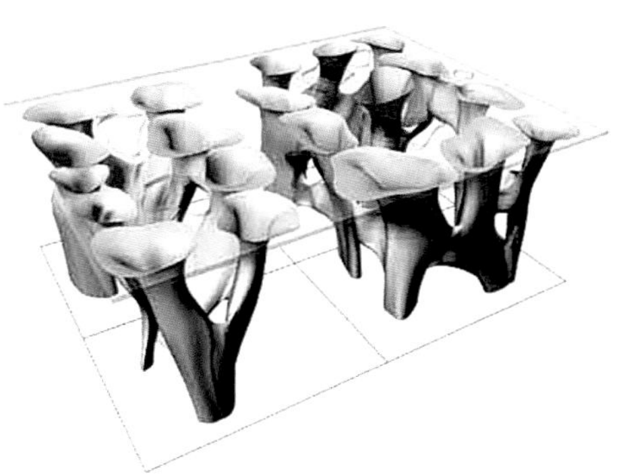

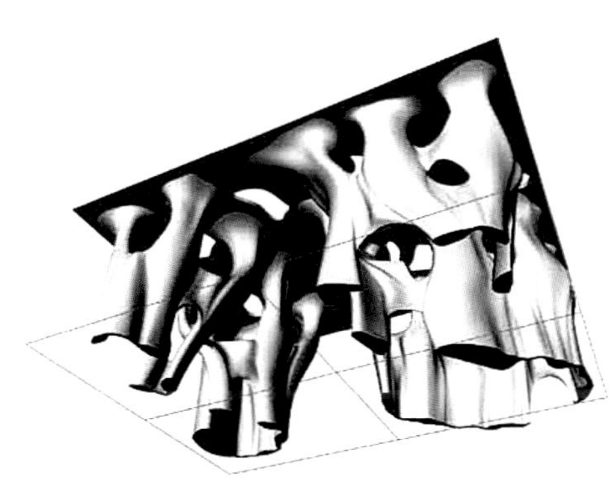

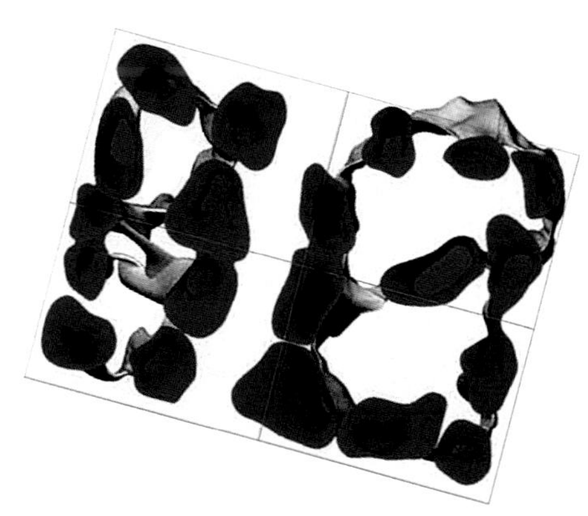

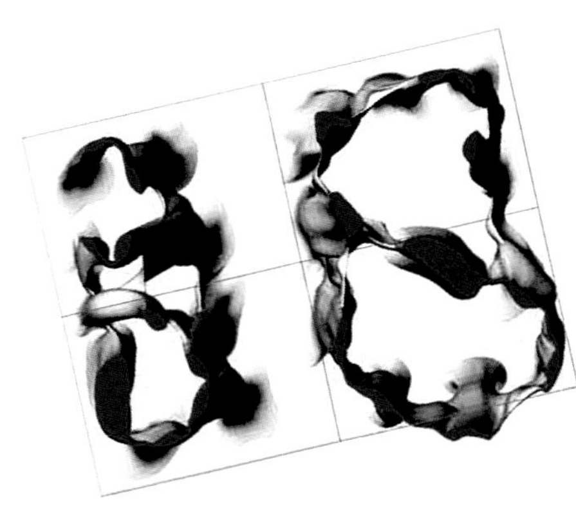

上海证大喜马拉雅艺术中心基地位于浦东石楠路以东，芳甸路以西，梅花路以南，樱花路以北，占地面积 28 893 m²。本工程基地不仅位于浦东的中心地区，且地处将来上海市三个会展商贸圈之一的区域内。

上海证大喜马拉雅艺术中心总建筑面积约为 155 000 m²，其中地上部分约 89 000 m²，地下部分约 65 000 m²，是一个集当代艺术中心、多功能演艺厅、五星级酒店、艺术家创作中心、商场为一体的复合设施。

整体功能分区可分为酒店、艺术家创作中心、当代艺术中心、商业设施、公共开放空间 5 个区域。

为了让这些多功能设施连接，本方案在水平与垂直方向上进行了三维立体的布置，让各种设施有效地相互连接，以形成特殊的复合型的高层设计。

为了确保西侧住宅区最大限度的日照，方案将五星级酒店配置在地基的北侧。五星级酒店为地上建筑面积约 50 000 m²，高度为最高限高 100 m 的高层建筑。相对在南侧配置的艺术家创作中心则为高度 67.5 m 的塔楼。这两个塔楼的中央连接部分为当代艺术中心，是本设施的核心，同时地上部分为向市民开放的公共空间。

同时本设计将设施在垂直标高 31.5 m 处分成上和下两个部分。北侧酒店客房部分配置在上方，酒店客房的塔楼为边长约 60 m 的明快的几何立方体。南侧设置的艺术家创作中心约为 11 500 m²。中央部分设有 60 m × 90 m 的空中庭园，配合地上的城市文化广场及地下的下沉文化开放广场，形成了立体的空间构成。在 31.5 m 以下的部分，配置了当代艺术中心、商业设施、多功能大厅、宴会厅、会议中心等功能区，这些功能区可延续邻接的新国际会展中心使用。

在办公空间性质多样化的今天，艺术家创作中心的设计不采用延续至今的均等的网格化办公空间，而是以基于一定条件配置的垂直核心筒为基础，将宽度约达 16 m 的办公空间通过模数、占地面积、安全疏散、采光等条件构筑出各种空间特征。本方案最终得出的是像中央空间围成内院那样的一种旋转上升的形态。

中央部分容纳的当代艺术中心和一个约 2000 人的多功能剧场为有机异形体。这种独特的有机造型运用了“进化论结构的最优化手法（ESO 方法）”的设计。一般从结构要素上的力的传导效率来看，与拉伸以及压缩的轴向力相比，弯力的传导效率极低。即在所有结构要素中均匀地分布没有弯力的应力结构，其力的传导效率是最高的，在这种情况下可实现用最小尺度的材料来组成结构体。使用电脑来生成这种形态的是 ESO 方法，根据所给出的力学条件及设计条件自动形成的造型，并在进行某种程度的调整后所得到的形态，将在被称为白色立方体的展示空间之外，以其崭新的空间形式被加以利用。

立方体形式在迄今为止的各个项目中频频亮相，最初的运用大约是在 20 世纪 70 年代设计的日本群马县近代美术馆中，那里把放置美术作品的地方诠释为“空洞”，作为对其的一种引喻，采用了立方体来构成美术馆。在喜玛拉雅艺术中心项目中，立方体被摆放在 31.5 m 标高的平台上方，作为酒店客房部分，在这里明快的几何造型作为都市的一种表象而体现。

在酒店以及艺术家创作中心下部的外墙上，采用了汉字造型的素材，且用更抽象化的手法制作 GRC 外墙幕，以双层幕墙的建筑立面形式覆盖在外墙上，起到了形象塑造和遮光的双重效果。

Shanghai Zendai Himalaya Art Center was based in Pudong with total area of 28,893 square meters. It's conneted to Shinan Road, Fangdian road, Meihua Road and Yinghua Road in four directions. The base is not only in the center of the Pudong area of Shanghai but within one of the three business exhibition cycles.

The total building area of Himalaya Art Center Shanghai is about 155,000 square meters,of which about 89,000 square meters overground and 65,000 square meters underground. It is an integrated establishment of contemporary art center, multi-functional performing arts room, five-star rate hotel, artists center as well as the shopping center. Overall functional area can be divided into five areas of the hotel, Artists Center, Contemporary Art Center,commercial facilities and public area.

This program conducted three-dimensional settings in the horizontal and vertical directions so that the variety of facilities can be well connected to each other to form a special high-level compound design.

To ensure maximum sunlight of the west residential area, the program set the five-star hotel in the north side, with the overground area around 50,000 square meters,as a high-rise building with the maximum height of 100 meters. While the south artists center is a 67.5 m-high tower building. The central connecting part of the two towers is the Contemporary Art Center, the core of the facility,the overground area of which is the public space open to the residents of the city.

At the same time we divided the buiding into two parts at the 31.5 m point of the vertical elevation. The north hotel rooms were set on the higher part. The hotel room tower is a lively cube with side length of 60 meters. The south Artists Center almost occupies 11,500 square meters. The central part was set the 60 m × 90 m air garden to form a three-dimensional space with the overground cultural square and the underground open cultural square,. In the lower areas of 31.5 m in vertical direction include the Contemporary Art Center, business facilities, multi-function hall, banquet hall, conference center and other functional areas, which can be used as the subsidiary space for the adjacent International Exhibition Center.

With the diverse features of the office space today, the design of the Artists Center gave up the even grids patterns, instead, based on certain vertical columns, construction areas, security evacuations, lighting and other spatial characteristics, in the 16 m wide space. The program eventually came to the spinning up form like a enclosed courtyard.

The central Contemporary Art Center and the multi-functional theater accommodating over 2000 persons are irregular organic body. These unique organic shapes were designed through "optimized method of evolutionary structural theory (ESO method) "Generally in terms of power transmission efficiency of the structural elements, the bending force is much lower than the axial tensile force or constringent force,that is, the even distribution of the stress structure without bending force produces optimized power transmission efficiency and in this case the structural body can be formed through minimal use of materials. The use of computers to generate this form is the ESO method. The shape self-obtained on the basis of the dynamic conditions and the design conditions and then reshaped through certain adjustment will be used as the brand new space form in the so-called white cube demonstrating space.

Cube forms have frequently appeared in the various projects till now , with the debut in the design of Museum of Modern Art in Gunma Prefecture, Japan in about 1970s, where the places of art exhibits were defined as the "empty hole." As a kind of allusion, the cube was used to form the museum. As for the project of Himalaya Art Center the cube was placed at the platform of 31.5 m high above the elevation. As the hotel room part, it showed itself as a city embodiment in lively geometric shape.

The lower part of the external walls of the hotel and the Artists Center takes the Kanji shaped element and the GRC curtain wall made through even more abstract approach. The double-layered curtain wall façade take both the image and shading effects.

上海交响乐团迁建工程

RELOCATION PROJECT SHANGHAI SYMPHONY ORCHESTRA

项目地点：中国 · 上海　**用地面积**：16 300 m^2　**建筑面积**：20 000 m^2
建筑设计：矶崎新工作室
建筑师：矶崎新，胡倩，高桥邦明，饭岛刚宗，李卓

LOCATION: Shanghai, China　SITE AREA: 16,300 m^2　BUILDING AREA: 20,000 m^2
DESIGN CORPORATION: AI&A
ARCHITECTS: Arata Isozaki, Hu Qian, Kuniaki Takahashi, Yoshitoki Iijima, Li Zhuo

本项目的设计旨在打造一座具有世界级高水准音响效果的音乐厅，以满足上海交响乐团的日常排练及高水平的交响乐公演。

作为上海交响乐团常驻场地，1200 人的大排演厅首要职能自然是要满足其高水平的日常排练需求。在此基础上作为公共文化设施的一部分向市民提供演出及招揽国际知名乐团来沪表演，为市民提供文化娱乐场地。由于其高品质的音响效果，该厅也可作为大型乐团的录音场地或在公演时录制、直播高水准的交响乐演出，借助媒体给更多的人带来艺术的享受。

为了使该项目的设计目标更好地得以实现，除矶崎新工作室外，团队还包括永田音响设计顾问公司、同济大学设计研究院及舞台、照明等众多优秀的设计团体。经团队的共同商讨与研究，确认了大排演厅的最佳净高、宽度及空气容积，更进一步确认了坐席的排布及反射板的位置关系。内部空间的功能化及形态美学自然是建筑设计中的关键，但是本项目的难点在于如何去协调建筑与声学的关系。经无数次的讨论、交流，借助先端的 3D 模型与音响模拟试验，同时依仗团队各部门的丰富经验及灵感将其确定为现在的形态。

该项目的大排演厅没有拘泥于传统的“鞋盒式”，而是采用了“Arena”式，即将舞台设置在中央，观众席分设在四周。这种形式的优势是观众席上的竖向墙体可以更利于声音的反射，且整体排布更加立体又具有亲和力。五块大型的侧反射板和天花反射板，经过周密的角度与形式的调整，不但对声音的反射至关重要，同时将照明等音乐厅功能需求统合于一身，还可利用高性能投影装置使其呈现影像。打破传统音乐厅的传统形式，丰富了视听，同时为现代交响乐演出形式提供了更多发挥的舞台。

可容纳400人的小排演厅首先可以适用于室内乐，而且可作为多种演出形式的高品质的录音棚。鉴于这两点要求，对小排演厅的净高、宽度进行了研究，确定了现在的小排演厅的体量比例关系。由于现代音乐多元化的发展及形式的日新月异，为了尽可能地满足各种形式的表演要求，将小排演厅的地板设置成了一个等标高的平台，且划分为12块，利用升降装置可以使任意一块或多块自由组合升起作为舞台。

建筑是人为活动、机能的空间容器，这一点对音乐厅的设计来说更加突出。面对音响近乎理想化的目标，克服基地的不利条件是该项目中我们最大的难题。对于音响的不利影响主要来源于两点：第一，本基地位于上海市衡山路风貌保护区，建筑体量受到高度的制约，音响所需空气容量难以保证；第二，新近建成的地铁从该基地内通过，噪声对音乐厅的影响是致命的。

解决上述两点唯一的方法就是将开挖深度加大，将音乐厅向地下延伸，首先保证音响设计所需的空间要求；解决来自地铁的噪声的方法是将整个音乐厅设计成一个像钢筋混凝土的盒子，并将其抬高架设在弹簧隔振装置上以达到隔振效果。

从整个建筑的平面布局来看，为突出其庄严又不失中国传统设计理念的元素，采用了左右对称的形式。大排演厅居中，小排演厅与入口大厅分居左右，以相似的体量遥相呼应。建筑与南侧的道路并列排布，借抽象化几何形态绿化带进行过渡，形成强烈的立体层次感。入口大厅与该抽象化几何形态绿化带相贯通，形成入口大厅的前庭空间，这样使入口大厅的功能从单一的功能转变为具有举办简易演奏会和宴会的多功能空间。

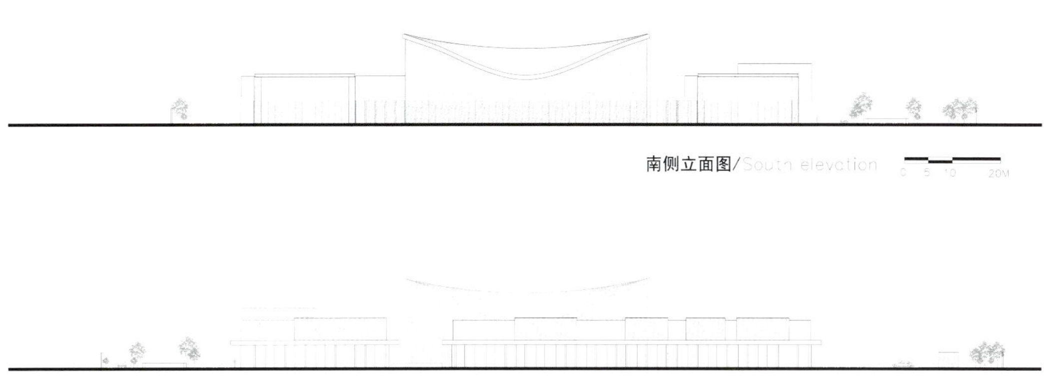

南侧立面图/South elevation

北侧立面图/North elevation

The design of the project aims to create a world-class concert hall of high sound quality to meet the needs of daily rehearsals and high level performance of Shanghai Orchestra.

As the permanent site of Shanghai Symphony Orchestra,the primary function of the large rehearsal hall accomadating 1200 people is to meet the needs of its high level daily rehearsals, and on this basis, as part of public cultural facilities it provides cultural and entertainment venue where world-renowned orchestras are also attracted here to give their shows. Because of its high-quality sound effect, the hall can also be used as the recording studio of large orchestras or a venue where high-level symphony performances can be recorded and broadcast live, which brings more people their enjoyment.

To achieve design goals better,besides the AI&A Studio the design team also includes Nagata Sound Design Consultant Company, Design Research Institute of Tongji University and and many other outstanding organizations of stage design and lighting design. After careful studying and discussion,the team made final agreeement on the clear height, width and air volume, and further confirmed the arrangement of seats and the relationship between the locations of the baffle-boards. The function of interior space and natural aesthetic form is the key to the building design, but the difficulty is how to coordinate the relationship between architecture and acoustics. Therefore the existing form is crystalization of numerous discussions, exchanges, and simulation tests with the help of 3D models on the basis of rich experience and inspirations of the departments of the whole team. Instead of the traditional "shoebox-style", the large rehearsal hall uses the Arena-style,to set the stage in the center and the auditorium peripheral. The advantage of the layout is that the vertical wall of the auditorium produces better reverberation as well as adds more affinity and space impression to the overall arrangement. Five large side baffle-boards and the ceiling baffle-board, through careful adjustment of the angles and forms, are not only essential to the sound reflection, but also assume lighting and other concert hall functions, and can also be used to render images through high-performance projection device. It broke the form of the traditional concert hall, enriched the audio visual sences and provide more space for modern symphony on its stage.

The smaller rehearsal hall accommodating 400 people first can be applied to chamber music. It can also be used as a high-quality recording studio for various perfrmances. Given these two requirements, the team studied on the clear height and the width of the small rehearsal hall and finally confirmed the body mass ratio. As the diversity of modern music and the rapid changing of its forms,the floor of the hall is designed as a platform of homogeneous elevation to meet requirements of a variety forms of performances. The floor is divided into 12 parts, each of which can freely lift or join to the others to form a stage.

Architecture is a human activity as well as a functionary spatial container, which is more significant to the design of the concert hall. In terms of the target of almost idealized sound effect,to overcome the disadvantages of the base is the biggest problem,which mainly come from two points: first, the base is located in Hengshan Road conservation area in Shanghai,where the building body mass is constrained in height and the needs of air volume for the sound is difficult to meet. Second, the newly built subway runs through within the base and the noise deadly affects the hall.

The only way to solve the problems is to increase the depth of excavation and extend the concert hall to the deeper underground, first to ensure necessary space requirement for sound quality; the way to remove the noise from the subway is to design the whole concert hall like a concrete box and rest it on the spring vibration isolator to avoid vibration.

The building plane layout takes symmetrical form to highlight the dignified yet traditional Chinese design elements. The large rehearsal hall lies in the middle, between the small rehearsal hall on the left and the entrance hall on the right echoing each other with similar body masses. The building stands parallel with the south side road and forms a strong space impression through the abstract geometric green belt ot which the entrance hall is conneted to form the vestibular space. Thus the entrance hall changes into a multi-functional space capable of holding simple concerts and banquets.

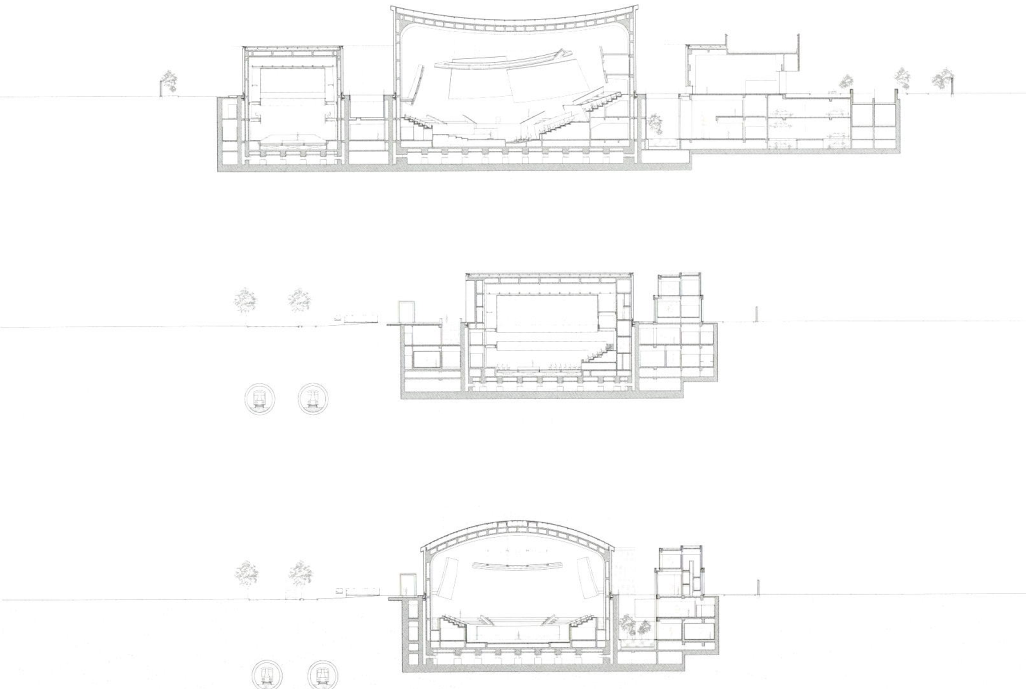

上海航天博物馆

SHANGHAI AEROSPACE MUSEUM

项目地点：中国 · 上海　用地面积：89 079 m^2　建筑面积：31 482 m^2

建筑设计：矶崎新工作室

建筑师：矶崎新，青木宏，胡倩，高桥邦明，福泽崇浩，曹振宇，佐野洋子，宫岛昇平，吉田凉子

LOCATION: Shanghai, China　SITE AREA: 89,079 m^2　BUILDING AREA: 31,482 m^2

DESIGN CORPORATION: AI&A

ARCHITECTS: Arata Isozaki, Hiroshi Aoki, Hu Qian, Kuniaki Takahashi, Takahiro Fukuzawa, Cao Zhenyu, Yoko Sano, Shohei Miyajima, Ryoko Yoshida

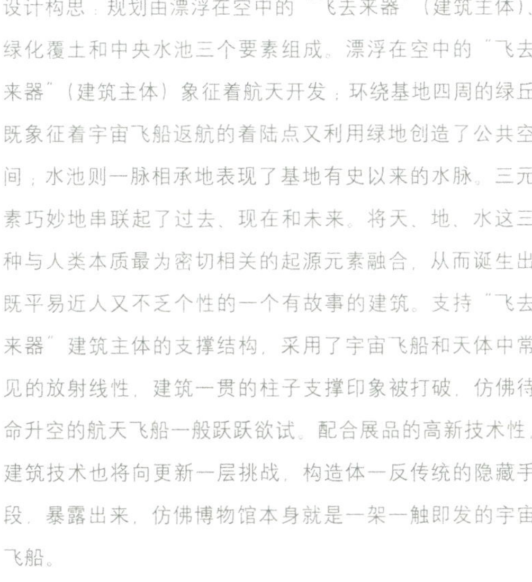

设计构思：规划由漂浮在空中的“飞去来器”（建筑主体）、绿化覆土和中央水池三个要素组成。漂浮在空中的“飞去来器”（建筑主体）象征着航天开发；环绕基地四周的绿丘既象征着宇宙飞船返航的着陆点又利用绿地创造了公共空间；水池则一脉相承地表现了基地有史以来的水脉。三元素巧妙地串联起了过去、现在和未来。将天、地、水这三种与人类本质最为密切相关的起源元素融合，从而诞生出既平易近人又不乏个性的一个有故事的建筑。支持“飞去来器”建筑主体的支撑结构，采用了宇宙飞船和天体中常见的放射线性，建筑一贯的柱子支撑印象被打破，仿佛待命升空的航天飞船一般跃跃欲试。配合展品的高新技术性，建筑技术也将向更新一层挑战，构造体一反传统的隐藏手段，暴露出来，仿佛博物馆本身就是一架一触即发的宇宙飞船。

功能分区：建筑体块大体分为地上的绿丘和漂浮在空中的“飞去来器”。地上部分的覆土绿丘，被引导河水的水渠、连接地铁站和入口的多媒信息桥形成了三块甜圈形的形态。南面的绿丘是三块绿丘中最大的，入口大厅、特别展厅、主题展厅、四维动态电影院、餐厅等主要的公共使用功能被安排在这里。西北的绿丘安排的是为南部部分提供后援的服务功能空间。南面和西北面的绿丘属于一期工程。东北面的属于二期，二期预设为科技馆。“漂浮”在空中的飞去来器形状的建筑空间内设置的是基本陈列展厅、科普互动室等功能空间。空间的公共开放性使参观者能够自由地体验空间。与包含了许多公共功能的“飞去来器”相连的是南部和西北部的绿丘，这样的体量组合使来馆人流和后勤流线都得到了恰当的连接。

Design Concept: the plan consists three elements of the "boomerang" floating in the air (the main building), green earthing and the central pool. The "boomerang" is a symbol of aerospace development; the green hills around the base both symbolize the landing point of the spacecraft and form the green public space; the pool expresses the water pulse in line with the history. The design tactfully uses the three elements to connect the past, the present and the future, and integates the heaven, the earth, and the water; the three natural elements are most closely related to human origin to bear a both approachable and individual building with stories. The structural support for the boomerang in the main building uses the radiation linearity commmon in spacecrafts and celestial bodies to break the conventional column structural support, as if it is a spacecraft ready to take off carrying its mission. To match the hi-tech of the product, the building technique also takes a big step to expose the structural body instead of traditionally hiding it, as if the museum itself is a touch-and-go spacecraft.

Functional Division: the building body mass can be divided into the floating "boomerang" in the air and the green hills on the ground. The earthing green hills were divided into three donut-shaped areas by the canal guiding the water from the river and the multi-media information bridge connecting the subway station and the entrance. The south side green hill

is the largest of the three,where the entrance hall, special exhibition hall, theme exhibition halls, four-dimensional dynamic theater, the restaurant and other major public facilities are arranged. The northwest green hill is the backup functional service space for southern part. The south and the northwest of green hills are the first stage of the project, and the northeast belongs to the second, which is planned to be the Science and Technology Museum. In the interior of the "boomerang" are the basic exhibition hall, interactive space of science popularization and other functional spaces. The openness of the spaces allows visitors free experiences in them. Connecting to the multi-functional "boomerang" are the south and northwest of the green hills. This combination of body mass provides proper connection for the visitor flow and the logistic streamline of the museum.

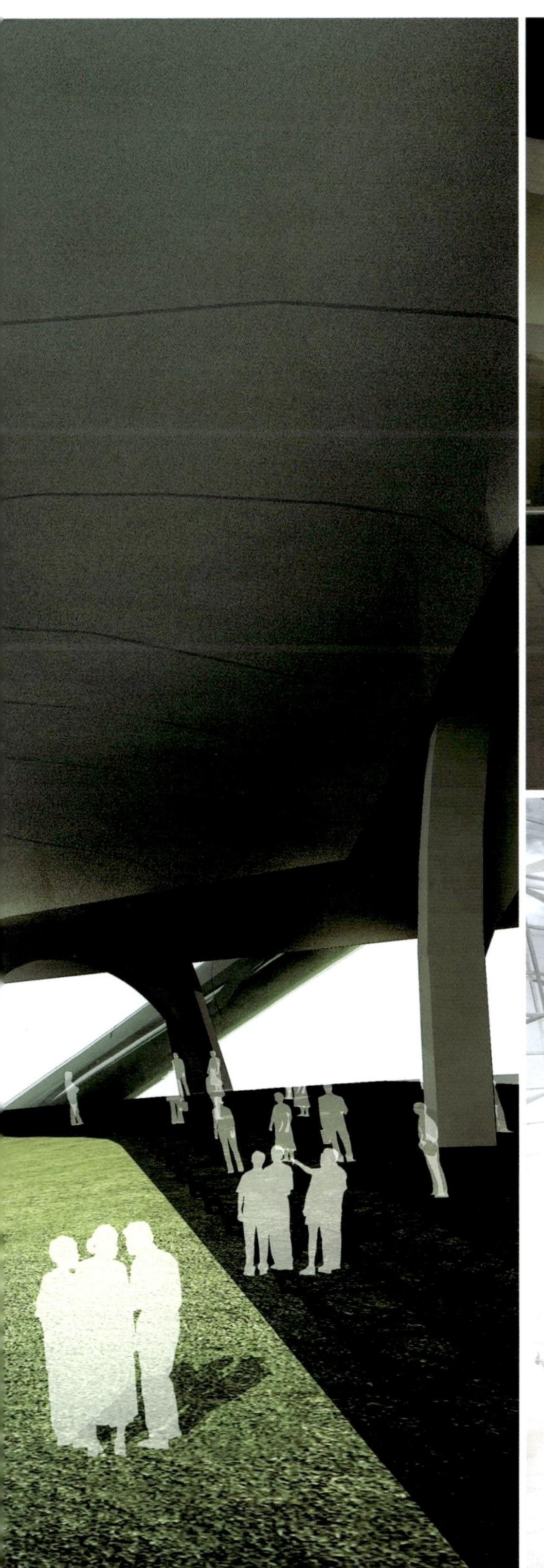

ARCHI-NATURE LLC

天津水木风光建筑工作室

王克宁

姚远疆

公司合伙人及主要设计师

- 王克宁 Wang Kening

1992年毕业于河北工业大学建筑系
曾先后在天津华汇建筑设计公司（1997—1999）
北京中外建天津分公司（2000）工作
现于天津城市建设学院建筑系教授建筑设计
2004年7月创建水木风光工作室

- 姚远疆 Yao Yuanjiang

1992年毕业于河北工业大学建筑系
曾先后在唐山房屋设计院（1992—1994）
新加坡建屋发展局（1994—1997）工作
现于河北理工大学建筑系教授建筑设计
2004年7月创建水木风光工作室

双港城市设计

SHUANG GANG URBAN DESIGN

项目地点：中国·天津　建筑面积：3 750 000 m^2
建筑设计：天津水木风光建筑工作室
合作单位：天津中天建都市建筑设计有限公司
建筑师：王克宁，徐璐，杨尧，高昊

LOCATION: Tianjin, China　BUILDING AREA: 3,750,000 m^2
DESIGN CORPORATION: ArchiNATURE LLC
PARTNER: Tianjin ZTJ Urban & Architecture Design Co., Ltd.
ARCHITECTS: Wang Kening, Xu Lu, Yang Yao, Gao Hao

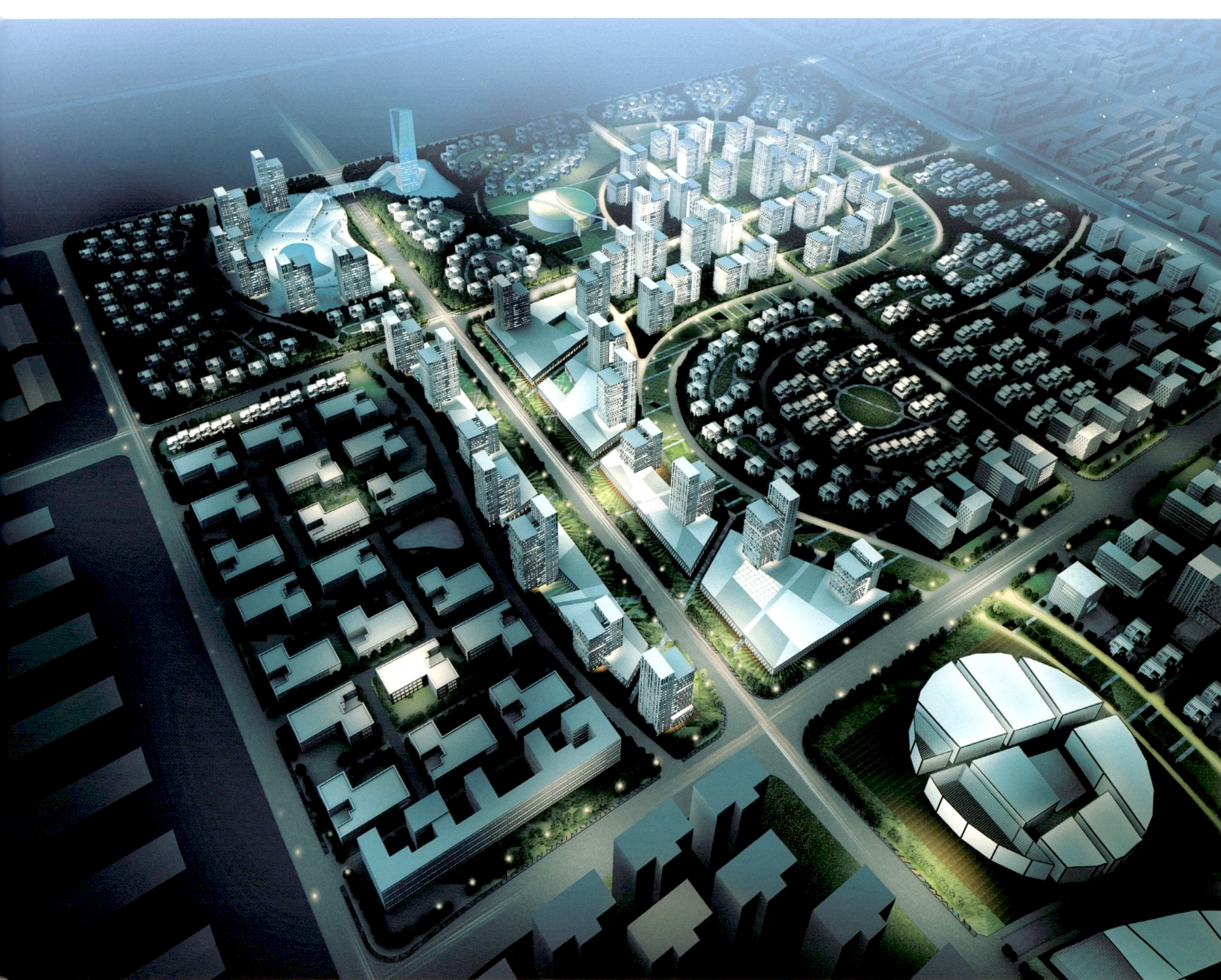

双港镇位于津南区西侧，处在市区和滨海新区中心区之间的黄金走廊上，是津南最靠近市区的产业发展基地。

紧邻外环线，双港充分依托市区资源，与空港、经济开发区互为补充、促进，发展潜力巨大，未来几年将成为环渤海经济区的又一个新兴的产业发展集群。

总占地 11 km^2，包含核心区、拓展区、提升区和居住区四个组团。北靠外环，南接北洋的地理位置使金港城核心区具有独特的发展潜力。OIT 不仅是双港的核心发展区域，也将成为带动津南发展的源动力。

规划理念

“四区域，五组团，一纽带”的“451”格局：

“4”即四个区域协调发展，共建金港；

“5”即五种功能组团，互为补充；

“1”即条纽带，服务周边，纵深发展，借势生长。

总体规划以景观绿轴为纽带，借鉴水的空间形态，通过绿带联系园区，串联起园区的各个公共空间，景观作为主线串联公共空间，形成新的城市形象。

城市新形象

金港城核心区通过建筑和景观的自然融合，为双港增添新的城市形象。沿外环线一侧的总部基地采用岛屿式组团的形式进行划分，便于开发运作。蜿蜒的景观带贯穿南北，由西向东随着路网和产品形式的转变，空间天际线也不断转变。主要道路两侧大厦采用高层办公、底层商业的模式，是主要的人员汇集场所。开放的公共空间、舒适的办公环境、高效的生产基地、宜人的自然景观、灯火辉煌的商业氛围都将成为金港城繁荣发展的起点，也将是双港美好未来的缩影。

The Shuang Gang town lies in the west of Jinnan district, and between the downtown area and the center of Binhai New district. It is the industrial development base closest to the urban area of Jinnan.

Close to the outer ring, fully relying on urban resources Shang Gang and Tianjin Airport Economic Area and the Economic Development Zone complement and promote each other and it possesses boundless development potential, Shuang Gang will be another emerging industry clusters Bohai Rim Economic Zone in the next few years.

It occupies total area of 11 km^2 including the core area, the development area, the promotion area and the residential area. With the north the outer ring and the south the coastal areas, the core area Jin Gang city has a unique potential for development. OIT is not only the core development area of Shuang Gang, but will also be the driving source to the development of Jinnan.

Planning Ideas

"4.5.1" pattern of "Four regions, five groups, one ligament":

"4" – four regions coordinatedly develop to construct Jin Gang city.

"5" – five functional groups complement each other.

"1" – one ligamnet serves the surrounding area, develops in great depth and seizes the chance to grow.

The master plan uses the green landscape as the axis , borrows ideas from the water form and connect the park area and the public spaces within it through the green belt zone. The landscape gets through the public spaces as the main line to form a new city image.

The new city image

The core area of Jin Gang city adds new image to the city through the natural integration of architecture and landscape. The base along one

side of the outer ring road is divided in island-style or in groups, easy to be developed.

The landscape winds from north to south, while the skyline constantly changes as network and product change from west to east. On both sides of the major roads are high-rise office buildings, in which the lower business office is the main people gathering site . Open public spaces, comfortable office environment, efficient production base as well as the pleasant landscape and the illuminated business atmosphere will become the starting point of prosperity and development of Jin Gang city, which is also the miniature of the future Shuang Gang city.

滨海临空产业园——EOD 总部港

COASTAL AIRPORT INDUSTRIAL PARK – EOD HQ PORT

项目地点：中国 · 天津　占地面积：442 000 m^2　建筑面积：500 000 m^2
建筑设计：天津水木风光建筑工作室　合作单位：天津中天建都市建筑设计有限公司
建筑师：王克宁，徐璐，杨尧，穆鹏，尹烁婧

LOCATION: Tianjin, China　PLOT AREA: 442,000 m^2　BUILDING AREA: 500,000 m^2
DESIGN CORPORATION: ArchiNATURE LLC
PARTNER: Tianjin ZTJ Urban & Architecture Design Co., Ltd.
ARCHITECTS: Wang Kening, Xu Lu, Yang Yao, Mu Peng, Yin Shuojing

滨海国际企业城紧邻建设中的航空城，是临空经济圈的重要组成部分。项目位于滨海新区与市区的交汇处，是集总部办公、技术研发、高科技生产、企业配套服务于一体的都市型现代工业产业集群。

四个方向的规划路把基地主要分成企业总部经济区、企业研发区、创意产业区、高科技产品生产加工区、配套服务区、员工公寓区六个组团。分期分批地进行建设。

(1) 企业之城：在这里建筑类型多种多样，展现着企业形象的多姿多彩。建筑或联排，或独栋，或小高层，或单元组合连接。多种办公户型可以满足各种规模各种类型公司的要求。

建筑的单体设计采用单元块叠加的模式，每个单元代表一个部门或企业。层层叠叠的公司通过通透的公共空间分隔并连接，各种新型的节能表皮凸显强烈的现代感。西南的配套服务区更是通过简洁流动的设计和节能的百叶赋予地块标志性的表现力量。

(2) 便捷之城：企业城规模宏大，交通系统更为重要。在已有规划道路基础上，各地块以环形路网为主，确保各类型的公司都能直接拥有一个靠近主要道路的便捷出入口。

(3) 生态之城：空港国际办公区旨在建成人性化的生态办公区（Ecological Office District），简称 EOD，是一种全新的商务办公方式，崇尚绿色环保、可持续发展理念，追求人与自然的和谐统一。基地南面是 50 m 左右的绿化景观带，结合西部的高压电缆绿化隔离区形成连续景观。组团内每个企业单元都拥有相对独立的院落，保证了企业的外部空间小环境。

(4) 科技之城：地块中部的主要规划路两侧的小高层公司总部，外立面以幕墙为主，结合铝板，铝制百叶遮阳，玻璃幕墙面与不透光铝板组合，各种现代材料的组合运用，使建筑具有新颖的形式和强烈的现代感。

在建筑屋顶设置太阳能板或绿色景观屋面，节能环保，新型技术将大量运用于进一步深化的设计之中，力求创造更高品质的高级企业环境。

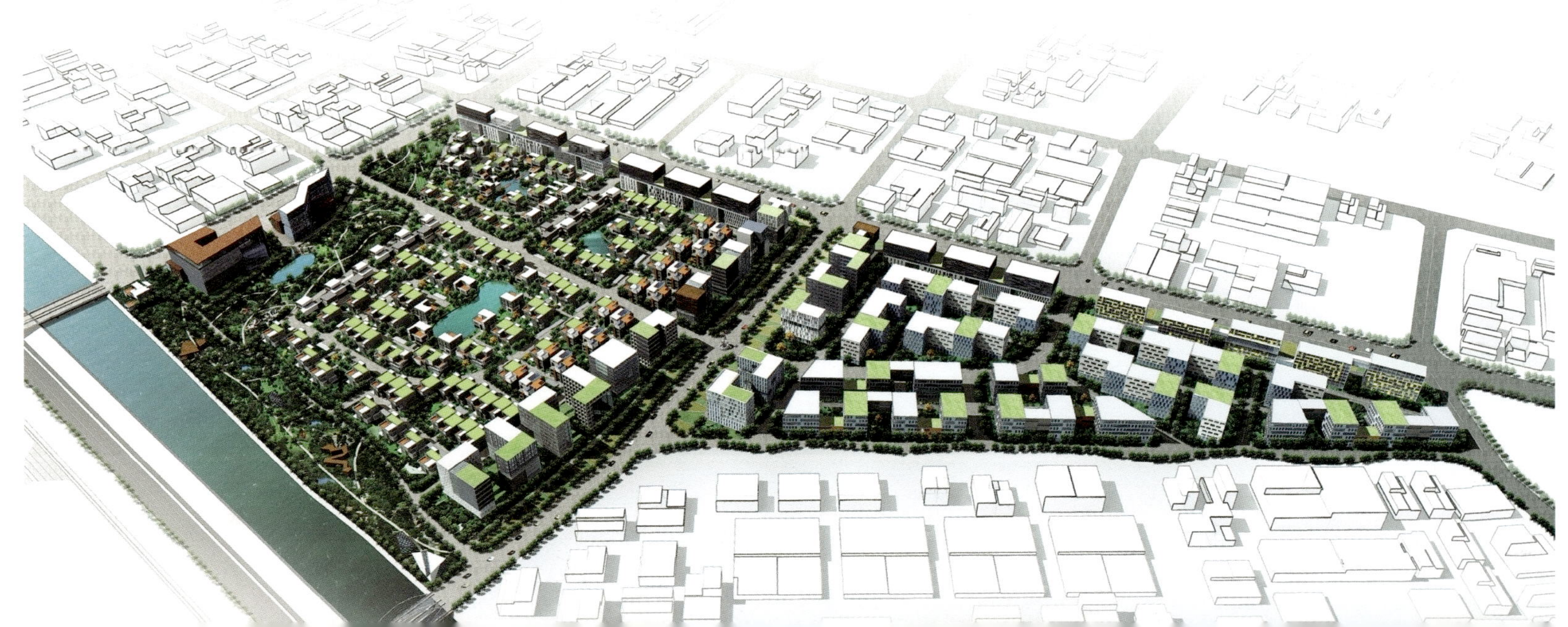

Binhai International, close to the aviation city under construction , is an important part of airport economic circle. The project is at the intersection of Binhai New Area and the downtown area, integrating headquarter office, R&D, high-tech manufacturing and business support and other services in the modern urban industrial clusters. The planning road in four directions separate the base into functional 6 groups including: the headquarter economic zone, R&D areas, creative industry area, high-tech production and processing zone, support service zone, the staff apartment, which are constructed in different stages.

(1) Enterprise City: Here buildings show their varied types as well as colorful images of the enterprise. They are terraced or detached, medium-rise or unit groups, meeting the demands of various types of companies.

Monomer building design takes superimposed cell blocks, each representing a department or an enterprise. Layers of companies are separated or connected by translucent public spaces. All kinds of new energy-saving skin highlights a strong modern ssense. The compact flowing design and energy saving shutters in the southwest supporting service area endow the block with powerful representability.

(2) Convenient City: Traffic system is more important to the large-scale enterprise city. On the basis of the existing planning of roads, each block, mainly with the ring road networks, ensures that all types of companies can have a direct and convenient road close to the main entrance.

(3) Ecological City: The office area of Airport International aims to build the Ecological Office District(EOD), with a new way of business working, advocating environmental protection, sustainable development, the harmony between human and nature. South of the Base is the 50 meters long greenbelt, combined with high voltage power lines greenbelt in the west area forming the continuous landscape greenbelt area. The independent courtyard of each business unit guarantees outside microenvironment.

(4) Science and Technology City: As for the small-scale high-rise headquarter in the middle of the block on both sides of the planning road, the large area of the glass curtain walls on the facade combined with opaque aluminum plate and shutters, endow the building with unique forms and strong modern sense.

The solar panels or green landscape roofs make energy efficiency and large use of new technology possible in further design, striving to create top-grade business environment.

	建筑面积	基地面积	容积率
A:配套服务区	25000平方米	24000平方米	1.04
B:微电子产业区（一期）	33580平方米	42800平方米	0.78
C:微电子产业区（二期）	103960平方米	88900平方米	1.17
D:精密仪器加工区	71340平方米	59250平方米	1.20
E:高科技产品生产加工区	226120平方米	135490平方米	1.67
F:员工公寓区	30240平方米	23510平方米	1.28
总:	490240平方米	444628平方米	1.10

渤海职业技术学院体育馆

GYMNASIUM OF BOHAI VOCATIONAL TECHNICAL COLLEGE

项目地点：中国·天津 建筑面积：8134 m^2

建筑设计：天津水木风光建筑工作室 合作单位：天津方标建筑设计有限公司

建筑师：王克宁，杨尧，姚远疆，徐璐，张文婷，唐涛

LOCATION: Tianjin, China BUILDING AREA: 8134 m^2

DESIGN CORPORATION: ArchiNATURE LLC

PARTNER: Tianjin Fangbiao Architecture Design Co., Ltd.

ARCHITECTS: Wang Kening, Yang Yao, Yao Yuanjiang, Xu Lu, Zhang wenting, Tang Tao

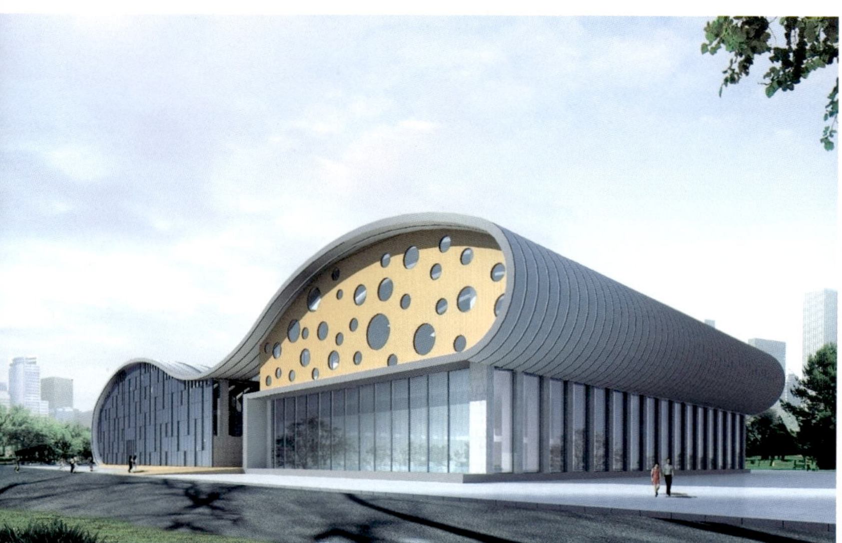

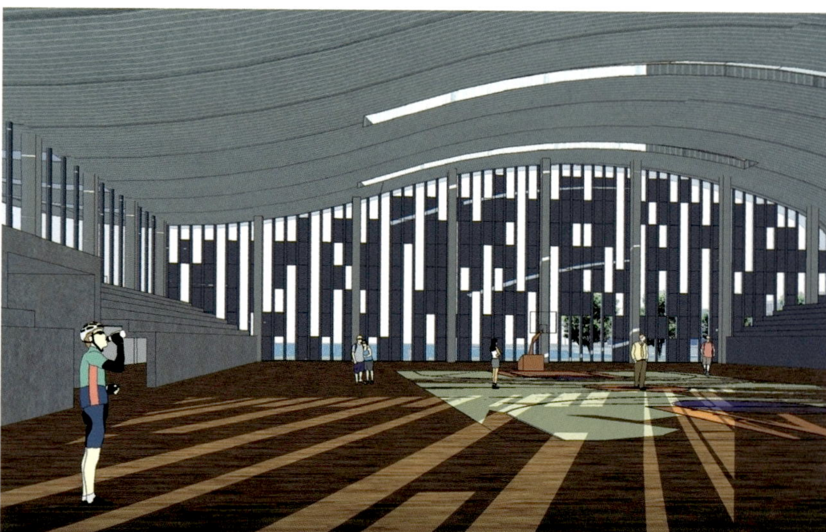

体育馆和培训中心位于西侧校园内湖和东侧体育场之间。建筑朝向两个方向，把西侧的湖和东侧的体育场的景色同时收入视野。其中，体育馆的体育设施靠近体育场，与体育场另一侧的看台遥相呼应，形成完整的运动区域。培训中心位于规划体育馆南侧，临近校园内湖，景色与朝向俱佳。
体育馆设置主馆、游泳馆及训练馆。主要入口设计成半开放式，从开放的平台上进入体育馆，整个建筑虚实结合，空间富有变化。建筑顶棚采用流线型设计，曲线自由、柔美，建筑外皮设计灵活，更好地体现了学校体育馆建筑的灵动活泼。

The gymnasium and the training center are set between the west lake and the east stadium on campus, seizing the scenery of the both.The sports facilities are near the stadium echoing with the grandstand at the opposite side to form the complete exercise area.The training center is located in the south of the gym, close to the campus lake with the good vantage of view..
The gym contains the main hall, swimming pool and the training center.The main entrance was set as the semi-open style. Entering the stadium through the open platform you will find the whole building a diversified space of the virtualization and the reality.The building roof takes the flowline design with free and soft curved lines.The flexible design of the construction better reflects the vitality of the gym.

海泰国际酒店综合体

HITECH INTERNATIONAL HOTEL COMPLEX

项目地点：中国 · 天津　占地面积：500 000 m^2　建筑面积：231 000 m^2
建筑设计：天津水木风光建筑工作室　合作单位：天津中天建都市建筑设计有限公司
建筑师：王克宁，张炼，徐璐，杨尧，穆鹏，李超

LOCATION: Tianjin, China　PLOT AREA: 500,000 m^2　BUILDING AREA: 231,000 m^2
DESIGN CORPORATION: ArchiNATURE LLC
PARTNER: Tianjin ZTJ Urban & Architecture Design Co., Ltd.
ARCHITECTS: Wang kening, Zhang Lian, Xu Lu, Yang Yao, Mu Peng, Li Chao

海泰国际酒店综合体涵盖了办公、酒店、酒店式公寓。

融入城市空间，和谐城市形象

建筑布局重点强调区域内的城市关系。产业园区内津静公路和海泰内环一路形成的两条主要轴线，酒店与办公各踞南北错开布置。北面朝向中心广场并与其呼应，南面朝向低层住宅区以及远处的高尔夫球场，视野开阔。

主体办公塔楼位于基地东北部，紧邻两条主要轴线，与园区主要的金融办公中心和广场遥相呼应；出租办公水平方向展开，沿街分布，底部沿街设置配套的商业服务设施。

南部的酒店靠近东侧海泰内环一路，面向海泰内环南路，环境相对安静，酒店的东南侧是城市绿地，酒店南侧是低层住宅，因此酒店南北视野开阔，可以直接远望北面的金融中心和西南的高尔夫球场，景色怡人。

酒店式公寓分布于基地西侧，与办公大楼和酒店紧密联系。低层低密度，保证充足的建筑间距，减少遮挡，也拓展了酒店及办公区的视野。灵活的户型设计提高了公寓的品质，使公寓更具有市场吸引力。

整合与划分

在设计初期，对综合体功能进行整体的划分，通过理性的分析量化各功能间的关系。各功能以单元模块的形式整合，排列，再划分，再整合，逐步形成最终的关系。

办公、酒店、公寓分区明确，又通过位于三层的超级楼层（Super Floor）联为一体，各个单元内部又由空中公共平台联系。单元划分与整合的概念贯穿设计始终。

立体交通，人车分流

基地内地下一层为地下车库，一条地下通道贯穿基地南北。汽车由南北两个方向直接进入地下车库，继而形成一条地下汽车街道。

西侧公寓设地面架空停车场，从二层平台进入公寓，利用空中平台使办公和生活流线分开。东侧办公与酒店部分则完全以步行交通为主。

通过立体交通及步行体系的建立，真正实现了人车分流。

立体生态景观与绿色高技术生态系统

地面绿化庭院、屋顶绿化平台、地面步行街、连廊等构成了多层次的立体景观环境。公共层的屋顶结合太阳能光电板布置绿化，丰富了屋顶空间，实现了向生态化、人性化的居住、办公环境过渡。

办公楼及酒店外表面以幕墙为主，结合铝板，铝制百叶遮阳，用水平与垂直的疏密百叶区分各个单元。玻璃幕墙面间以不透光铝板。屋顶及局部墙体表皮采用太阳能光电板集能，环保节能。建筑的新颖形式和强烈的现代感体现了高品质的设计理念和办公环境，恰好反映了海泰工业园区高科技、国际化、现代化的时代特征。

单体设计

(1) 办公楼：立方单元的整体设计概念符合企业和写字楼相互依存提升品牌形象的要求。几个立方体通过交通核和公共层组合到一起，立面用不同的表皮区分，每个立方单元代表一个单元或是一类公司。单元体竖向发展形成集团办公大楼，楼体高大的现代气质突出企业形象。单元体水平发展，不同企业可以自由选择符合企业形象的立方单元，灵活组合，形成群体的公司园。

(2) 酒店按四星以上标准设计，各种活动娱乐设施配置齐全，同时满足酒店、办公及生活的要求。大型公共服务空间位于基地中心，是生活与办公半径穿插重叠的部分，为各区域使用提供方便。

(3) 公寓的多种户型设计增加了可选择空间，每三层配有独立的共享空间，朝向不同景观，在此休憩或远望皆可体会高品质生活。

(4) 综合体的配套服务设施主要沿街，环境优越，交通便利，易通达，极具吸引力。

海泰国际酒店综合体是城市中办公、居住、餐饮、会议、文娱、交通等城市生活空间的组合，是一个多功能、高效率、复杂而统一的综合体，是丰富人们生活的城市舞台。

它蕴涵着现代的建筑理论，创造着和谐的城市环境，由它带动全新塑造的新的城市形象也将慢慢浮现。

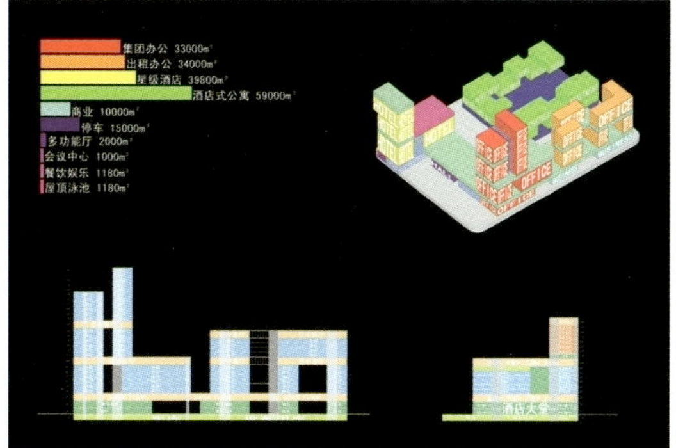

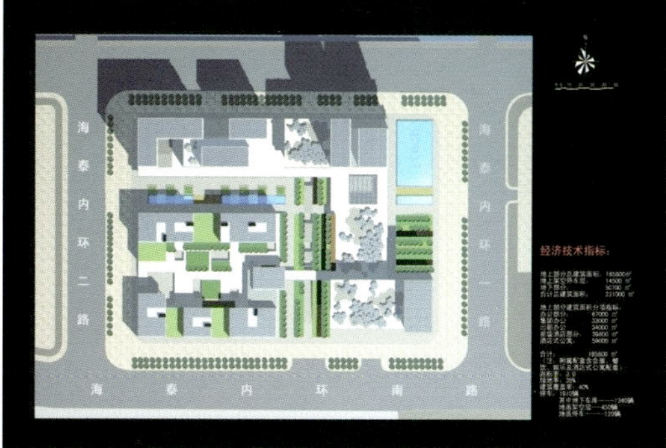

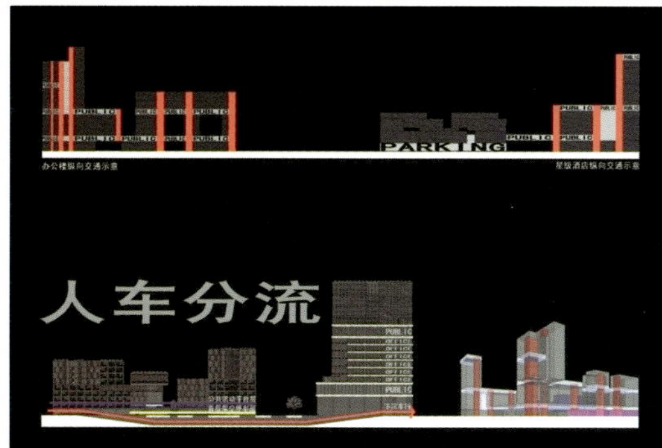

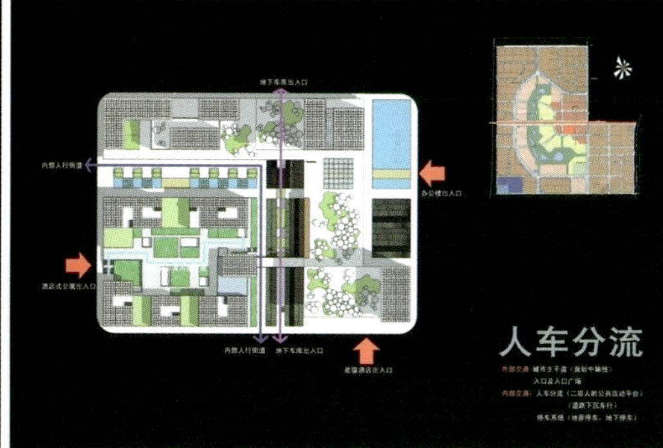

Haitai International Hotel Complex includes office areas, hotel, hotel-style apartments.

To fuse into the urban space and harmonize the city image

The architectural layout mainly emphasizes on the regional relations within the city. The two main axes of Jin Jing road and Haitai inner ring road in the Industrial Park hold the hotel and the offices separately in the north and the south. The north faces and echoes the central square, and the south takes the broad view of the low-rise residential areas and the distant golf course.

The main office tower stands in the northeast, close to the two main axes and echoing the major financial office center and the square; the rental and office areas unfold in the horizontal direction, distributed in the street, supporting commercial service facilities are set at the bottom.

The south hotel is close to Haitai inner ring road in east and facing the south part of it with the environment relatively quiet. The southeast of the hotel is the greenland, and the south is the low-rise residence. So the hotel has a full view, directly overlooking the pleasant view of the north financial center and the south-west golf course.

The apartment is located in the west of the base, closely related to the office buildings and the hotel. The minimum density and elevation ensure adequate space and reduce blocking between buildings, while expanding the vision of the hotel and office. The flexible design improves the apartment quality and makes it more competent in the market.

Integration and division

The functional division to the overall complex in the early design was given rational analysis to quantifying the relationship of the functions. Each function is integrated, ranked, divided and reintegrated in modules to gradually form the final relationship.

Office, hotel, apartment areas are clearly laid out and integrated through the third Super Floor, with each area having its public platform to connect with others. The concepts of unit division and integration gets through the whole design.

Three-dimensional transport and separation of people and vehicles

B1 is the garage with an underground passage or the car street from north to south, through either of which the car directly enters into the garage.

The west apartment area was facilitated with superterranean parking ground entering the apartment from the second floor platform, through which streamlines of office and living are separated. The offices and hotels in the east are completely dominated by pedestrian traffics.

Through the establishment of the three-dimensional transport and walking system the separation of the people and vehicles is realized.

Three-dimensional ecological landscape and green high-tech eco-system

The green courtyard, green roof platforms, the walking street as well as the corridor constitute a multi-layered three-dimensional landscape environment. The public floor roof combined with the solar panels makes the whole space green and activated, and thus achieves the ecological and humanized residential and office environment transition.

The outer surface of the office buildings and hotels mainly use the curtain walls, combined with aluminum plates and shutters. The horizontal and vertical shutters ,thick or dense,distinguish each unit. The glass curtain walls are partitioned with opaque aluminum plates. Roof and partial wall surface use solar panels to achieve environment conservation and energy saving. The unique forms and strong

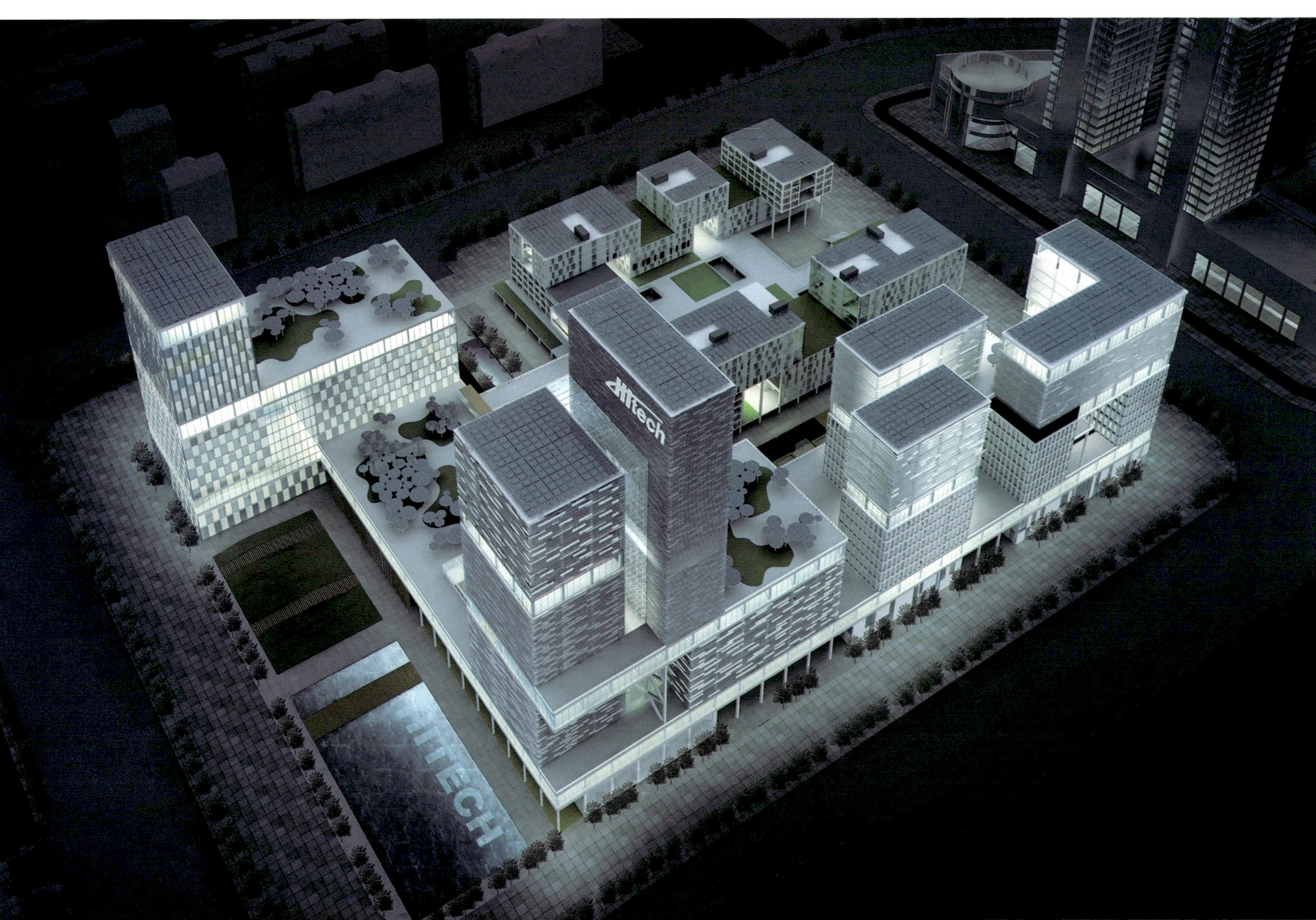

modern sense reflecting the high quality design concepts and office environment fully embody the high-tech, international and modernized characters of Haitai Industrial Park.

Single building design

(1) Office buildings: the overall design of cubic units is consistent with the demand of interdependence between business and office to enhance brand image. Several cubic bodies are combined through the transport core and public floor. The facades are distinguished by various skins, each cube representing a unit or a class of companies. The units vertically growing form groups of office buildings, are tall and large to highlight modern corporate image. The units horizontally expanding allow different companies to freely choose their cubes which are proper to the company images, which are flexibly combined to form park companies.

(2) The hotel is designed by four-star rate standard or above with various entertainment facilities while meeting the hotel, office and living demands. Large-scale public service spaces lie at the center of the base, also the overlapping part of the office and the living areas, which is available for each area.

(3) Multiple-type design of the apartment increases the choice of spaces, every three storeys are equipped with independent shared space facing different views, where one can rest or look out to experience the high-quality life.

(4) The supporting facilities of the complex are attractive and accessible with advantageous environment and convenient transportation mainly along the street.

Haitai International Hotel complex is the integration of office, living, dining, conference, entertainment, traffic and other urban lives. It's a multifunctional and efficient complex of stage which displays the rich life of the citizens.

It implies the modern architecture theory, creates the harmonious urban environment, and will gradually lead to the emerging of the new city image.

AS. ARCHITECT-URE -STUDIO

法国AS建筑工作室

1973年在巴黎创办的法国AS建筑工作室本着开放性的原则和团队合作的工作哲学，形成了一支专业的建筑设计团队，并且在不断扩大中。

法国AS建筑工作室在事务所12位建筑师合伙人的周围团结了一支由25个不同国籍的优秀建筑师、城市规划家、室内设计师及成本评估师等组成的专业团队。

2005年，法国AS建筑工作室在中国上海成立了永久事务所，更加强了它在国际市场的影响力。2007年北京分公司也宣布成立。在中国的50多位建筑师延续了法国AS工作室的设计理念及工作方式，并与巴黎总部紧密联合,将中法两国的精神理念和文化特点融合在一起。

法国AS建筑工作室将建筑和城市规划定义为：与社会密切关联的艺术，同时也是建设人类生活的构架。最根本的就是建立在团队合作、分享经验之上，通过团队成员的沟通和交流，将个人的感知转化成更强大的集体潜在创造力。

法国AS建筑工作室在2005年通过了ISO9001认证，同时凭借在法国乃至全球的技术优势、理财以及特殊顾问等领域的合作网络实现了各种不同的项目：医院、办公楼、公司总部、酒店、住宅楼、体育综合性项目、商业中心、学校项目(小学、初中、高中、大学)、研究院、文化项目(博物馆、剧院、歌剧院) 以及城市规划项目等。

Set up in 1973 in Paris, AS.Architecture-Studio has formed a professional team in architecture design, which is also enlarged continuously with the principle of openness and the philosophy of team work.

AS.Architecture-Studio has established a professional team consisting of excellent architects, urban planners, interior designers and cost appraisers from 25 different countries around 12 partners.

In 2005, AS.Architecture-Studio set up Yongjiu (Forever) Office in Shanghai, China to enhance its influence in the international market.In 2007, it set up its Bejing Subdivision. The more than 50 architects in China use the design concept and working modes of AS.Architecture-Studio, and through close contacts with its headquarters in Paris,merge together the spiritual concept and the cultural characteristics of both China and France.

AS.Architecture-Studio defines architecture and urban planning as an art closely related to society and a framework to build up human life.

The most fundamental principle is to establish team work, sharing experience, communication and exchange via team members so as to transform personal perception into more powerful potential creative force of the team.

AS.Architecture-Studio passed ISO9001 certification in 2005. Meanwhile, thanks to the cooperative network concerning such fields as technology, finance management and special consultation both in France and the world, various projects have been realized, including hospitals, office buildings, company headquarters, hotels, residential buildings, sports complexes, commercial centers, schools (primary, middle and high schools and universities), research institutes, cultural projects (museums, theatres, opera houses) and urban planning projects.

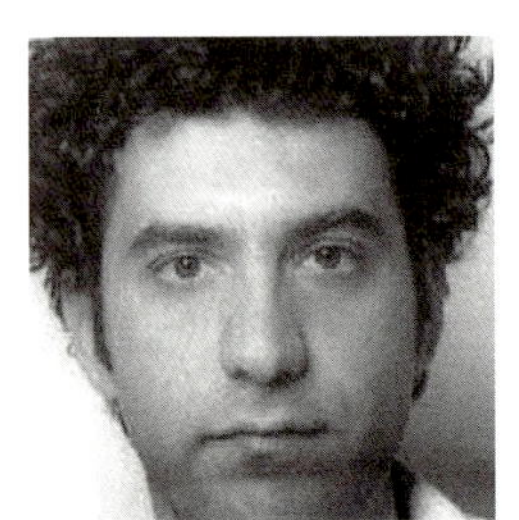

自左至右：

Martin Robain, Rodo Tisnado, Jean-François Bonne

Alain Bretagnolle, René-Henri Arnaud, Laurent- Marc Fischer

Marc Lehmann, Roueïda Ayache, Gaspard Joly

Marica Piot, Mariano Efròn, Amar Sabeh El Leil

太原市钟楼鼓楼地区修建性详细规划及建筑设计

DETAILED CONSTRUCTIVE PLAN AND DESIGN OF BELL TOWER AND DRUM TOWER DISTRICT IN TAIYUAN

项目地点：中国 · 山西　占地面积：57 ha

建筑设计：法国 AS 建筑工作室

LOCATION: Shanxi, China　SITE AREA: 57 ha

DESIGN CORPORATION: AS.Architecture-Studio

设计任务：本地区拟建设成为以商业、文化娱乐、服务为一体的城市传统历史文化街区，市级商业中心，将钟楼街建设成为传统商业格局的步行街。

项目位于山西省太原市钟楼鼓楼地区，用地范围内以商业金融和居住用地为主。原有的钟楼与古楼均被拆毁，现存文物古迹的状况令人担忧，另外，内部道路交通系统混乱。

“两点一带三区”是本方案的规划结构体系——“两点”即整个基地的发源点：钟楼、鼓楼；“一带”指的是绿色文化商业带；三区分别为北侧高层办公区、中心区以及南侧居住区。

采用对比的手法对待鼓楼与钟楼，我们建议原址原貌重建鼓楼，而钟楼采取原址新建的手法，建成具有时代内涵的新地标。

沿着太原市历史传统发展轴，一条连接钟楼鼓楼，有一系列以特色商业功能为依托的开放式广场组成的绿化景观步行带，贯穿整个基地，成为本方案城市空间布局的重要轴线，同时力图使其演变为一条具有历史人文内涵的绿色走廊，也为太原市增添一条亮丽的风景带。这条绿色人为走廊一直延伸到东西两侧的山脉，将自然生态景观引入城市，使得城市与自然融为一体。

Design task: the district is planned to be a traditional historical and cultural area that covers commerce, cultural entertainment and services. It'll be an urban commercial center of this city, while the street will be built as a walking street in traditional commercial pattern.

The project is located in Bell Tower and Drum Tower area of Taiyuan in Shanxi, which is mainly used as a residential as well as commmercial and financial area. The original Bell Tower and Drum Tower had been demolished. The ancient cultral relics and historic site are in danger, moreover the traffic system of inside road is disordered.

"Two points,one belt and three areas" is planning structural system

of the program – "two points" refers to the origin points of the entire base: Bell Tower and Drum Tower; "one belt" refers to the green cultural and commercial belt; "three areas" refers to the northern high-rise office area, central area and the southern residential area.

To treat Bell Tower and Drum Tower with contrastive techniques,we recommended reconstruction to the origiani Drum Tower and rebuilding to the Bell Tower to make it a new landmark of times connotation.

Along the historic development axis a green landscape walking belt connecting Drum Tower and Bell Tower and containing a series of open squares that are dependent on unique commercial functions runs through the whole base and becomes the key axis of the urban space layout in the program. In the meantime we strove to change it into a green corridor that carries historical and huaman cultural connotation to increase a bright scenery belt to the city. The green cultural corridor extends directly to the east and west mountains to invite the natural ecological landscape into the city, and makes it integrated to nature.

巴林麦纳麦国家大剧院

BAHRAIN AL-MANAMA NATIONAL GRAND THEATRE

项目地点：巴林　用地面积：10 370 m^2
建筑设计：法国 AS 建筑工作室

LOCATION: The Kingdom of Bahrain　SITE AREA: 10,370 m^2
DESIGN CORPORATION: AS.Architecture-Studio

剧院拥有 750 座的大剧院厅、150 座的小剧院厅和展览厅。巴林，"两个海洋间的国家"，带给剧院岛屿的景色，那里海天一线。在麦纳麦，地面和海洋一样也成水平线，这使景色给人一种延展的印象，在平面上看像没有边际一样。我们设计的建筑在海天之间，用巴林与海洋的结合来演绎一道水平线的风景。

国家大剧院同样显现出阿拉伯世界的文化特征。它的组织形式是阿拉伯宫殿式的，周围有一个中央空场，宫殿中有庭院，里面有大厅。在这个雄伟的空间的中心，就是大剧院的剧场，就像一件珍宝躺在首饰盒里一样。因为，在阿拉伯建筑风格中，水和光影是建筑整体的一部分。

国家大剧院将成为巴林的一个文化标志建筑，准备迎接各项国内国际活动。它将在区域间以及在全球范围显示王国的艺术和文化。它也将成为城市文化的窗口，成为艺术家们创造和排演的地方，而艺术家们在建筑里的创作活动又会给建筑本身带来艺术生活的气息。

The theatre has 750 Grand halls and 150 small theater halls and exhibition halls.

Bahrain, known as "the country between oceans" brings the theater with island landscape, where the sea and the sky are connected. In Al-Manama, the ground and sea are also at the same level, impressing people with the landscape extending and boundless when seen from the horizon. We placed the building between the sky and the sea to interprete the horizontal landscape with the combination of Bahrain and the ocean.

The National Grand Theatre also reveals the characteristics of the Arabian culture. Its organization is the Arabian palace form with a vacant auditoria in it. There's a courtyard in the palace and a hall in the courtyard. In center of this magnificent space is just the large playhouse of the theatre, like a treasure lying in a box. Because, in the Arabian architectural style, water and lighting is an integral part.

The Bahrain National Grand Theatre will be a cultural landmark in preparation for the national and international activities. It will show the arts and culture of the Kingdom both in the local area and around the globe. It will also be a window of urban culture, as a creational and rehearsal place for artists whose creative activities in the building will bring artistic life atmosphere to the construction itself.

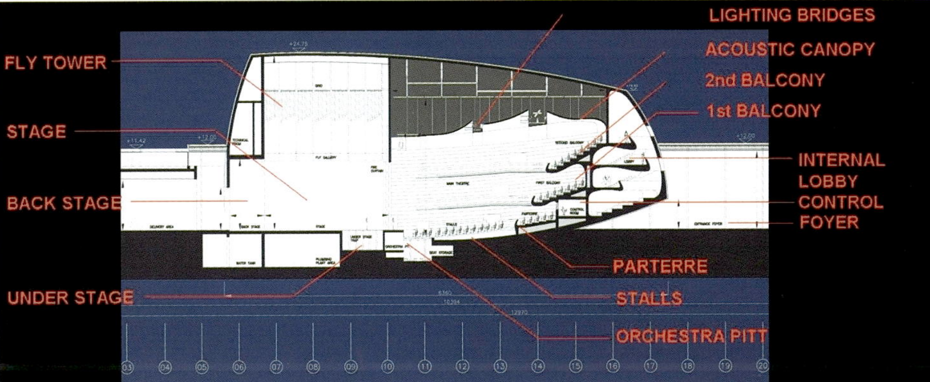

CA'ASI 艺术展览馆

CA'ASI ART GALLERY

项目地点：威尼斯　面积：400 m²
建筑设计：法国 AS 建筑工作室
摄影：法国 AS 建筑工作室

LOCATION:Venice　AREA: 400 m²
DESIGN CORPORATION:AS.Architecture-Studio
PHOTOGRAPHER:AS.Architecture-Studio

2009 年 11 月 21 日，法国 AS 建筑师工作室的 12 位合伙人在威尼斯举行 CA'ASI 艺术展览馆的揭幕仪式，其灵感来自法国 AS 建筑师工作室著名的公社结构。

威尼斯公国的独特焰式建筑诞生于狂热的商人、银行家和建筑师大行其道的年代，但也保留了大量的自然元素，其诞生对原先的建筑构成了极大的挑战，在人类的城市建筑史上留下了浓墨重彩的一笔。她体现着马可波罗的光辉形象，随着时光推移，来自商业和航海的财富孕育并吸引了各种才华横溢的人才到此发展：从提香（Titien）到丁托列托（Tintoret），从罗托（Lotto）到维洛奈思（Véronèse），从提埃坡罗（Tiepolo）到卡纳列托（Canaletto），从威尔第（Monteverdi）到维瓦尔第（Vivaldi），从帕拉弟奥（Palladio）到斯卡帕（Scarpa）！

归入新生的意大利版图的前道奇共和国直到 20 世纪才恢复其原先的艺术使命：首先通过影展（世界上第一个电影展）邀请电影人，然后举办当代艺术和建筑双年展。依靠新的文学和艺术事业的资助者（从佩吉 - 古根海姆到弗朗索瓦 - 皮诺）及其基金会，建筑大师安藤忠雄和伦佐·皮亚诺挥笔重建了皇宫和历史古迹，再次丰富了威尼斯的文化遗产。

以马可波罗为榜样，法国 AS 建筑师工作室踏上了寻找建筑和城市表达及试验的新大陆之旅，足迹遍布中东和近东（土耳其、亚美尼亚、黎巴嫩、约旦、叙利亚、阿联酋、卡塔尔、巴林、阿曼、沙特阿拉伯）以及亚洲（阿富汗、中国、日本）。面对全球经济一体化造成的城市群膨胀以及气候失常，新世纪伊始，建筑界的思辨变得极其重要。法国 AS 建筑师工作室的基本原则是宣扬矛盾的对话和文化现实的共享，以催生建筑的活力。基于这一原则，法国 AS 建筑师工作室希望围绕当代建筑构造一个集会场所。

之所以选择威尼斯，一是因为她有着颇为神秘的背景：不断上升的海平面让这座城市变得悲情而易逝；二是因为她是吸引着全世界的创作人和游客的伟大的文化殿堂。集中在这里的不再是香料或其他物质财富，而是我们的世界十分需要的思想、理念和其他非物质的价值。

里亚托桥不远处马可波罗故居旁的 CA'ASI 艺术展览馆是一座三层高的美丽的中世纪公馆，矗立在卡那雷周的中心。

艺术展览馆占地 400 m²，以现实精神进行了合议式的翻新：一楼是展览馆，面对坎皮耶洛圣母新堂广场，二楼是宽敞的会客室。在细长的赭红色隔墙后，有几间带有卫生间、厨房和附属设备的房间，以接待逗留于此的宾客。其中最幸运的人还将享受到位于顶楼的两个房间的小卧室。

作为对当代建筑爱好者开放的俱乐部，CA'ASI 艺术展览馆将成为城市规划、建筑和艺术反思的普通试验室，便于举办展览，使更多的人可以接触到职业辩论。

建筑师、城市规划师、风景画家、设计师、造型艺术家、作家或哲学家们不仅可以在此找到隔墙、屏幕和其他分享他们的思想和经验的讲坛，还可以找到发表期间的宿营地。

On November 21, 2009, 12 partners of AS (Architecture-Studio) held the opening ceremony of CA'ASI art gallery in Venice. CA'ASI. It was inspired by the famous commune structure of French AS Architect Studio.

The unique flame construction of Venice was born in the era when the fanatical businessmen, bankers and architects were quite popular, but many of the natural elements remained in the construction. Its emerging was a big challenge to the buildings existing then and was a monument in architectural history. It embodies the splendid image of Marco Polo. As time went by more and more talents were produced or attracted here by the wealth from bussinesss or navigation,including Titien, Tintoret, Lotto, Véronèse, Tiepolo, Canaletto, Monteverdi, Vivaldi, Palladio and Scarpa!

The former Doge Republic, which has been included in the new Italian territory,had not returned to its art mission untill 20th century: at first it invited film people through the world first film exhibition, and then it held the Biennale of contemporary artistic and architectural exhibition. Dependent on the new subsidizers(from Peggy-Guggenheim to Froncois-Pinault) and foundations to literature and art, the architectural masters Ando Tadao and Renzo Piano magnificently rebuilt the royal palace and the ancient historical sites and enriched the Venetian cultural heritage once more.

Following the example of Marco Polo the AS. Architects Studio started the exploration of urban architectural expressions and trials. Their steps spread the Middle East and the Near East,including Turkey, Armenia, Lebanon, Jordan, Syria, United Arab Emirates, Qatar, Bahrain, Oman and Saudi Arabia as well as Asia, including Afghanistan, China, Japan. Facing the mass urban inflation as well as the deterioration of the climate condition resulted from Economic Integration of the globe, intellectual enquiries of architectural world has become extremely important at the start of the new century. The basic philosophy of AS is to advocate the coexistence of the paradoxical dialogue and the cultural reality to facilcitate architectural vitality. Therefore AS aimed to create a gathering place with the theme of contemporary architecture. There are two reasons for the selection of Venice:one is its mysterious background, i.e. the continuously rising horizon makes the city in pathos and easy to pass away;the other is that it's a great cultural sanctuary to attract the world creaters and tourists. It's no longer the spices or other material wealth that gather here but the ideas, eidios and the immaterial values that are essential to our planet.

The CA'ASI art gallery which is near the former residence of Marco Polo and not far from Rialto Bridge is a beautiful 3-storey Medieval Manor standing at the center of Cannaregio. The gallery occupies an area of 400 square meters and was rennovated in collegial style with realistic spirit. The first floor is the exhibition hall in face of Campiello Holy Virgin New Palace Square. The second floor is the spacious reception room. Behind the long and narrow soil-red partition wall are several rooms equipped with bathrooms, kitchens and other accessary equipment to accomadate guests sojourning here, among whom the luckiest ones will can enjoy the small bedrooms of the attic.

As a club that is open for contemporary architectural enthusiasts CA'ASI art gallery will become the ordinary laboratory for urban planning, architecture and art reflection. It can also be convinient for exhibition and make professional debate accesssible to more people. Architects, urban planners, landscape painters, designers, modeling artists, writers and philosopher can not only find the partition wall, the screen and other platforms where they can share their ideas and experience but also the camping site during the delivering.

法国广播大厦改建工程

FRENCH BROADCAST BUILDING RENOVATION

项目地点：法国　面积：110 000 m^2
建筑设计：法国 AS 建筑工作室

LOCATION: France　AREA: 110,000 m^2
DESIGN CORPORATION: AS.Architecture-Studio

法国 AS 建筑工作室成功成为法国广播大厦改建工程的中标设计方。该项目预算为 2.4 亿欧元，即 24 亿人民币。

此项目无疑是法国现今最大的建筑工程，在保留其原有风格的基础上进行翻新改建，不仅给文化传播注入了新的活力，同时亦象征了法国广播台对公众开放自由民主的精神。从今年直到 2012 年，巴黎翘首期待多年的大音乐厅将落成，与同时汇聚了博物馆、现代展览厅、餐厅、咖啡休闲厅、商场等的荣誉大厅交相辉映。

法国 AS 建筑工作室保留了 Henry Bernard 设计的法国广播台原有建筑的标志性形象，在此基础上重新定义了城市与建筑的关系，并对室内作了一系列的创意设计，使得其使用价值以及运作效率更适合当今广播传播的需要。

建筑的主入口重新装饰后，入口前的广场面向一直延伸到塞纳河的肯尼迪总统大街。地下车库的创意设计，实现了法国广播台围绕于花园中的规划，并大大提升了人员进出的便利性。

建立于大厦正中心的中庭，已经成为了建筑物新的核心，它解决了广播大厦内部定位与长廊人员流动的问题。以罗盘形排列的玻璃制长廊将中庭及环形建筑物内侧有序地连接起来，整体设计为改建的大厦提供了纯自然的光线、无障碍的疏通以及一个开放的中心花园。中庭内镶嵌着感光板的彩玻璃，也同时被赋予了一种能量循环再生的功能。

AS successively won the bidding to design for the French Broadcast Building Renovation. The budget of the project is 204 million euros, i.e.2400 million RMB. It's undoubtedly the grandest project in France so far. The renovation on the basis of its origianal style not only injects new blood to culture but also symbolizes liberal and democratic spirit carried by the building. By 2012 the odeum in public expection will be completed in Paris. It will add radiance and beauty to the Glory Hall which includes the museum, the modern exhibition hall, the resturant, the coffee bar, and the market etc.

AS Architecture Studio has remained its origiani landmark image designed by Henry Bernard. They redefined the relation between the city and the architecture and added a series of new ideas to the interior design to better meet the needs of the modern broadcasting in use value and working efficiency .

After the redecoration of the main entrance, the square faces to the President Kennedy Avenue by Seine. The creative design of the underground garage realized the plan of the building being surrounded by the garden, and more practically it enhanced the convinience for population flow.

The patio built in the center has been the new core of the construction which solves the problems of interior position and the corridor population moving. The glass corridor like compass orderly connects the patio and the inner side of the building. The overall design provides pure natural light, good ventilation and an open central garden for the building. Photographic color glass set in the patio has the function of energy regeneration.

金桥科技园

JINQIAO OFFICE PARK, SHANGHAI

项目地点：中国·上海　用地面积：41 300 m^2
建筑设计：法国 AS 建筑工作室

LOCATION: Shanghai, China　SITE AREA: 41,300 m^2
DESIGN CORPORATION: AS.Architecture-Studio

基本构思是要体现一种前后的连续性和互补性。整个区域的核心部分是中间空阔的湖面，建筑则是沿湖面的周围布置，创造出一种和谐的空间。

建筑的几何外形特色比较鲜明，建筑的外观呈果壳形，中间被多个层面割断，从而避免了建筑受到过多光线和噪音的干扰。建筑的布置方式尊重了园区的总体规划，提供了一个有活力的、现代的、柔和的、与环境和谐共处的空间。

建筑实际上成了被光线穿透的岩石。各层面之间互相穿插、互相层叠，有助于组织各种形式的立面及工作空间。

Basic concept of our design is to show a kind of continuity and complementarity between the past and future. Core part of this area is the broad lake in the center. While architecture is arranged around the lake in order to create a harmonious space.

Geometry shape of the architecture bears distinct characteristics. Its appearance is shell-shaped, and is cut off by several levels, which avoids interference of excessive light and noises. Arrangement of architecture respects overall planning of the park zone, and provides a dynamic, modern, soft space that is harmonious with the environment.

Architecture actually becomes rock that is penetrated by the ray. Various levels interlude and cascade to each other, which contributes to organize various façade and work spaces.

世博中心室内设计

THE EXPO CENTER INTERIOR DESIGN

项目地点：中国 · 上海　面积：45 020 m^2
建筑设计：法国 AS 建筑工作室
摄影：法国 AS 建筑工作室

LOCATION: Shanghai, China　AREA: 45,020 m^2
DESIGN CORPORATION: AS.Architecture-Studio
PHOTOGRAPHER: AS.Architecture-Studio

建筑从总体来看就如同玻璃盒中的巨蛋，所以在这个方案的设计中我们把会议大厅整个体量用木板条包覆，球体的表面分为两层，并在其背后设置了灯光，使其成为整个内部空间的中心点，也成为整个建筑体量内的一个发光点。行走在其内部的各层空间可透过球体上的空洞享受到穿透进大堂的自然光线，同时也自然地形成了一个舒适的休息空间。

会议大厅作为整个建筑体量的中心点，同时也作为几大功能区中最重要的场所，我们在整个会议大厅的设计中注入了更为细致的元素，利用乐器中皮膜与声音的共振原理我们以表皮的肌理效果作为概念，运用在墙面的设计，同时结合灯光及声学的各项要求，巧妙地运用了不同的材质和光效营造出一个可同时满足会议和演出的功能空间。

整体空间以金属灰为主，风格完全智能化，突出了科技焦点，设计以塑造科技、智能形象为基点，完整地体现了一个超越时空的会议功能区域，生成新的设计理念和因素，进而辐射开去，成为一个聚焦点。

Seen as a whole it's just like a huge egg in the glass box, so in the design of this peoject we made the whole conference hall wrapped with the wooden laths.The surface of the egg including two layers.The light at the back side makes it a luminous point of the whole bulding. Walking in the different layers of the inside space you can enjoy the natural light penetrating into the hall which also naturally provides a comfortable rest area.

As the central point of the whole construction and the most important functional area, it was injected with more delicate elements. Taking use of the resonance of skin and sound in the musical instrument we applied the notion of the texture effect of the surface to the wall surface design, and combining the requirements of optics and the acoustics ingeniously used different materials and lighting effect to create a functional space to meet the needs of conferences and shows.

The overall space uses the metalic gray as the main color line. The style is totally intellectualized and highlights the technological point.The cardinal point of the design is to create a high-tech and intellectualized image. It completely shows a conference area that transcends space and time , which is the new design concept and element that further radiates and finaly becomes a new focal point.

传播媒介图书馆、艺术研究与试映电影院

COMMUNICATIONAL MEDIUM LIBRARY, ARTISTIC RESEARCH AND PREVIEW CINEMA

项目地点：法国 · 圣马洛　**面积**：6000 m^2

建筑设计：法国 AS 建筑工作室

LOCATION: Saint Malo, France　AREA: 6000 m^2

DESIGN CORPORATION: AS.Architecture-Studio

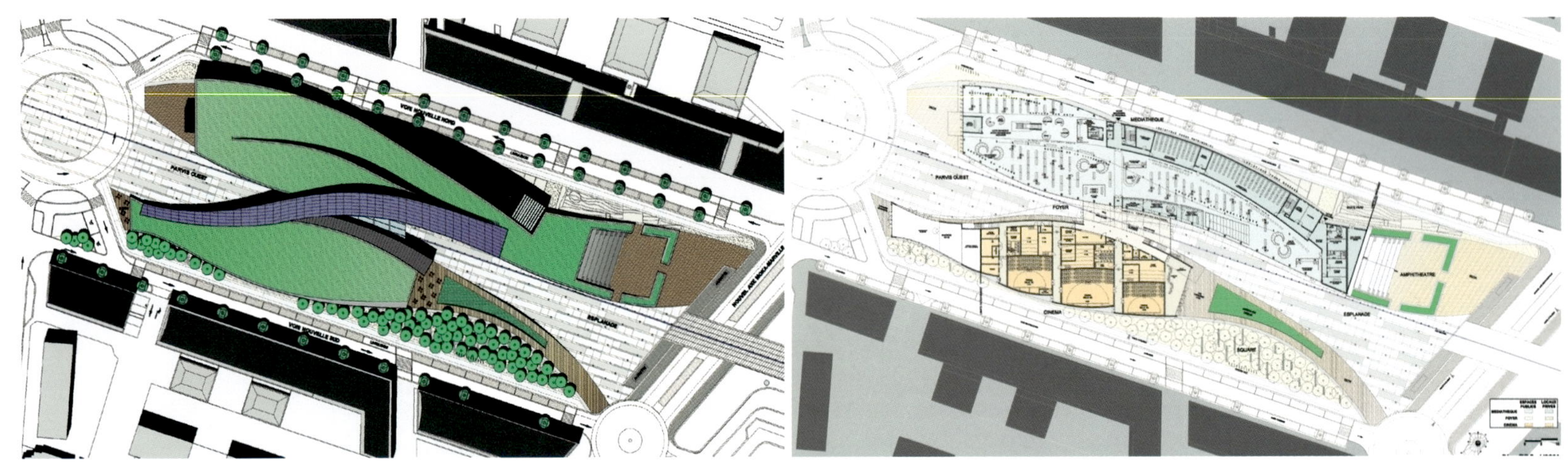

Coupe A 1/200°

此项目包括了举办城市圣马洛为实现与新的文化设施相结合的一个传播媒介图书馆、三个艺术研究与试映电影院、一座公寓、一家餐厅以及一个咖啡馆。该项目的总面积将为 6000 m^2，预计费用为 1500 万欧元。此项工程预计在 2010 年年底开始并于 2012 年竣工。

两座被设计为波浪形的建筑物的正前方便是火车站广场，而它们浓厚的建筑艺术气息更是突出了这座城市一个新的大门。圣马洛城市的中心轴，一个象征着朝向大海的开放型城市的标志，贯穿于连接传播媒介图书馆和电影院之间的玻璃通道上。这两种标志性的文化建筑的屋顶都将被绿化植被覆盖，并且由一条长条形太阳能光板架空其上，而地热系统设备将被安置在地下室。该项目不仅持有法国 NF 的第三产业建筑高环境质量可持续发展认证，还有 THPE ENR 的使用标志。这些都是圣马洛城市一个全新的文化标志 。

This project includes a communicational medium library that shows the perfect integration of the host city Saint Malo and its cultural facilities,three art and preview cinemas, a mansion, a resturant and a coffee bar. It occupies totally area of 6000 m² with the budget 15million euros. It's expected to start in the end of 2010 and be completed in 2012.

In front of the two wave-shape buildings is just the railway station square. The thick architectural flavor highlights the new gate of the city. The central axis of Saint Malo, a symbol of an open city toward the sea, runs through the glass passageway connecting the library and the cinema. The roofs of these two landmark buildings will be covered by green vegetation and a strip of solar power light panel set up on it, while the underground heat system will be set in the basement. The project not only got the high environmental quality and sustainable development certificate in tertiary industry architecture, but also had the logo of THPE ENR. They are all the brand-new cultural symbols of Saint Malo.

BLVD. INTER-NATIONAL INC.

毕路德国际

国际企业

毕路德国际（Blvd. International Inc.）是由几位有志于中国建筑市场的加拿大建筑师、室内设计师于2001年10月在加拿大安省注册的。毕路德国际的主要设计师在加拿大继续从业活动的同时，以毕路德国际的形式在中国进行建筑设计、规划设计、室内设计和景观设计。

国内三地

（加拿大）毕路德国际在中国的载体是北京毕路德建筑顾问有限公司，于2001年12月在北京注册。2002年北京毕路德建筑顾问有限公司深圳分公司在深圳注册，目前在深圳办公室工作的设计人员有近120人，是（加拿大）毕路德国际在中国的主要设计基地。2005年北京毕路德建筑顾问有限公司上海分公司在上海注册。

整合资源

运用先进的网络通信手段，毕路德公司实现了多伦多办公室、深圳办公室、北京办公室的联合设计，使得几地的国际、国内设计师可以共同交流，多方合作，完成同一设计作品。充分的人力资源和信息资源整合，使毕路德公司在保持高效率的同时，具有更高的设计质量。

竞争优势

毕路德公司具有从规划、建筑、室内到景观设计的完整团队，致力于为业主创造一体化的建筑环境。设计范围相互渗透，从大环境到微环境、从室外到室内的完整设计，提高了完成项目的综合品质。

重视策划、注重创意

公司倡导设计师们相互交流，思想碰撞，强调设计原创性，为客户提供更好的综合解决方案，确保项目的独特价值观。项目的创意不能脱离市场的需求，最大限度地满足市场的需求和适当地引导消费同样重要，用设计的眼光看待策划，根据用地情况寻找市场切入点。

Blvd. International Inc. was cooperated in Ontario in Canada at Oct. 2001. It was formed by several Canadian architects and interior designers who have the common interest in the Chinese market. By continuing practicing in Canada, Blvd started business in architecture design, urban planning, interior design and landscape design in China. By Dec. 2001 Beijing Blvd Architecture Consultant Co., Ltd. was registered in Beijing. And in 2002 Beijing Blvd Construction consultant Co., Ltd. registered a branch in Shenzhen. There are almost 110 designers in Shenzhen office now, and it is the design center for Blvd in China. In 2005 Beijing Blvd Architecture Consultant Co., Ltd. registered another branch in Shanghai.

By advanced communication technology, Blvd realized the cooperation design between Toronto, Shenzhen and Beijing offices. The designers located in different offices can work on the same project at the same time. With the advantages of a wider human resources and information resources pool, Blvd has enhanced its efficiency and design quality.

Blvd offers full scope of design services ranging from architecture design, interior design, planning and landscape design. As the result Blvd can create a comprehensive built environment for the client. From macro to micro, from outside to inside, the effort of merging the different scopes of design improves the quality of the design product. The designers specialized in different fields, work and communicate together, create better understanding and solution to the project.

Innovation is what Blvd can contribute to the project. For the specified client, project and market, we deliver a custom made solution. From the function layout to the massing form, creativeness is the soul of design procedure. Every designer in the team is putting all their effort on creating a "one of a kind" design solution. Putting effort on the"market" , to create built environment for the final purchaser/user, is also important for Blvd design team. To help the clients on their successfulness on the market also helps blvd to win more clients.

As for today, Blvd has finished many successful urban planning, architecture design, interior design and landscape design projects in China, and has won many international and domestic awards. blvd has become a rising star in the design industry of China.

Thanks to all the clients for the past seven yeas to give us support, understanding and most importantly trust.

遂宁河东新区滨江景观带规划

RIVERSIDE LANDSCAPE PLANNING IN THE NEW AREA IN SUINING EAST RIVER

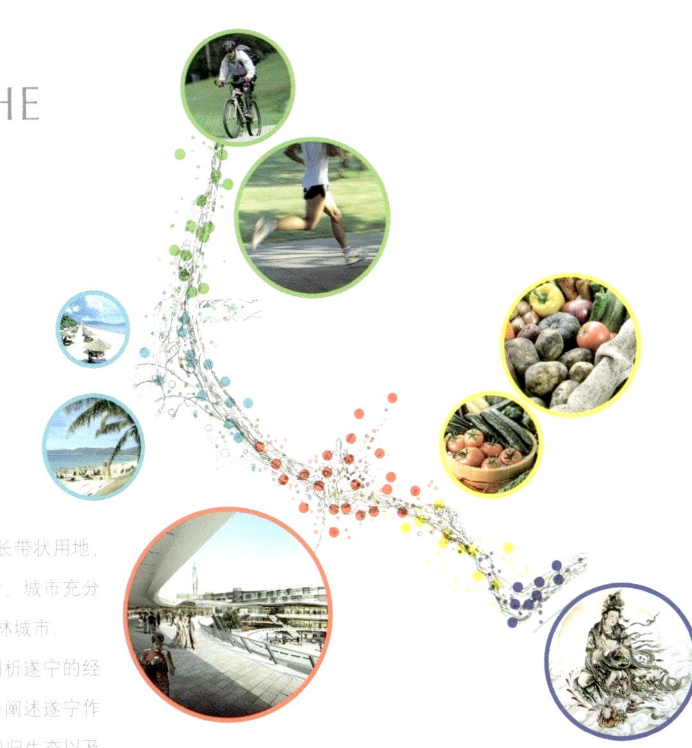

项目地点：中国 · 四川　规划面积：370 000 m^2

建筑设计：深圳毕路德建筑顾问有限公司

建筑师：杜昀

LOCATION: Sichuan, China　PLANNING AREA: 370,000 m^2

DESIGN CORPORATION: Shenzhen Blvd Architectural Consultant Co., Ltd.

ARCHITECT: Du Yun

项目基地位于遂宁市河东新区西南面，紧邻涪江，与老城区隔江相望，是联系新老城区的纽带。基地为一条狭长带状用地，东临建设中的河东新区，西面为观音湖，基地与水域之间有高出基地约 3 m 的防洪堤以及大量的原生河道滩涂。城市充分利用“两面临山水、中间一座城”的山水特点，突出生态园林特色，将遂宁建成一个现代化的国际旅游山水园林城市。汲取“舞动”的灵魂含义，使用“张扬”的设计手法，去寻求文化、商业、空间、生态的突破与回归。深刻剖析遂宁的经济态势与历史文化，通过设计为项目与城市之间找到了一个嫁接兴奋点。立足于项目的优势，借用新生地带去阐述遂宁作为一个千年城市的“城市复兴”。作为城市复兴计划的一部分，我们从设计内容上通过融合商业、传承文化、回归生态以及创造体验空间与旅游度假的多功能复合型城市模式，把遂宁五彩滨江景观带打造成一处现代、时尚且具有品牌效应的旅游购物天堂，使之成为遂宁市新的城市中心、休闲体验中心、购物旅游中心、滨水观光中心，促进遂宁经济的发展。从设计形式上打破传统的规划模式，传承东方园林中“象外之象、景外之景”的高度融合意境，运用流动的线条与聚集的小圆点所产生的韵律，创造一处建筑、景观、自然协调统一，具有极强艺术感染力的景观体验场所。

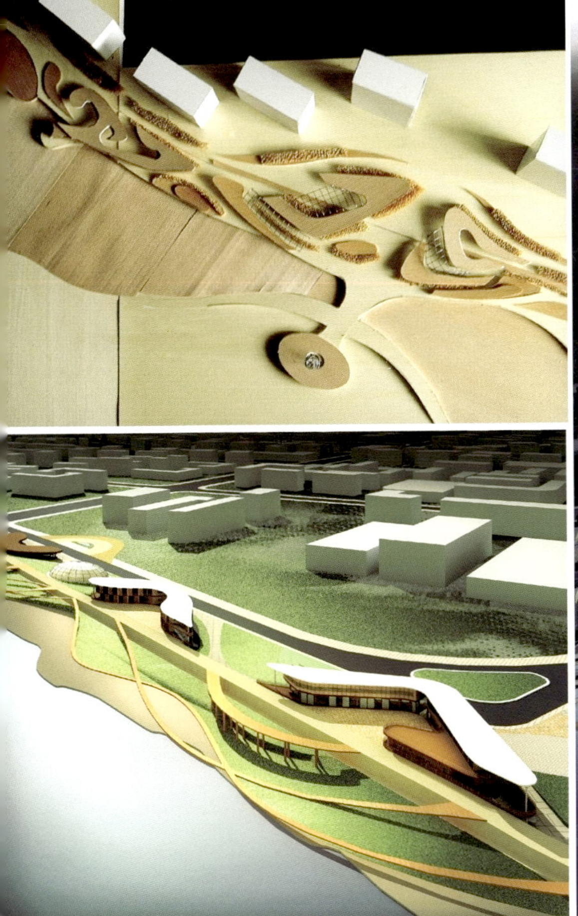

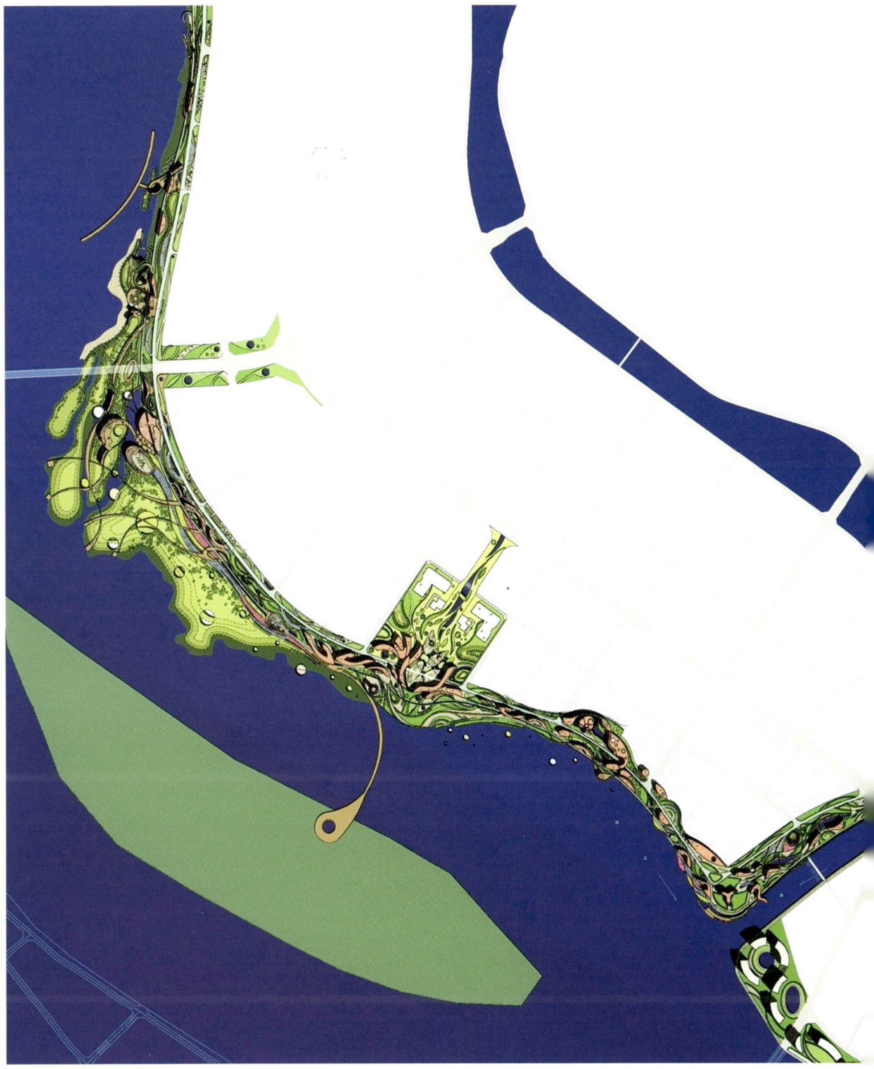

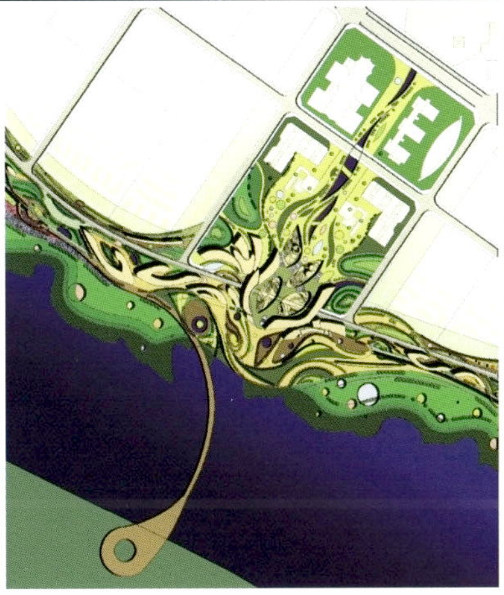

The project base is located in the southwest of New Area in Suining East River, close to the Fujiang River. and across the river is the old town. It is the bond beween the old and the new cities. The base is a long and narrow strap-shaped site with the east to the East River Area in development,the west to Guanyin Lake. Between the waters and the base is the floodwall about 3 meters high and the large area of native river beach. We took full advantage of the landscape, " a city between mountains and waters " to highlight the eco-garden features and built Suining into a modern international tourist Garden City.

We absorb the connotation of the "dancing soul", using a "publisizing" design technique to seek for the breakthrough and recurrence of culture, commerce, space and ecology. Through deep analysis of the ecology and history of Suining, the design gives a thrilling grafting point between the city and the project. Based on the preponderance, the project interprets the"city renaissance" as a thousand-year old city by the newly born zone. As part of urban regeneration scheme, we created the riverside landscape zone into a modern, stylish and

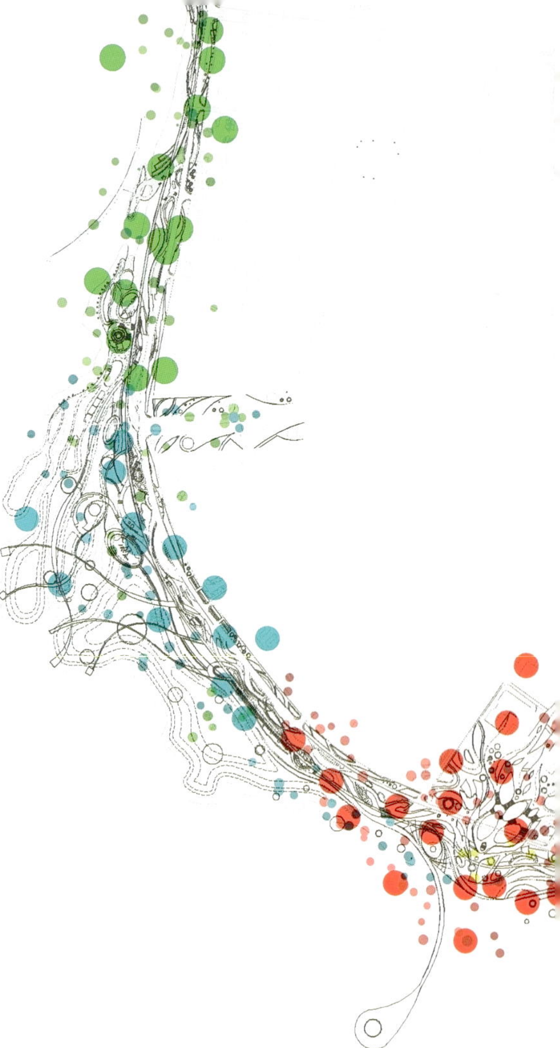

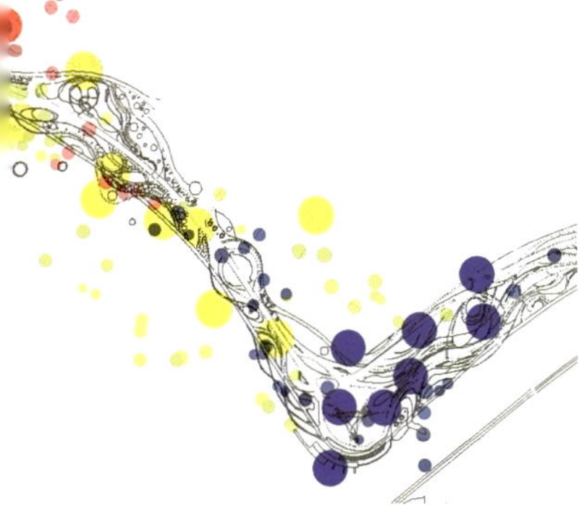

tourist shopping paradise of brand effect with the design content of commercial integration, cultural heritage, ecological regression and the creation of space and vacation experience of a compound & multi-functional city model, so that it will become the new center of Suining for leisure experience, shopping, riverside site seeing to promote economcal development of Suining. In design form we broke the traditional planning style and inherited the artistic conception of east garden" to see one scene outside the other". It creates a sight seeing area which is an harmonious integration of architecture, landscape and nature through the rhythms produced by the flowing lines and the little dots.

AGENCE C&P ARCHITECT-URE

法国C&P(喜邦)建筑设计公司

法国C&P（喜邦）建筑设计公司是由法国国家建筑师R.Chareyre及J.L.Pagnier合作创立的一家专门从事工业与民用建筑设计、规划及工程咨询的专业设计公司，总部设于法国第二大城市里昂市。多年以来C&P（喜邦）设计公司设计了诸多工程项目，涉及居住、医院、学校、办公、商业及文化体育等各种类型的建筑。C&P（喜邦）公司在设计中十分注重将西方先进的设计理念与中国深厚的文化底蕴相结合，同时也积极探索新技术、新工艺、新材料的应用。丰富的设计实践和经验形成了C&P（喜邦）独特的设计风格，公司业务范围不断扩展。

随着改革开放力度逐年加大，中国经济与社会有了巨大的变化与发展，尤其中国加入世界贸易组织后，吸引了世界各国建筑师的目光。法国C&P（喜邦）建筑设计公司自1995年开始介入中国建筑市场，目前已在中国北京、厦门、天津及南京设立了分支机构，并于2009年设立美国分部（明尼苏达州）。公司在积极参与中国发展建设的同时与中国相关建筑设计单位建立了长期稳定的合作关系，充分发挥各自的技术优势，利用法国C&P（喜邦）建筑设计公司先进的建筑设计理念、丰富的建筑设计经验及国际化的建筑技术能力，结合中国国内建筑设计单位所具备的满足中国建筑市场的工程设计、工程承包及工程监理方面雄厚的设计力量以及对中国建筑市场的深刻了解，开拓中国建筑大市场。目前已在中国设计了诸多工程项目，在中国建筑设计市场的国际招投标中多次荣获一等奖，取得中国同行与业主的广泛好评，并获2009年中国城市规划与建筑设计行业"最佳设计机构"称号。

在新的世纪里，法国C&P（喜邦）建筑设计公司必将努力开拓，积极进取，充分发挥中西方文化交流之桥梁作用，为中国的进步与发展贡献力量。

Agence C&P Architecture headquartered in Lyon, the second largest city of France, is a professional design company co-founded by French architects R. Chareyre and J.L.Pagnier,specializing in industrial and civil design as well as planning and engineering consulting. Over the years, C&P designed many projects involving housing, hospitals, schools, office, commercial,cultural, sports and other types of buildings. C&P attaches great importance to the combination of advanced western design ideas and Chinese cultural heritage. In addition it actively explores the use of new techniques, and materials. Rich experience in design practice leaves the unique design style of C&P, which leads to the continous expansion of the scope of our business.

With the enhancement of the reform and opening up policy year after year, China has got tremendous change both in economy and society, particularly after its access to the WTO,it has drawn the attention of architects around the world. C&P has been involved in China's construction market since 1995 and established branches in Beijing, Xiamen,Tianjin and Nanjing. In 2009 it set offic in the U.S (Minnesota). C&P actively engages itself in China's development and construction, and on the other it has established long-term cooperative relations with related building design units in China. They fully take their technological advantages and cooperate to develop the huge Chinese construction market with the advanced architectural design concepts, rich experience and international technical capacity of C&P and the powerful engineering design, contracting, supervision and the deep understanding of Chinese construction market of the domestic building design units. Currently C&P has designed many projects in China. It has won some first prizes in the international biddings for architectural design market in China and has been widely acclaimed by Chinese counterparts and the owners. In 2009 it got the title as one of "the top design agencies" in Chinese Urban Planning and Design industry.

In the new century, the Agence C&P Architecture will strive to make more progress. It will fully play its role as a media between Chinese and western cultural exchange, and contribute to China's development.

上海沿海·丽水馨庭二期

THE SECOND STAGE OF LISHUI XIN TING PROJECT, ALONG THE COAST IN SHANGHAI

项目地点：中国·上海　总占地面积：150 503.81 m^2　建筑面积：240 497.05 m^2
建筑设计：法国 C&P(喜邦)建筑设计公司
建筑师：朱劲松，黎明

LOCATION: Shanghai, China　THE TOTAL AREA: 150,503.81 m^2
BUILDING AREA: 240,497.05 m^2
DESIGN CORPORATION: Agence C&P Architecture
ARCHITECTS: Zhu Jinsong, Li Ming

项目位于上海市区西南的闵行区与松江区交界处的新闵经济园西侧，是上海成熟的别墅区之一。基地位于明中路北侧，东为晶苑四季御庭项目，北至砖新河，西至一号桥河，小茜浦泾河呈南北走向贯穿基地中央。

二期规划用地面积为 150 503.81 m^2，地上建筑面积为 160 311.97 m^2，地下为 80 185.08 m^2，户型为联排别墅、双拼别墅和小高层公寓式住宅。

本案充分利用丰富的天然水系资源，建设自然近人的居住环境。同时注重邻里交往空间的设计，为改善当今都市人群冷漠的交往现状做出一点努力。

建筑设计：新中式合院式 townhouse，独具现代美学风格的中式合院型别墅。丽水馨庭产品在传承中华院落精髓的基础上，融合西方现代建筑思想与艺术装饰风格。一个个小单元的“四合院”式联排，一改传统 townhouse 兵营式排布的单调布局，创造出重庭叠院的人居韵味，街、弄、院、园、宅、庭六维空间，私家庭院、中央景观带、邻里共享空间、会所空间、生态空间层层递进，为主人带来强烈的归属感与领域感，更富生活情趣。

合院别墅，私家车库，匹配主人优越生活，几户家庭围合

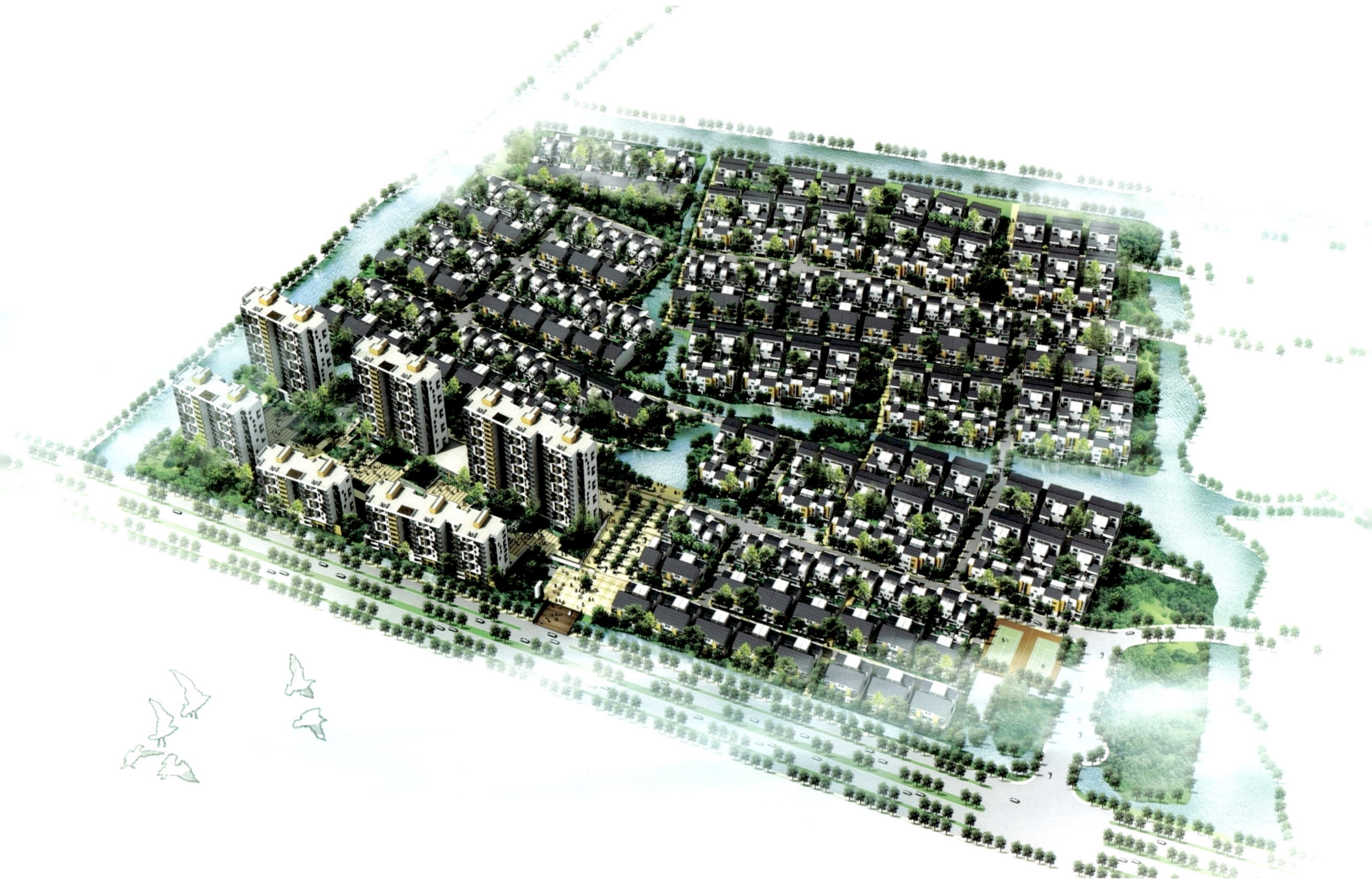

共享中庭，让每一座合院都是一个社交圈，每一栋别墅都是一个私享王国。无论是大院社交、会所社交，还是庭院沉思，房间休憩，均营造了兼具开放性与私密性的最佳动静分离。组团围合式的联体别墅，新颖独特，扩展庭院视野，让人们回归邻里温情。

高层住宅区院落空间绿地向各个组团内部渗透。中心绿地、体育活动场地、休闲步道、组团绿地、自家院落等多层次网络状的绿化景观系统使每户住宅都能享受到优美的绿化环境，即景观的均好性。借助错落的矮墙和不同的地面铺装，将大院落分割成具有不同功能、富有情趣、宜人的活动空间。

The project is located at the junction of Minhang District and Songjiang District in the west of the new Min Park, southwest of Shanghai. It's one of the mature villa area in Shanghai. The base is in the north of the Ming Zhong Road, to the Four-Season Crystal Royal Court project in the east, north to the Zhuan Xin River, and the west to the NO. 1 Bridge Creek,and from north to south it's inserted by Xiao Xi Pu Jing River. The second planning land area: 150,503.81m^2,with overground area 160,311.97 m^2 and underground: 80,185.08

m^2, the household types are townhouses, semi-detached villas and small high-rise apartments.

We took full advantage of rich natural water resources to construct living environment close to nature. At the same time we focuse on the neighborhood association space design to improve current status of indifferent socializing of urban people.

Architectural Design: as a new townhouse and a unique modern aesthetic villa of Chinese courtyard , Lishui Xin Ting not only inherited the essence of traditional Chinese compound but also integrated the modern Western design ideas and Art Deco style. The small units of "courtyard" type houses creatively changed the traditional townhouse monotonous barracks-style layout and built home charm of overlapping courts and yards. The six-dimensional space of street, alley, yard, park, home and court,the private courtyard, the central landscape area, the neighborhood space, the club space as well as the ecological space, bring the owners strong sense of belonging and domain layer by layer, adding more pleasure to life.

The courtyard villa and the private garage well match the occupants superior life style; several houses share one enclosed courtyard which resembles a social circle and each villa is a private Kingdom. Whether the social compound and the social club, or the meditation garden and the sitting room display the best separations between dynamic and tranquility that carry both privacy and openness. The uniquely structured joint villas extend the courtyard view, allowing people to return to the neighborhood warmth.

The greens of the high-rise residential courtyard penetrates to the internal space of each block. Every household can enjoy the beautiful multi-layered green landscape networks of the central green space, sports venues, leisure trails, block greens, home-like compound, etc. that is the all-good quality of the landscape. The scattered low walls and different ground pavements divide the yard into different functional areas full of fun and pleasure.

COBBLE-STONE DESIGN CANADA/ COBBLSTONE URBANISTS + ARCHITECTS WORKSHOP SHANGHAI

加拿大考斯顿设计
上海考斯顿建筑规划设计咨询有限公司

今天，信息技术日新月异并影响着人类环境的各个方面。作为营建人造景观的建筑及其相关领域毫无疑问地应当有所回应。考斯顿清晰地以城市主义为价值导则，以对城市生活和事件的深刻关注为设计起点去完成规划、城市设计、建筑和景观的综合解答。以城市专家、建筑师、景观建筑师和研究人员为研发设计组合的考斯顿团队致力于在信息化、多元化的今天，提供超越图纸空间的全方位的设计服务：对前期市场研发产品定位、中期产品设计、后期适时应对市场变化的动态调整。孜孜不倦寻求设计的依据和价值，并努力斡旋于平衡人的需求和所有的外部压力。重实效、敏锐、多元化、创造性和多选择的设计服务，以及对文化和生态环境关注将持续贡献我们的努力于当代的人文景观。我们的专业服务涵盖商业办公、星级酒店、酒店式公寓、大型城市混合体、居住区、工业园区、城市街区的策划、规划、建筑设计、景观和室内设计全过程。

Nowadays, information and technology are changing with each passing day and impacting every aspect of the human environment. As an artificial scene, architecture associated with other design disciplinary should undoubtedly respond to this change. Cobblestone comprehensively meets the issues of planning, urban design, architecture and landscape with clearly understanding for urban life and urban events. The Cobblestone team combined with urbanists, architects, landscape architects and annalists will be devoted in providing overall services beyond limited the paper space, which are pre-design study, project design, and post-design revision. We mediate between mankind and all external forces to create a balance. Pragmatic, sensitive, plural, innovative and optional design excellence to our culture and mankind environment would contribute to the contemporary human landscape. Our services include commercial, office, hotel, service apartment, urban complex, residential quarter, industrial park and urban block with programming, planning, architectural design, landscaping and interior design.

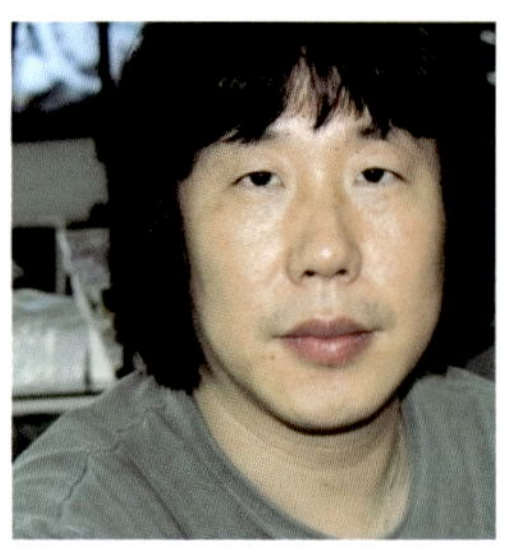

公司合伙人及主要设计师

- 刘廷杰　Tingjie (Peter) Liu
设计总监

加拿大马尼托巴大学专业建筑硕士
加拿大皇家建筑学会会士
中国注册建筑师、高级建筑师
20年中加两地经验
曾任职于加拿大温哥华IBI Group，温哥华 DGBK事务所，
温尼泊Raymond Wan事务所，深圳大学设计院等

获奖作品

- 2000.4 入围OTIS国际建筑学校住宅设计竞赛
- 2001.6 入围加拿大不列巅省建筑学会青年建筑师作品展
- 2004 主创设计Multi-Tenant Facility of Smart Park in University of Manitoba，获2004 年度加拿大总督奖提名（Ray Wan Architects）
- 2006.12 中国创新90中小套型住宅设计竞赛命题类鼓励奖
- 2008.12 黄石“东风路地块”中国人居典范建筑规划设计方案竞赛金奖
- 2009.10 上海松江“新弘国际”获建筑学会中国人居奖综合设计大奖
- 2009.10 上海新江湾城“05-2地块”获建筑学会中国人居奖规划和方案设计大奖

新弘国际社区，商业街设计

XINHONG INTERNATIONAL COMMUNITY, SHOPPING STREET DESIGN

项目地点：中国 · 上海　**建筑面积**：290 000 m²　**商业面积**：36 000 m²
建筑师：刘廷杰

LOCATION: Shanghai, China　BUILDING AREA: 290,000 m²　COMMERCIAL AREA: 36,000 m²
ARCHITECT: Tingjie (Peter) Liu

1 前庭广场
2 低层住宅区入口
3 滨水广场
4 商业内庭广场
5 游泳池
6 网球场
7 商业区入口广场
8 商业内街
9 商业步行街
10 树阵广场
11 紧急消防出入口
12 生态内河
13 围合下沉式公共庭院
14 湿地公园
15 公寓区入口
16 喷泉跌水池
17 大型跌水石景喷泉
18 亲水平台
19 泵房
20 组团绿化一
21 动态活动区
22 动态活动区
23 组团绿化二
24 组团绿化三
25 动态活动区
26 滨水散步道
27 滨水散步道

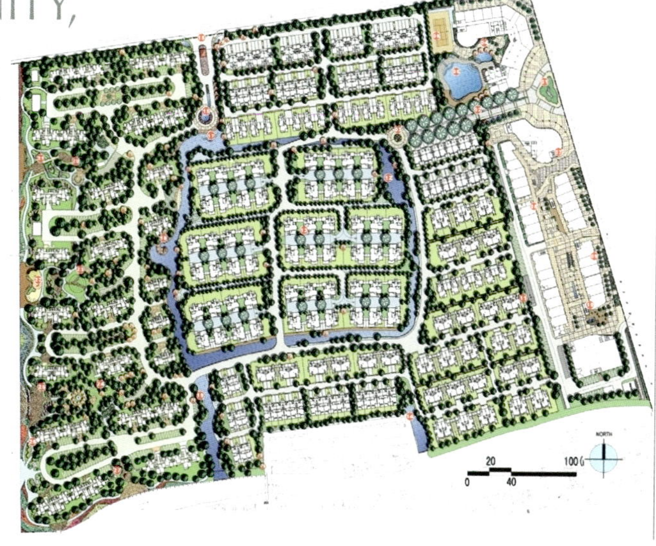

新弘国际既营造国际化的居住社区，更通过城市复合型商业的业态布局和空间塑造，改变松江新桥地区高端零售商业缺失的现状，从而推动区域和本社区品质的双赢。建筑特征以上海地域文化的红砖作为衍生载体，融合当代的形式语言，使商业群态兼具商业功效和文化品质的双重性。

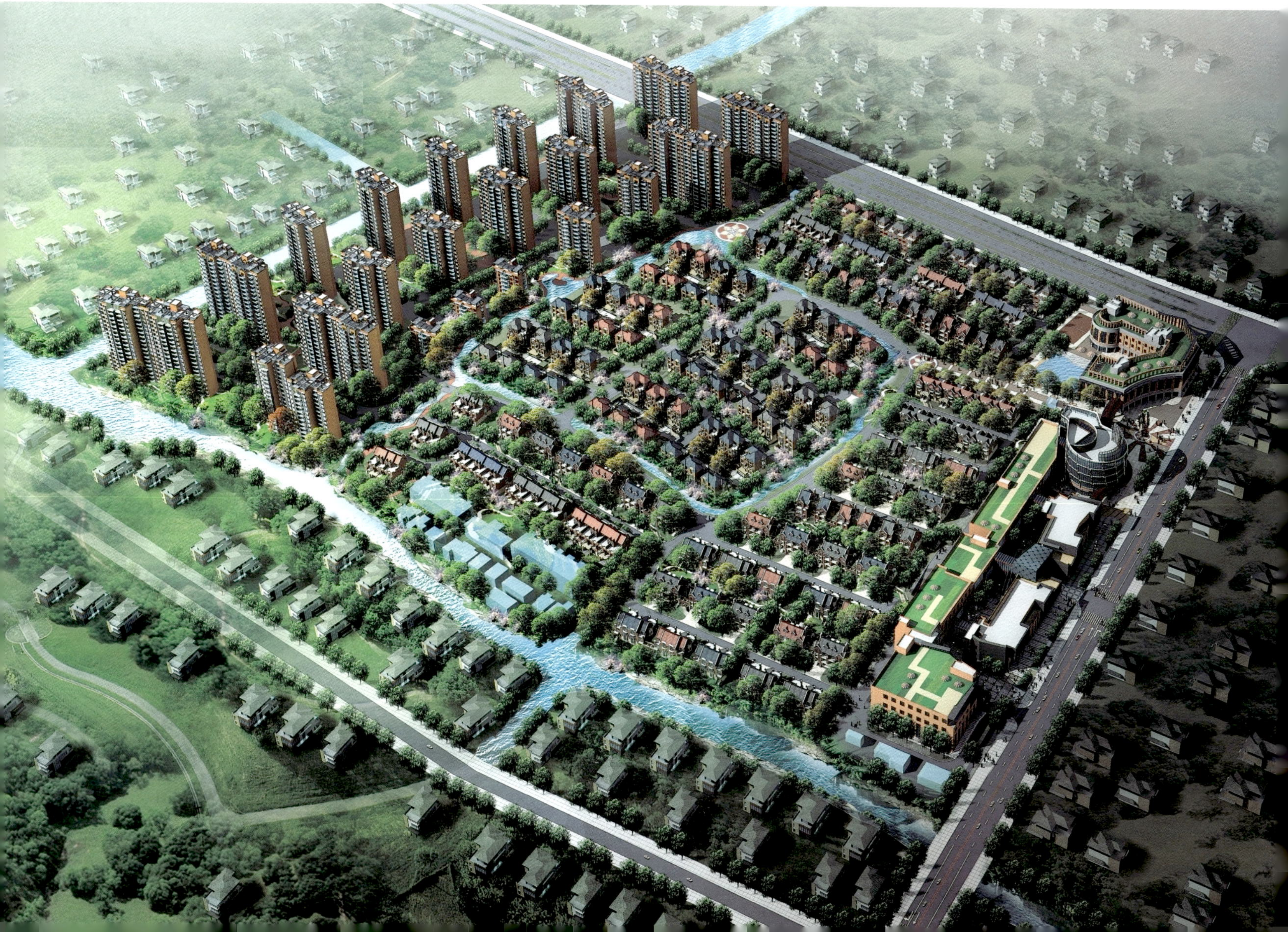

Xinhong International not only provides international residential neighborhood, but also aims to change the status of lacking high-end retail business in Songjiang New Bridge district through the layout and space creation of the city complex commercial industry, so as to enhance both the quality of the region and the community. The building features use the red brick as a derivative carrier, which is the symbol of Shanghai local culture, and fused into the contemporary language froms, which makes the commercial groups assume the dual natures of business effectiveness and cultural quality.

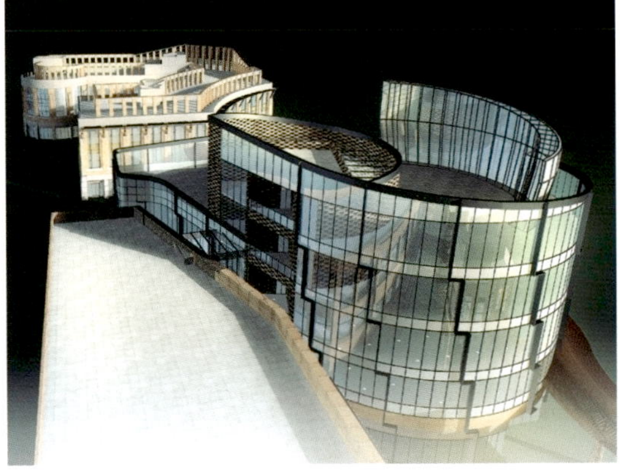

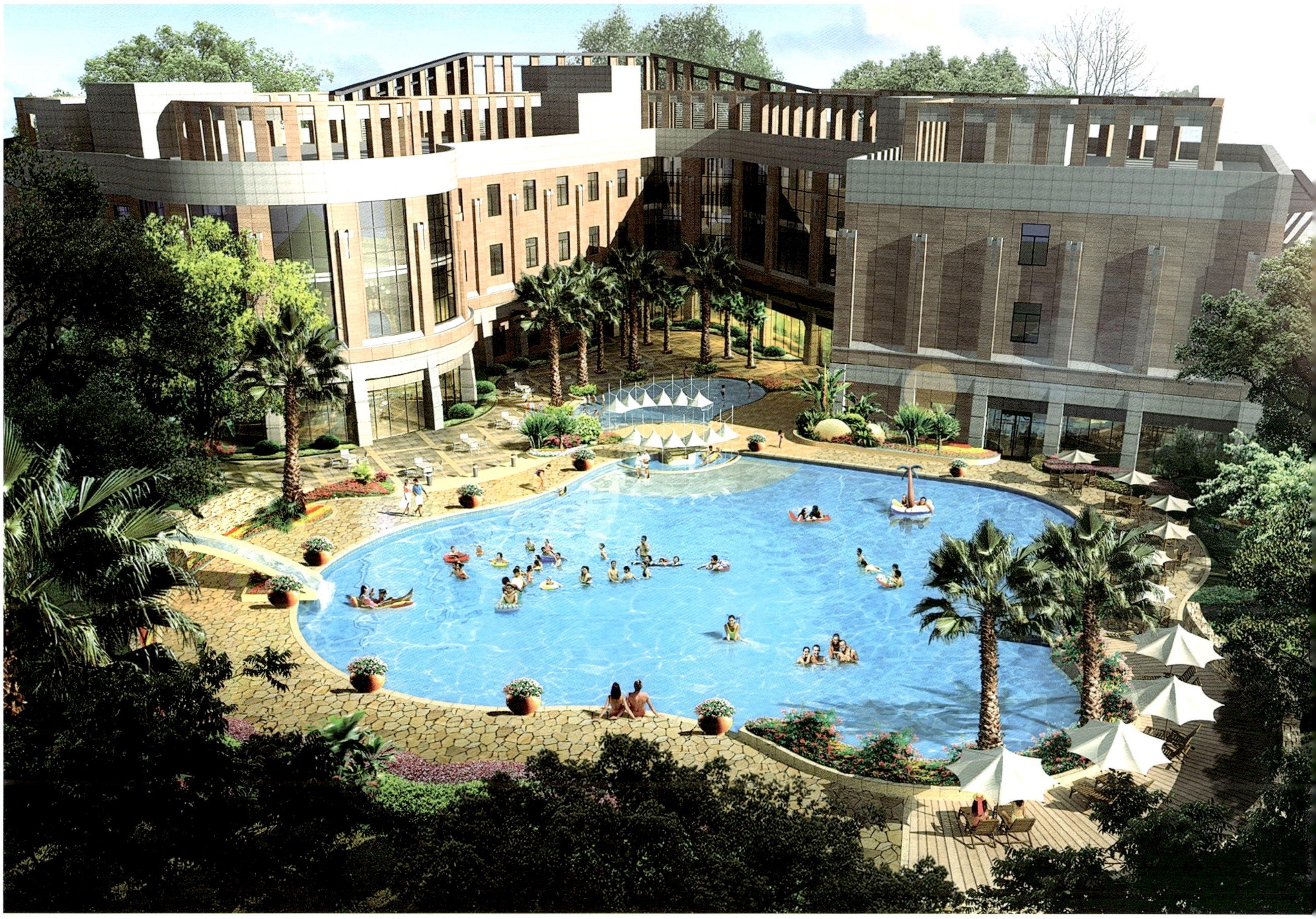

1. 网球场
2. 休憩座椅区
3. 树阵广场
4. 景观树池
5. 儿童池
6. 按摩池
7. 张拉膜阳伞
8. 洗脚池
9. 景墙
10. 落水梯
11. 微地形
12. 木平台
13. 花池
14. 景墙

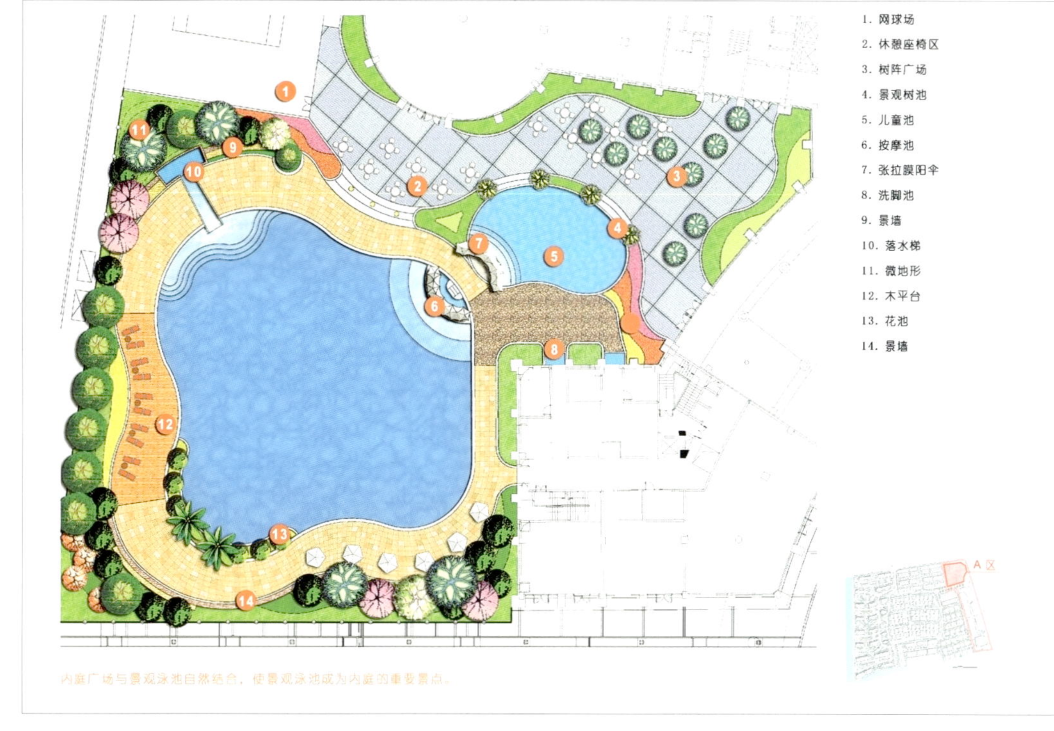

内庭广场与景观泳池自然结合，使景观泳池成为内庭的重要景点。

DATRANS ARCHITEC-TURE OFFICE

德默营造建筑事务所

德默营造建筑事务所正式成立于2004年，由陈旭东和吴洁创建，前身是2001年于德国柏林组建的都市建筑艺术设计和研究机构DAtrans，2005年初德默营造入驻上海莫干山50号创意园区的办公场地至今。

从创立伊始，德默营造关注全球化环境下的当代都市和建筑的变化与发展，并以积极的态度和专题概念，深度介入当代中国环境更新和城市改造；以多学科、跨领域、国际化的兼容并蓄的操作，严谨实务与专注研究的互动工作方法，广泛地参与本土建筑设计实践。同时，还坚持在新材料应用和数字技术实验的基础上，努力探索未来新建筑美学的可能性。

现有城市设计师、建筑师、景观设计师、平面设计师和艺术及经济管理专家组成的多专业复合型团队，同国际多家设计咨询、技术支持和研究机构形成长期稳定的交流与合作网络。

德默营造重要的设计项目有Art Deco凹凸家具库、杭州滨江双塔、M50／莫干山50号总体规划及建筑改造、开封火车站前城市设计、与艺术家艾未未合作的杭州西湖江南会、广东美术馆人文图书馆以及瑞典斯德哥尔摩公共图书馆竞赛等。同时，获得多项建筑竞赛的奖项。

德默营造的项目已被《纽约时报》、美国《建筑实录》、意大利《Domus》和德国《建筑世界》等国内外知名的大众和专业媒体报道和收录，并出版有研究和作品集《二手摩登：M50／莫干山50号的城市营造》。

德默营造也活跃在如伦敦中国发电站艺术展、深圳建筑双年展等国内外重要专业和艺术展览文化活动上。因其具有代表性的、对当代城市建筑的思考和结合中国文化气质的独特实践，受到荷兰国家建筑研究所、比利时布鲁塞尔建筑中心的邀请，参加当代中国建筑展，引起了更加广泛的关注和讨论。

DAtrans was founded as an international research association dedicated to architecture and urbanism in 2001 in Berlin. And in 2004 DAtrans Architecture Office was established by the partners CHEN Xudong and WU Jie in Shanghai and moved into the current working space in M50 by Suzhou Creek.

From the establishing on, DAtrans focuses on the evolution and mutation of the contemporary cities under the globalization's context, involved in the environmental transformation and urban renewal in contemporary China with the positive attitude and the critical argument. With the multidisciplinary, pluralized and international collaboration we also devote to the local architectural practice. We try to follow the interactive working method with the rigorous research and the dynamic strategy. As the final approach we try to discover and define a new architectural aesthetics according to the application of new materials and digital technology.

The best known projects of DAtrans Architecture Office are M50 urban renovation, Art Deco Furniture Gallery, Jiangnanhui Resort West Lake, Twin Towers Hangzhou and the Library of Guangdong Art Museum (GDMOA), competition for Train Station Commercial Area Kaifeng , Asplund Library in Stockholm and Theme garden for Expo 2010 Shanghai etc.

And as one of the most young and creative architecture offices the works have been presented in the important exhibitions of NAI, CIVA and RAM and introduced in many well-known international public and professional medias such as *The New York Times, Architectural Record, Domus and Bauwelt* etc. Meanwhile, DAtrans could also be discovered in many important art exhibitions and cultural events such as Chinese Power Stations London, Shenzhen/ Hongkong Biennale of Urbanism and Architecture and etc.

“M50 艺术园区”建筑改造

"M50 ART PARK," BUILDDING TRANSFORMATION

项目地点：中国 · 上海　基地面积：23 600 m^2　建筑面积：42 000 m^2

建筑设计：德默营造建筑事务所

建筑师：陈旭东 ,Max Mueller, 鲍伟，杨光，陈启春，左欢，蒋迪 ,Bojra Trujillo, 严梦菲 ,Lucy Schofield, 陈婉云，陈杨

摄影：张宗眉，德默营造建筑事务所

LOCATION: Shanghai, China　SITE AREA: 23,600 m^2　BUILDING AREA: 42,000 m^2

DESIGN CORPORATION: DAtrans Architecture Office

ARCHITECTS: Chen Xudong, Max Mueller, Bao Wei, Yang Guang, Chen Qichun, Zuo Huan, Jiang Di, Bojra Trujillo, Yan Meng fei, Lucy Schofield, Chen Wanyun, Chen Yang

PHOTOGRAPHERS: Zhang Zongmei, DAtrans Architecture Office

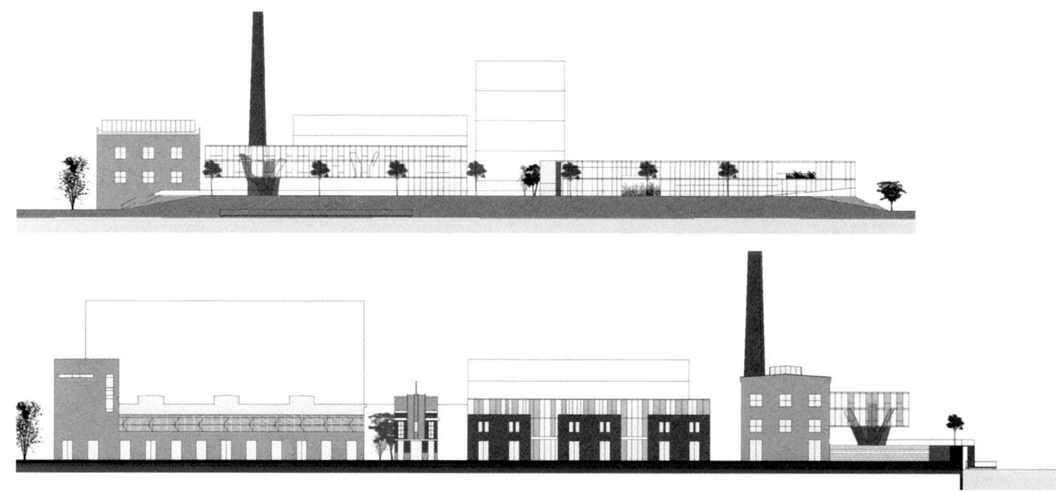

上海莫干山路地区毗邻上海火车站，三面被苏州河环绕，区内既有传统尺度的里弄街坊，又有快速建成的高层住宅。莫干山工业区曾是 20 世纪 30 年代上海民族工业的发源地，莫干山路 50 号原为 1938 年正式开工生产的英商信和纱厂，后屡经发展和结构调整，作为国营纺织企业更名为上海春明粗纺厂。

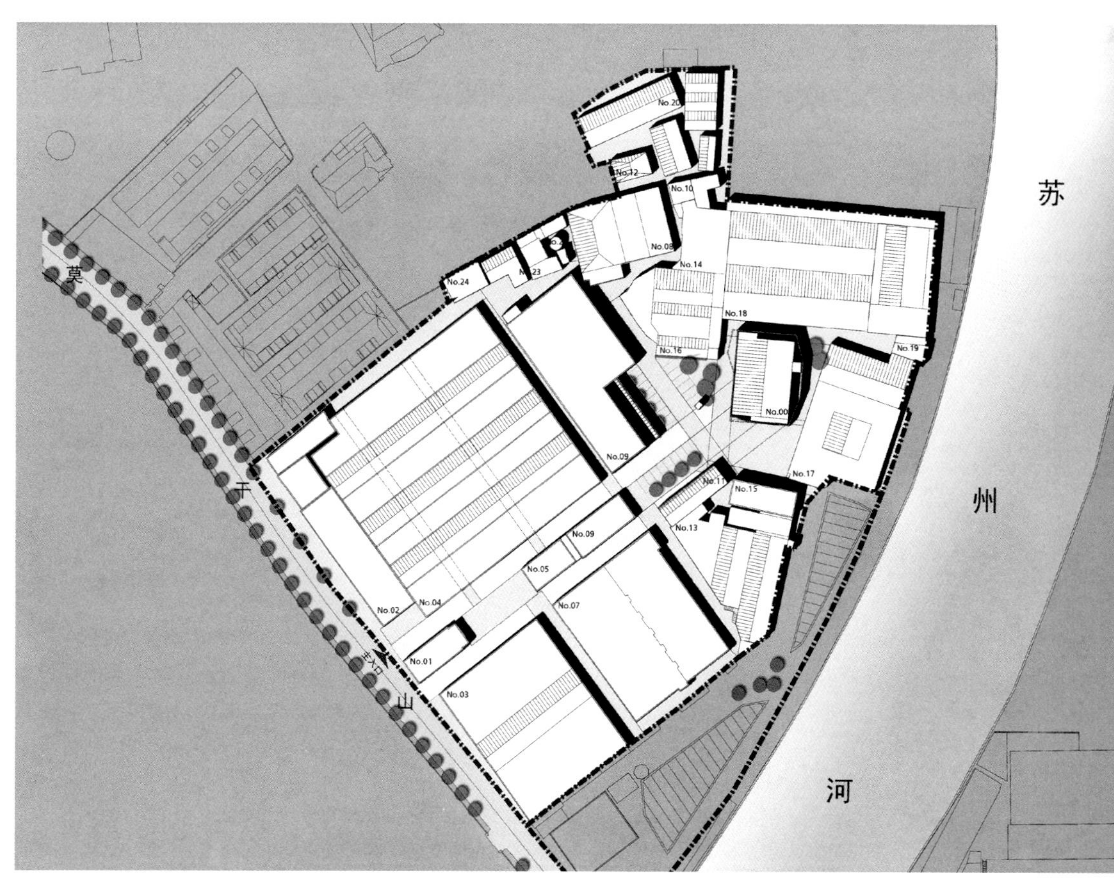

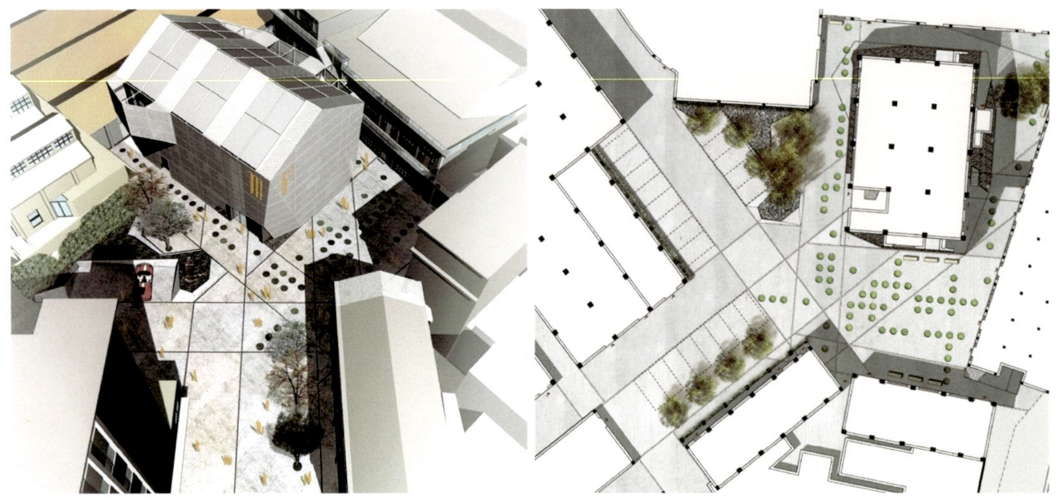

20 世纪 90 年代末，随着众多艺术家和文化机构进驻春明粗纺厂的旧厂房，这里逐渐演绎成为上海当代文化和城市生活的重要地标。2005 年作为国内首批创意产业聚集区之一，被上海市经委命名为"M50 创意园"。随后，M50 被美国《时代周刊》列为上海全球"推荐参观之地"，成为国内乃至亚洲最有影响力和代表性的都市艺术文化社区。

从 2005 年初开始，针对如何进行现状更新，设计师与业主、用户和公众之间进行了大量的互动式讨论，最终与业主、社区代表共同确定了改造的总体纲要，即确保在转型和改造的整个过程中，妥善地延续原有的区域特征，并完整地保护工业建筑风貌，使整个改造过程具有可持续性和灵活性。

同时，我们以此为框架发展出整体改造设计方案，围绕多功能的文化艺术社区、均质化的网络自治体系、开放的滨河公共的空间结构和个性化视觉效果的设计目标，进行了功能调整、环境改造和建筑更新等一系列改造项目。

2005 年我们完成了园区入口广场和信息中心的建筑改造，随后陆续完成了包括凹凸家具库、东八书仓和爱普生影艺坊等内部空间的改造。2009 年开始整个园区的二期改造的设计和施工，完成了入口标志塔、中心广场和滨河仓库等项目。

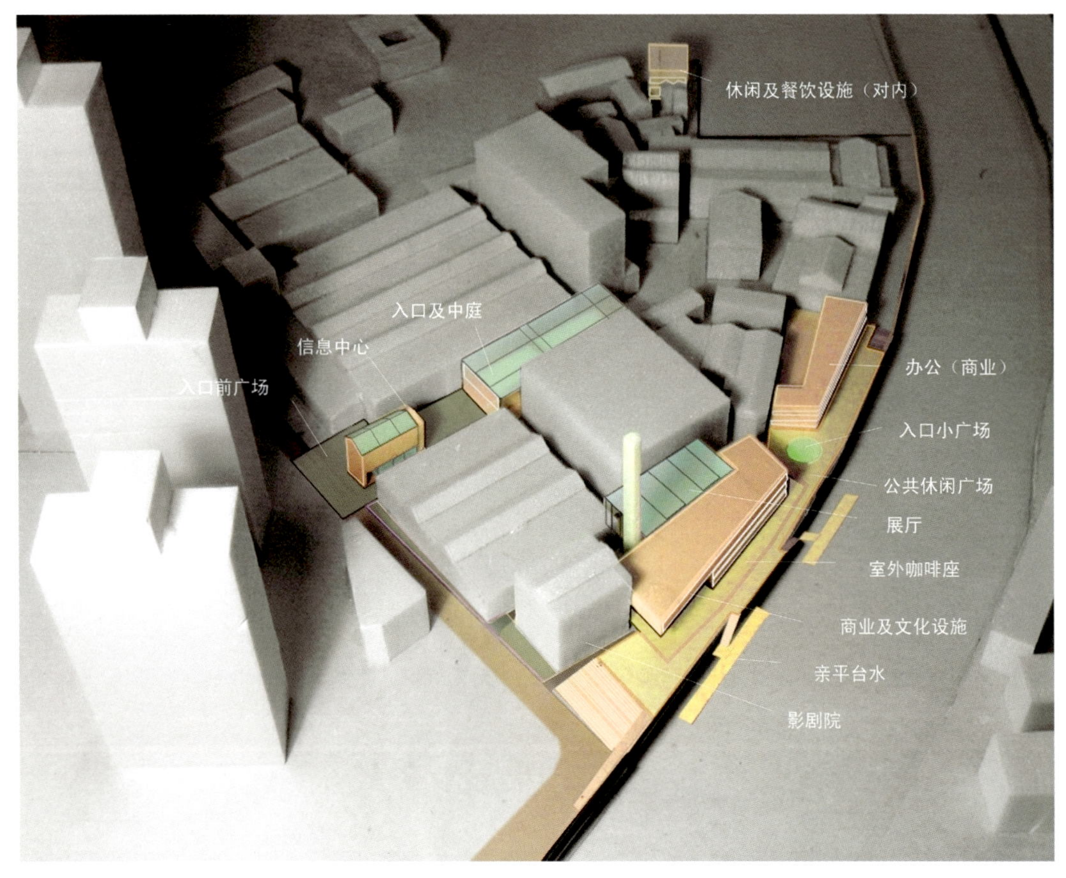

The Moganshan Road area is located in the peninsula that is close to Shanghai Central Station and surrounded by Suzhou Creek. The urban contrast could be discovered obviously between the traditional residential blocks of Lilong and the modern high-rise compounds to be built rapidly.

The national textile factory which has the brilliance history more than 70 years, located in Moganshan Road 50, is one of the so little survivals during the modern urbanism's crisis. After the entering of avant-courier artists and culture institutions since 2000, the factory is becoming an important and influenced artist's enclave in the urban landscape. And following its renaming M50 has been growing as one of the leading creative parks since 2005 in China and Asia.

After the winning of the competition we have continually involved in the master plan as well as several interior projects within the warehouses and the urban study on the surrounding area. Under the general guideline decided by the client, renters and the public we define the futural M50 as the multi functional cultural community, the equal and self organising network, the public urban space and the attractive unique landmark.

The three frames on the development of our project are the diagrammatic, the urban and the architectural. The diagrammatic frame takes as a point of departure the actual events once affecting the site of the industrial warehouses that partly generate the oppositional and programmatic aspects of M50. The urban and the architectural frames provide the stage for the diagrammatic. Both the urban and architectural frames are approached as a delayering of the contexts that constitute the buildings. At the architectural level we focus on the two issues: the relationship between preservation and development in the process of modernization, and the question of nostalgia as one of the forms through which the relationship between the past and the present arbitrates presence in the physical, social and cultural context of M50.

Since 2005 the key projects we've finished are the entrance, center square and warehouse by riverside. And we also rebuild the current elevation, visual system and landscape to form an integrated environment.

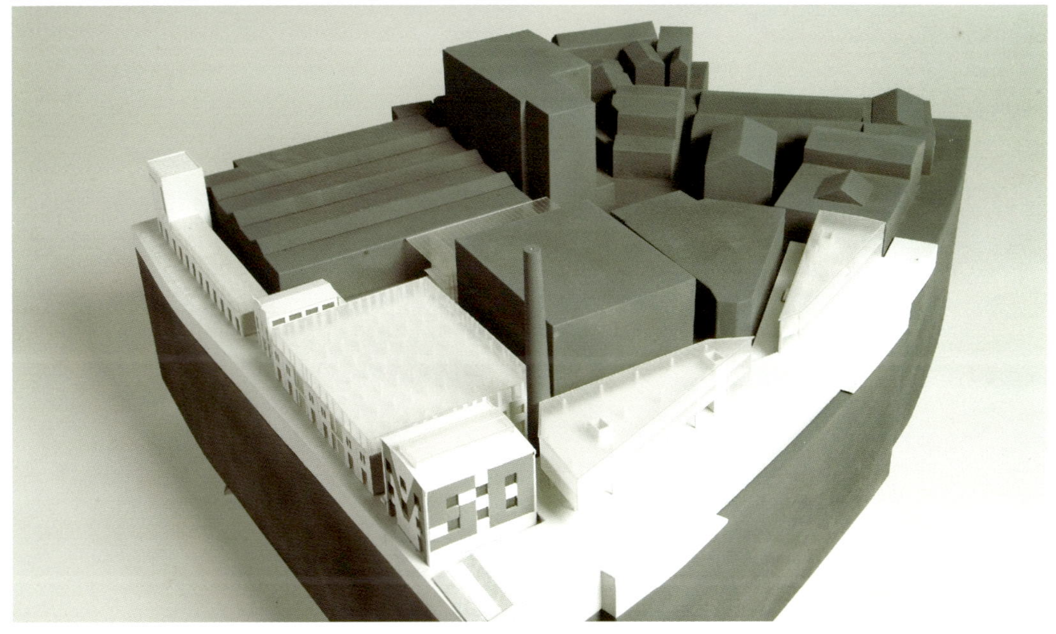

北京天地合院精品酒店

COURTYARD BOUTIQUE HOTEL BEIJING

项目地点：中国 · 北京　基地面积：3200 m^2　建筑面积：6900 m^2
建筑设计：德默营造建筑事务所
建筑师：陈旭东，严梦菲，陈杨

LOCATION: Beijing, China　SITE AREA: 3200 m^2　BUILDING AREA: 6900 m^2
DESIGN CORPORATION: DAtrans Architecture Office
ARCHITECTS: Chen Xudong, Yan Mengfei, Chen Yang

项目基地位于北京东城金宝街金宝汇南侧，东临文化保护建筑沈从文故居。对于现代建筑而言，本次设计就是一次对传统四合院空间特征的探索和学习过程：四合院的结构布局和空间特征体现了传统中国对外温厚封闭、内部有序和谐的大家庭生活形态。

由于基地形状特点和酒店的功能划分需要，设计师以一个两进合院为参照，把建筑分为两个空间性质不同的区域——南为“客厅”，空间开放、气氛活跃，用作大堂、餐厅和酒吧；北为“居室”，空间私密、气氛恬静，用作酒店卧房和spa会所。南北各有一内院，一动一静，是南北两区空间性格的集中体现。布局上建筑体量尽可能满铺整个基地，尽量压低建筑高度以接近传统合院适宜居住的建筑尺度，并预留出一定面积的室外庭院。

在规划许可的高度控制范围内，为了削弱由连续立面所带来的沉重体量感，设计师对建筑屋顶采用了褶皱的处理手法，褶皱屋脊也在一定的模数前提下呈现出高低错落的生动形态。屋顶西北高东南低，除了日照采光功能的需要外，也是向东侧院落曾经的主人——沈从文先生——致敬。北部屋顶局部开口作为跃层客房的私密庭院，重现内向的精品气质。

建筑外部材料的运用，在与周边传统四合院气质协调的同时，展现了强烈的当代精神：深灰色的锌板从外部包裹了整个建筑主体，而红色丝网印玻璃在内院展开，演绎着缤纷绚丽的都市生活。

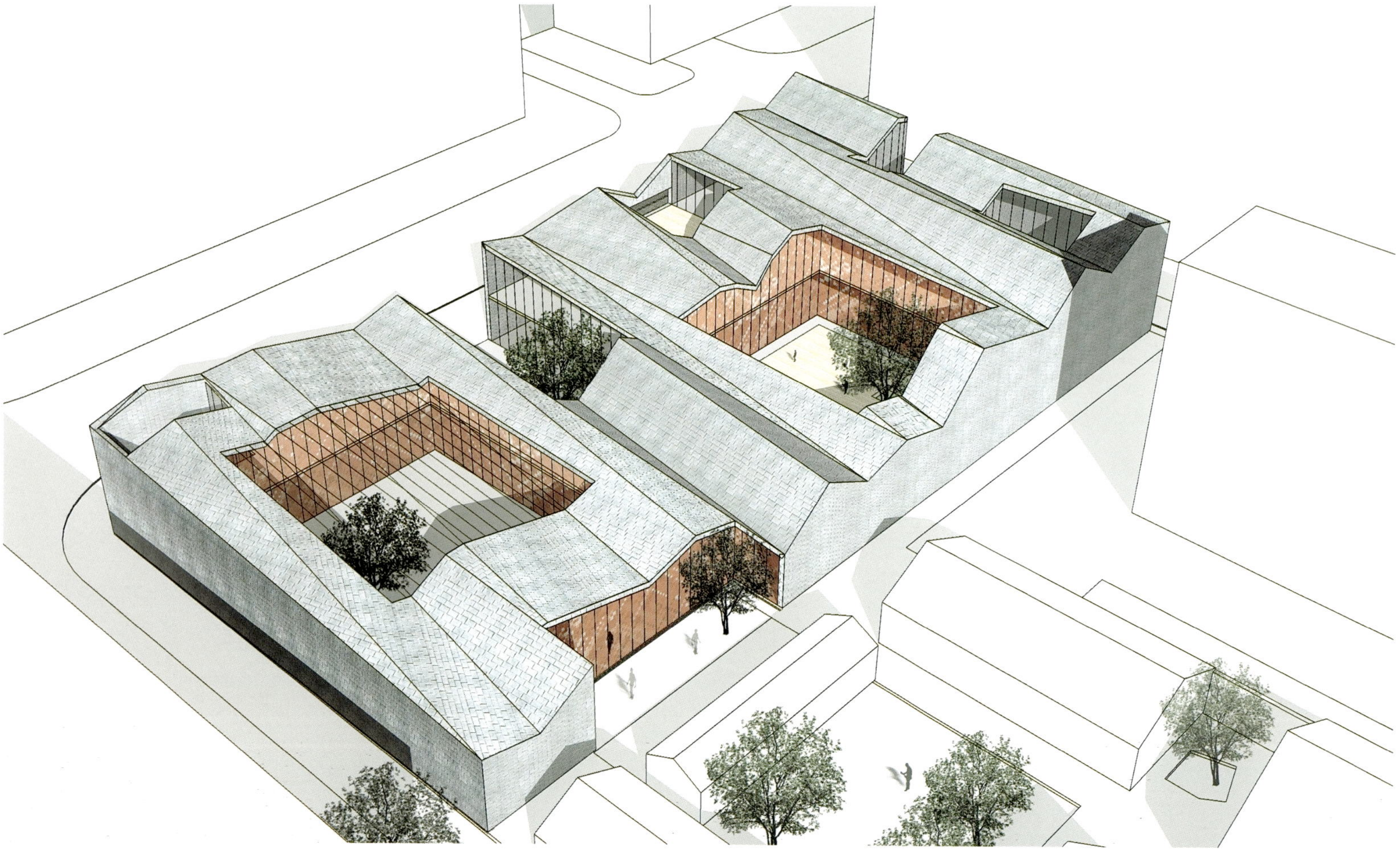

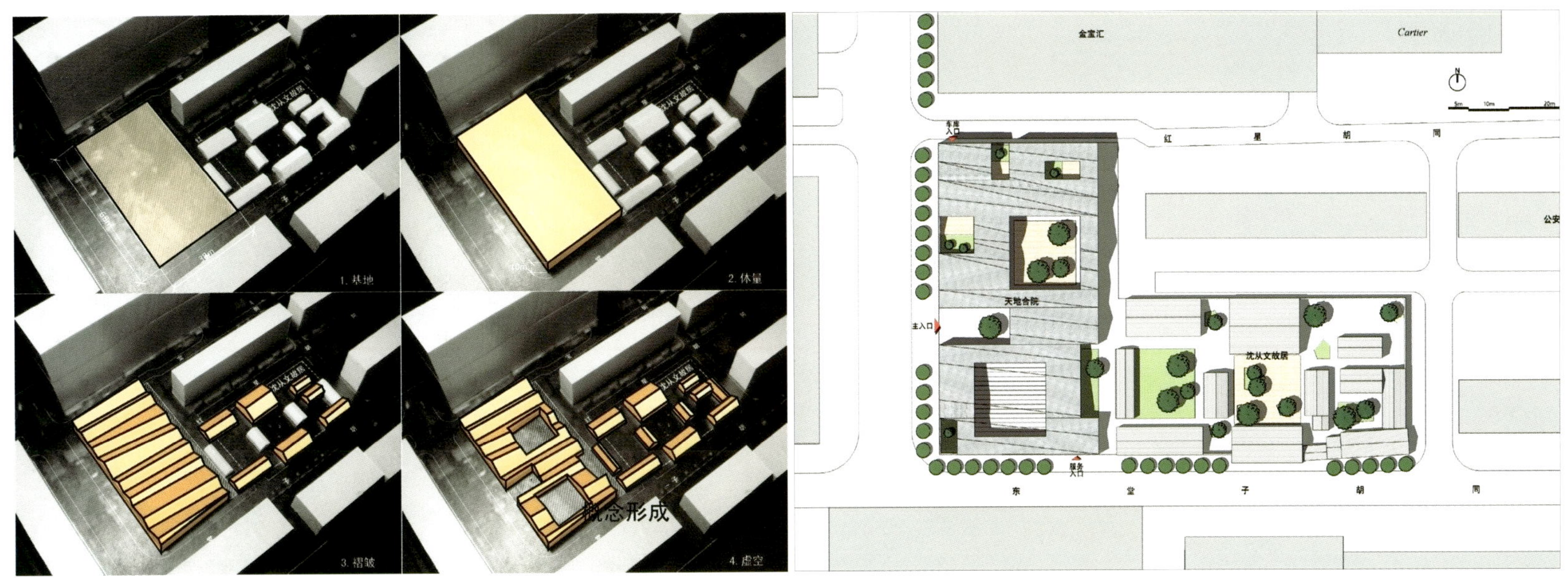

The project's site length is 40 m x 80 m, and is situated on Jinbao Street in Beijing. The south is approximately opposite the boutique store complex and neighbored to the east by a protected heritage building, the Former Residence of famous writer Shen Congwen.

Based on research into courtyard houses, Chinese traditional residences, this design is in itself a discovery process demonstrating how a modern building can derive the spatial characteristics of the traditional courtyard house. The layout and spatial characteristics of courtyard houses reflects the Chinese' traditional conservative and closed attitude to the outside world, with an open and communal attitude within, where family life was ordered by seniority. The characteristics of these spaces are what a busy, down-town complex featuring boutiques and restaurants need; privacy from the outside, the preservation of an inner space for sharing and exchange, and clear divisions of the zoning.

Due to the site length and width and the requirements of the division of the hotel's feature zones, we have taken a two-entry courtyard house as a reference. Hence, this boutique hotel is divided into two spaces with different properties along the north-south axis. The south area is "a living room", an open space evoking a lively atmosphere, to be used as a restaurant and bar. The north area is "a bedroom", a private space with quiet and peaceful atmosphere, to be used as a hotel bedrooms and spa club. The North and South share an inner courtyard, which is alternately lively and tranquil, thus embodying succinctly the characteristics of the North and South spaces. In layout, the entire base of the building is carpeted as far as possible. Thus in accordance with a pre-designated building area, the building height will be lowered as far as possible. This will bring the building closer in size to the size appropriate to the residences of the traditional courtyard houses whilst reserving an area sufficient for an inner and outer courtyard.

Due to the substitution of the small-scale group of buildings of the traditional courtyard house by one whole building, in addition to the height restriction, for the building's roof, we have chosen a folding approach. Weakening the connecting facades brings a sense of body volume. The folding roof takes different heights. Under the limit of the modules, this creates a dynamic sense of scattered highs and lows. The North-West roof is high and the South-East roof is low. In addition to the feature requirements, this also pays homage to the heritage building of the Former Residence of Shen Congwen on the east-side. Some small pits in the North area of the roof opens as the hotel bedroom's independent private courtyard.

The material of the building's outside is grey zinc-plate and red screenprinting glass. The screenprinting glass pattern is taken from the contemporary art work. This fuses the fashionable material and the spirit of contemporary art with the grey-and-red colouring of Beijing' traditional courtyard houses.

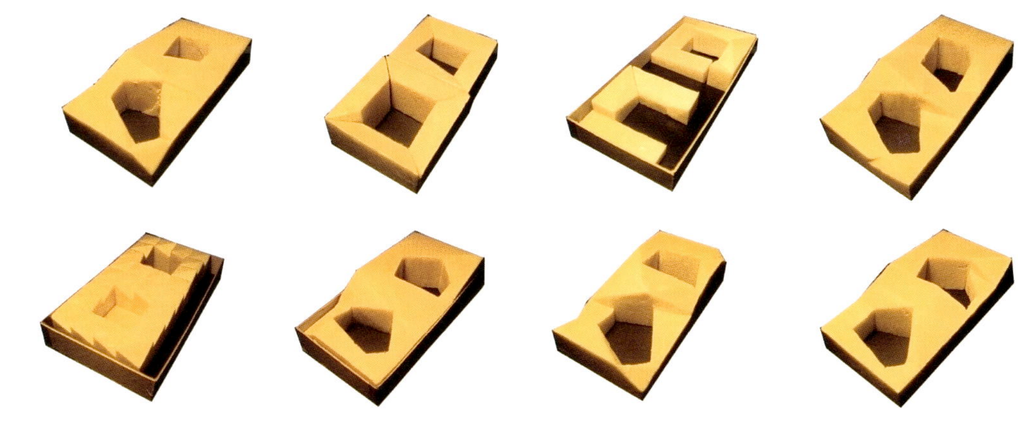

千叶院：徐州彭城一号展示中心

GINGKO PAVILION XUZHOU

项目地点：中国 · 江苏　建筑面积：1000 m^2

建筑设计：德默营造建筑事务所

建筑师：陈旭东，Matej Dobis，Paulo de Araujo，徐鑫，沙少磊

LOCATION: Jiangsu, China　BUILDING AREA: 1000 m^2

DESIGN CORPORATION: DAtrans Architecture Office

ARCHITECTS: Chen Xudong, Matej Dobis, Paulo de Araujo, Xu Xin, Sha Shaolei

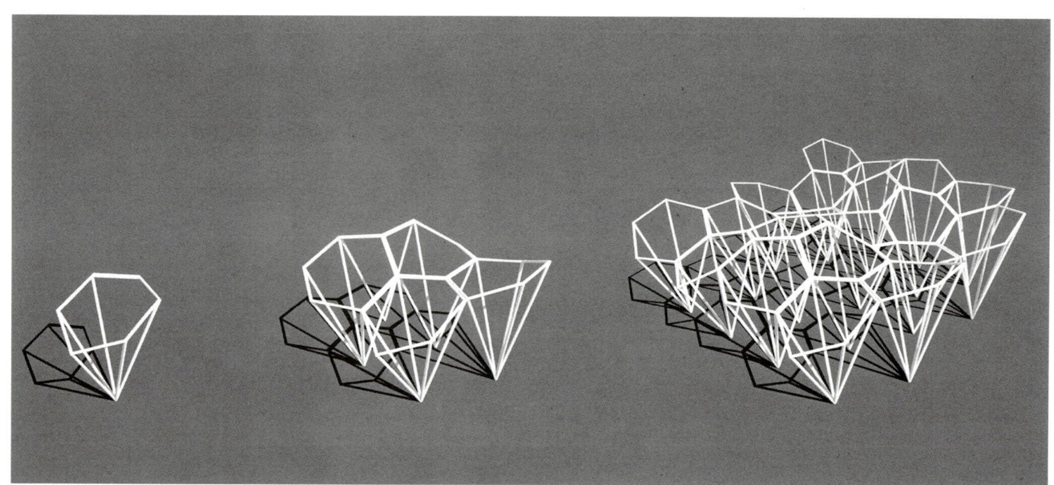

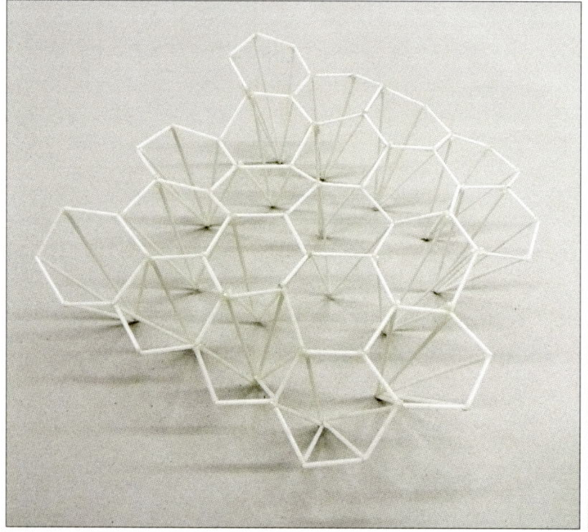

项目基地位于徐州市鼓楼区彭城一号时尚休闲商业街，地处中心商圈核心区。“彭城一号”作为城建重点工程项目和中心商圈的十大版块之一，业主希望我们能在广场入口处设计一个标志性建筑，并兼备展示中心的功能以吸引和服务商业人流。

设计从自然中提取灵感，以徐州市树银杏树作为形式的母体——外露颀长的圆形钢柱和动态起伏的曲形屋面犹如一片在微风吹拂下缓缓摇曳的银杏树林，室外耀眼的阳光透过银杏叶状的半透膜，投下斑驳影像，更给人们一种林中漫步的氛围。我们将若干个外接直径为 6 m 的正六边形按蜂窝状排列，形成了建筑的基本平面。然后利用数字造型软件设计出起伏的三维曲面，并通过将基本平面投影至该曲面，最终得到了富有韵律感的顶面主梁结构。每个六边形主梁作为一个模数单元被弧形的次梁划分为三片“银杏叶”；由一根细长的钢柱连接每片“银杏叶”恰似依托住树叶的叶柄，“叶柄”斜撑至地面与剩余的两根钢柱交汇于六边形投影的中心点。连续相交的斜柱与主梁形成一个个三角形框架，为结构的稳定性提供了保障。

我们利用这个模型单元将自然界创造的银杏树叶以建筑的语言诠释，通过增加模数单元实现建筑的自由生长，借助建筑流动的形体和柔软的轮廓，带给人们关于自然的联想，希望通过简单的几何结构的衍生及组合，创造出非几何的脱胎于自然的有机体。

建筑的主入口公共空间以开放友好的姿态迎接着由南端富庶街进入广场的主人流。建筑底层局部架空，使空间尽可能地开敞、流动；镂空的两片六边形屋顶下栽种了几株银杏树，配合建筑边缘一片水景，让整个底层空间与自然交融；二层为一个完整的室内展厅，可根据需要举办各类公关活动，比如展览、开幕式、推广会或商务派对等。

建筑主体结构和膜以白色为主。建筑灯光可以根据需要设置种色彩方案，根据使用需要和举办活动的主题呈现不同色彩。

The project is based at the fashion, leisure and commerce center in Gulou District, Xuzhou, located at the heart of downtown. No.1 Pengcheng is a key urban construction project and one of the top city's business districts. The proprietor would like us to design a symbolic building at the entrance area to the entrance square and an exhibition centre which has the potential to attract and cater to the needs of the public.

The design takes its inspiration from nature. Adopting Xuzhou's gingko tree as the motif, the design unveils tall, rounded steel columns and dynamic, undulating, curved layers like a gingko forest swaying slowly in the breeze. In the design process, we arrange a number of hexagonal honeycomb forms with an external diameter of 6 meter, forming the building's basic plan. Afterwards, we will use computer modelling software to design an undulating, wave-like 3D curved surface. Each hexagonal girder comprises a model with three inward-facing curved sub-beams in the shape of three "gingko leaves". Each leaf is connected by a long thin steel pole, like the stalk which supports the leaves. This stalk structure inclines towards the floor and the two remaining steel rods intersect with the central point of the hexagonal projections. The continuously intersecting columns and the girder form a triangular frame ensuring the stability of the structure.

Using the modelling unit, we can interpret the natural creation of the leaves of the Gingko tree through the language of construction. Through increasing these units we can make the building's independent growth a reality, and drawing on the construction's fluid shapes and soft outline, bring people into association with nature. We hope that through the derivation and assembly of the simple geometrical structure, to create a non-geometric body born out of natural organisms.

In the first floor we have installed a small café and second floor is a complete indoor exhibition area, catering to the basic requirements of a range of public-related events including exhibitions, openings, promotions and networking events etc.

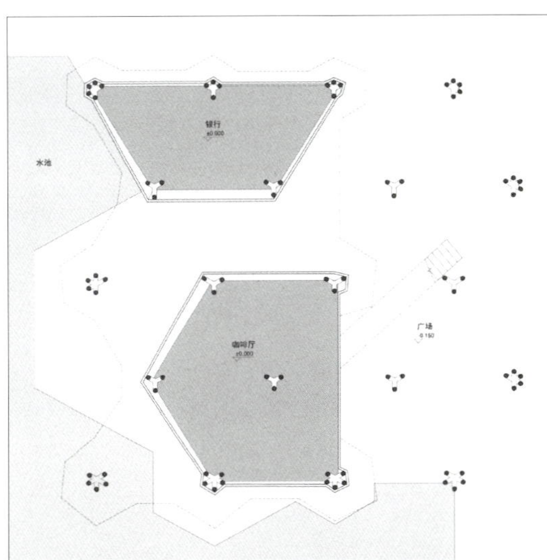

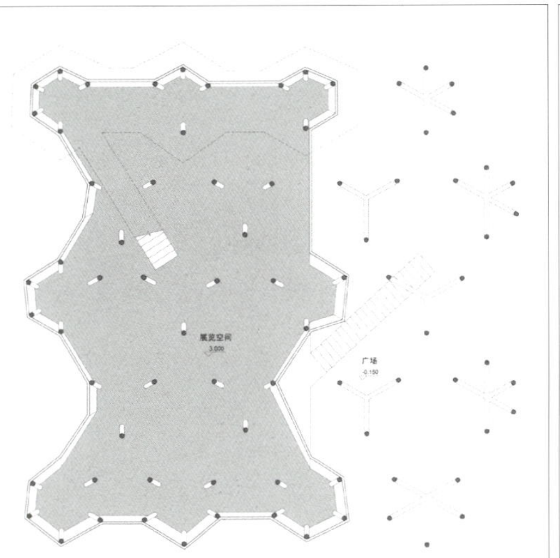

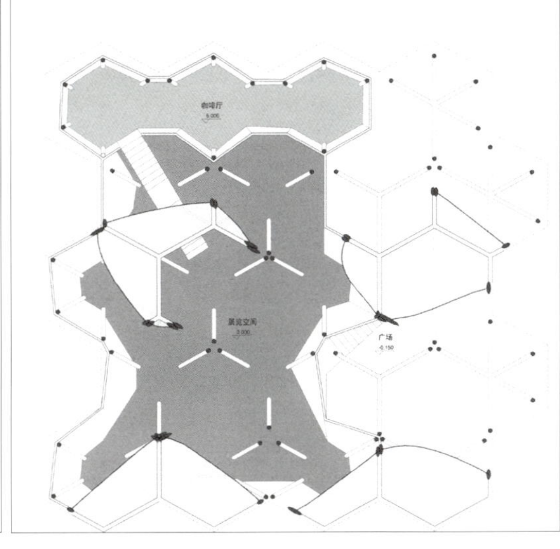

平刚

崔哲

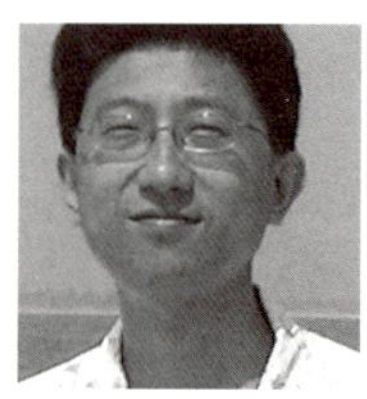
揭涌

万江蛟

公司合伙人及主要设计师

- 平刚　Ping Gang　执行董事 首席建筑师

平刚 先生毕业于中国南京东南大学建筑系，取得建筑学士学位。之后在上海、新加坡、澳大利亚工作与学习。1998年在新加坡创立DC国际，遵从单纯的自我实现与社会实现的双重建筑理想展开从"未建成"到"建成"的建筑实践活动，努力以国际性的视野，解决地域性问题。

- 崔哲　Cui Zhe　设计总监

崔哲 先生毕业于西安建筑科技大学建筑系，取得建筑学士学位。1999年从西安到上海学习与工作，取得同济大学建筑硕士学位。崔哲先生的优势在于他成熟优雅的审美观、快速的理解力，擅长处理复杂的公共建筑体系。崔哲先生致力于转变大众对公共和私有空间领域的思维方式和体验。

- 揭涌　Jie Yong　技术总监 中国一级注册建筑师 中国注册规划师

揭涌 先生毕业于西安建筑科技大学建筑系，取得建筑学士学位。揭涌先生擅长大规模的社区规划，高档别墅设计，领导过诸多综合性、跨领域、不同尺度的设计项目，揭涌先生丰富的项目管理与设计经验充分保证了其所负责项目的成功实施。

- 万江蛟　Wan Jiangjiao　项目总监 中国一级注册建筑师

万江蛟 先生毕业于东南大学建筑系，取得建筑学士学位。万江蛟先生在各类项目的总体规划和建筑设计方面具有丰富实践经验和专业知识。他主张每一个项目都应该有自己独特的风格和视觉效果，并对传统建筑文化如何与现代生活融合进行认真的探索。

DC国际（DC Alliance,Singapore）致力于提供专业的城市规划、城市设计和建筑设计服务。

1998年DC国际在新加坡注册成立，2001年在上海设立事务所开始拓展中国市场，同时以各种形式参与设计业务。DC国际旨在以精湛的专业技术为业主提供高标准的服务，与此同时以优秀的作品强化我们的建筑环境。经过努力，DC国际的工作得到了中国乃至世界的认可，2005年10月，宁波东部新城中小学及广西柳工集团总部工程入选美国纽约建筑周参展作品。2008年5月，宁波东部新城安置区项目获得美国《商业周刊》和《建筑实录》联合颁发的"Good design is good business"设计大奖。

2008年8月，DC国际被《设计新潮》评为中国"长三角地区最具潜力公司"。由于DC国际团队的多重文化背景，我们在中国所努力追求的便是一种既能结合中国国情又能反映国际建筑思潮的建筑。达到这一建筑途径是依靠集体的智慧，发掘当地的文化特质、生活情态、建筑施工条件、技术特长，以此为基础借鉴西方的先进建筑技术。只有这样，西方现代建筑才会积极地融入中国，而中国现代建筑也才会被世界建筑文化认同。

DC国际的建筑实践基于以下三个根本原则。

首先，全面细致地了解业主的需求。以高度的责任心及丰富的经验为业主带来最佳的经济效益，以敏感的设计和精良的作品来表达业主的精神追求。通过与业主的密切合作建立相互的尊重及信任。

其二，为业主提供最健全的服务。DC国际既注重结果又强调过程，在设计过程中各阶段始终组织得有条不紊，同时努力实现建筑设计到建成作品的高完成度。

最后，DC国际团队采用最新的设计手段。设计人员对各种计算机辅助设计程序及三维模型程序应用娴熟，搭建流畅的思维与设计平台，追求完整的设计体验。

DC国际设计作品涵盖各种类型的公共建筑、文化建筑、住宅建筑和城市设计。无论项目的规模大小，事务所一直坚持向业主提供高端优质的设计服务，并为复兴中国现代建筑做出自己的努力。

DC Alliance commits to offering professional service in city planning, urban design, architecture design as well as landscape design. DC Alliance was established in Singapore in 1998. During the following 9 years, DC Alliance has been expanding quickly in China's Mainland: it has, in various ways, participated in design projects of many cities such as Shanghai, Nanjing, Ningbo, Qingdao, Nanning, Beijing, Wuxi, and Xi'an. In 2001, DC Alliance set up its chinese design office in Shanghai, with its goal as providing excellent service to clients through its exquisite expertise, and improving our architectural environment through outstanding works. DC Alliance, through its industrious work, has achieved acknowledgment worldwide and two of its projects were selected in New York Architecture Week in America in October, 2005: one was the design of a primary and high school in Eastern New City of Ningbo, the other was that of the general office of Guangxi Liugong Group. The project Ningbo Eastern New City Economical Housing has won The Best Residential Projects in the 2nd Business Week/Architectural Record China Awards in 2008.In August 2008, DC Alliance was selected by Design Trend as the Most Promising Company in Yangtze Delta of China.

With its multicultural background, DC Alliance commits itself to designing such architecture that reflects international trends while being in accordance with the Chinese present situation. To achieve this, we rely on teamwork to exploit local specific cultural temperament, life style, conditions for construction, and technological strength, and based on all these, we draw on the advanced technology of the west. Only through this, can the western modern architecture be well introduced to China and China's modern architecture be recognized by the international architectural culture.

The architectural practice of DC Alliance has been based on the following three principles:

First, we try to understand clients' needs well. We try to bring about as much economic benefit as possible for clients with our strong sense of responsibility and rich experience; we convey clients' beliefs through exquisite design and excellent works; and we establish a mutual respect and trust with clients through close cooperation.

Second, we provide quality service for clients. DC Alliance values both the results and the course of projects. During different stages of design, we are well-organized and try our best to realise high-level accomplishment from our designs to finished projects .

Third, DC Alliance adopts the most advanced design technology. Designers are proficient at computer-aided design programs and three-dimensional model program, and we set up an efficient platform for thoughts and design for designers to have an all-around experience of design.

DC Alliance has its works covering various types of public, cultural, and residential architecture, and city planning projects. We always dedicate ourselves to providing high-end and quality design service for our clients whether a project is big or small-sized, and we strive for the revival of China's modern architecture.

宁波科技园区均胜总部基地

JUNSHENG HEADQUARTER BASE OF NINGBO SCIENCE AND TECHNOLOGY PARK

项目地点：中国 · 宁波　用地面积：44 000 m²　建筑面积：130 000 m²
建筑设计：DC 国际
建筑师：崔哲，王雷鸣

LOCATION: Ningbo, China　SITE AREA: 44,000 m²　BUILDING AREA: 130,000 m²
DESIGN CORPORATION: DC Alliance
ARCHITECTS: Cui Zhe, Wang Leiming

基地位于宁波中央商务区以南，会展中心以东，呈不规则形用地，集办公、商业、会议、公寓等多种功能于一身，总用地面积 44 000 m²。整个地块具有多方向性特点，因此，各建筑应都具有较强的标志性。

项目努力提供全方位服务的社区，对人才具有最大限度的包容性，理想中的园区不再是单一的生产、办公空间，而会是集居住、生活、交流、休憩、运动等功能于一体的综合社区。

通过景观步行道的设置，将各功能空间及广场串联起来，形成一个系统的街区，给人提供了更多生活交往状态的可能性。在系统性的前提下，各功能空间和广场又具有各自不同的特点，丰富了街区的层次和内容，满足了各种生活行为的需要。

作为后工业社会的今天，工业园的内涵已经发生了一系列的重要变化，21 世纪的工业园（孵化器）应适应后工业（信息）社会的发展，找到新的发展模式，应成为以高附加值的智慧型上游产业为主体，结合生产、研发、办公、生活等多种功能和谐共融，以正生态理念为标准进行建设的混合产业社区。

生产、生活、生态三者在工业时代两两对立，进入后工业时代走向三者合一，这是后工业园建立的基础，唯有这样才能使之成为吸引人才的高地。

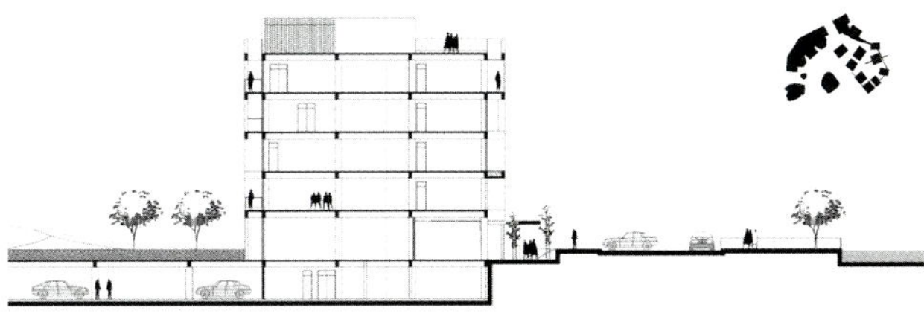

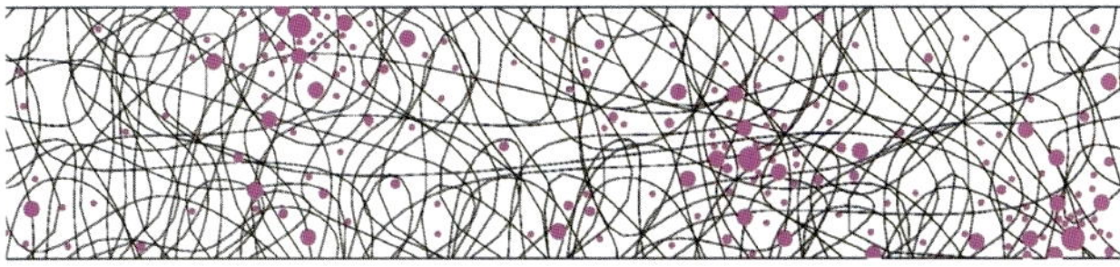

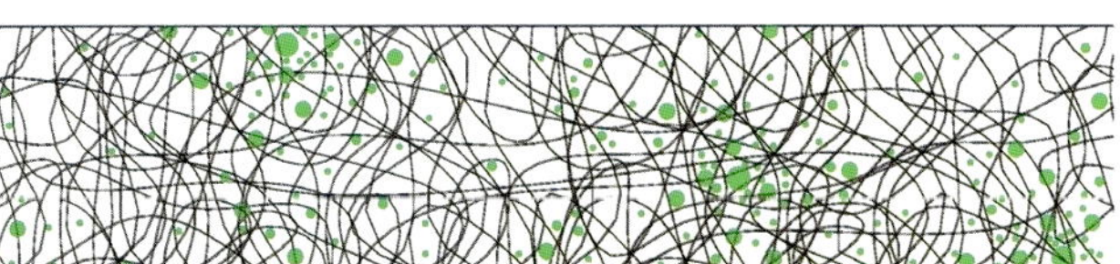

The base is located in the south of the central business district in Ningbo,the east of the Convention and Exhibition Center. It's an irregular shaped land which integrated official,commercial, conference , apartments and other functions in a total land area of 44,000 m^2. The entire area has a multi-directional characteristic, so each building should have a strong symbolic feature.

The project attempts to provide full-service community and shows maximum inclusiveness to talents. The ideal park area is no longer a single production and office space, but will be an integrated community of living and residential, leisure and communicational as well as sports and other functions.

The functional spaces and the square are linked together to form a systematic block through the pedestrain landscape setting, providing more possibility of contacts in life. In the context of systematicness, the function spaces and the squares also have their own different characteristics,which has enriched the levels and content of the neighborhood to meet the needs of the various acts of life.a

As a post-industrial society,the meaning of the industrial park has undergone a series of important changes today. The 21st Century Industrial Park (incubator) should keep in pace with the post-industrial (information) development,to find a new development model,and become a mixed industrial community which focuses on the upstream intelligent industry with high added value,combining with harmonious coexistence of the production, R & D, office, living and many other functions and be constructed in the ecological concept standard.

Production, living and ecology are opposed to each other in the industrial age, but have been integrated in the post-industrial age,which is the basis for the establishment of the post-industrial park,and only this can it be the highland to attract talents.

大红鹰学院慈溪分院新校区

NEW CAMPUS OF CIXI BRANCH OF THE RED HAWK COLLEGE

项目地点：中国 · 宁波　用地面积：300 000 m^2　建筑面积：180 000 m^2
建筑设计：DC 国际
建筑师：崔哲，王雷鸣，朱立

LOCATION: Ningbo, China　SITE AREA: 300,000 m^2　BUILDING AREA: 180,000 m^2
DESIGN CORPORATION: DC Alliance
ARCHITECTS: Cui Zhe, Wang Leiming, Zhu Li

教学综合体
科研学术交流综合体
实验实训综合体
实验行政图书综合体
体艺综合体
生活服务综合体
教师公寓
专家楼

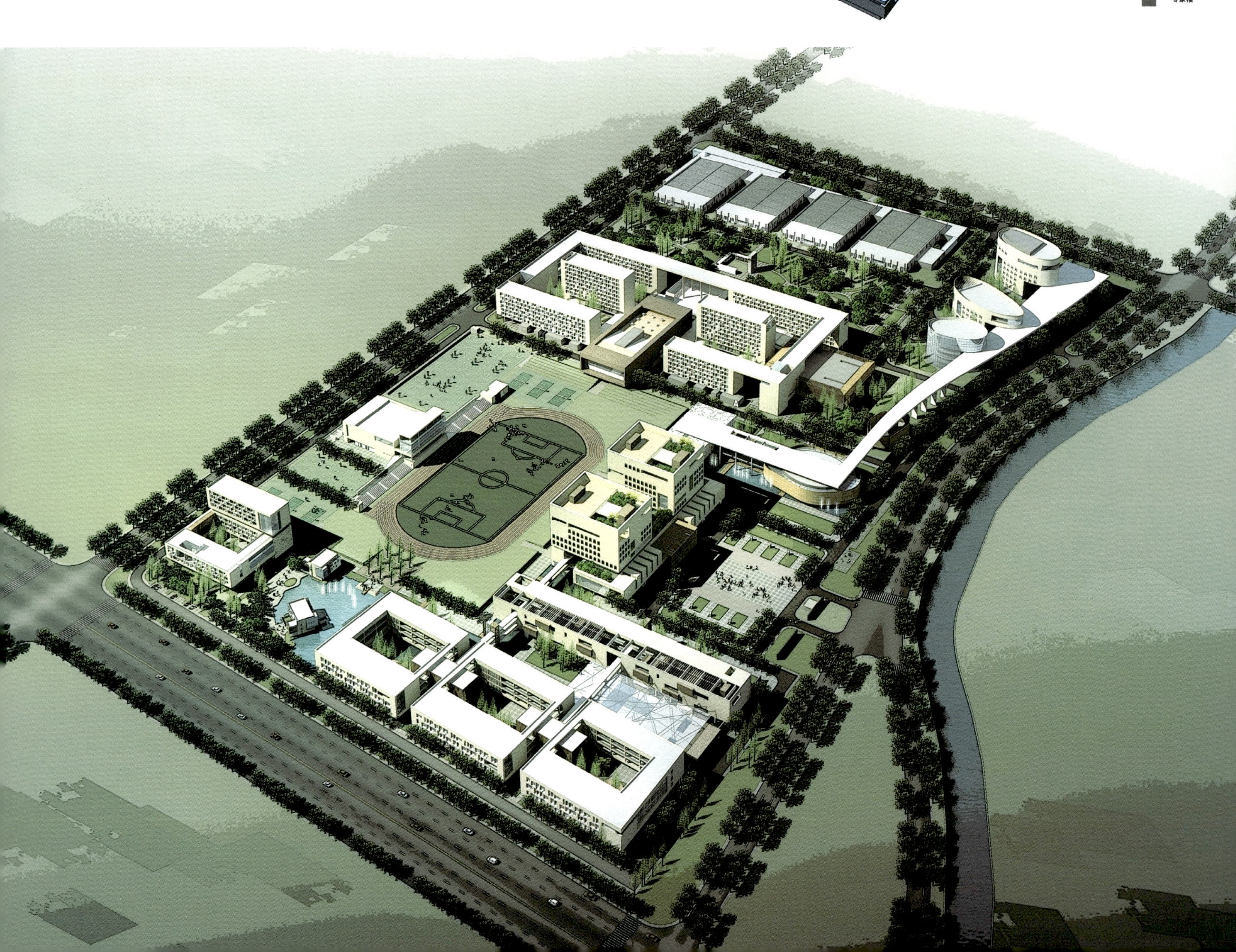

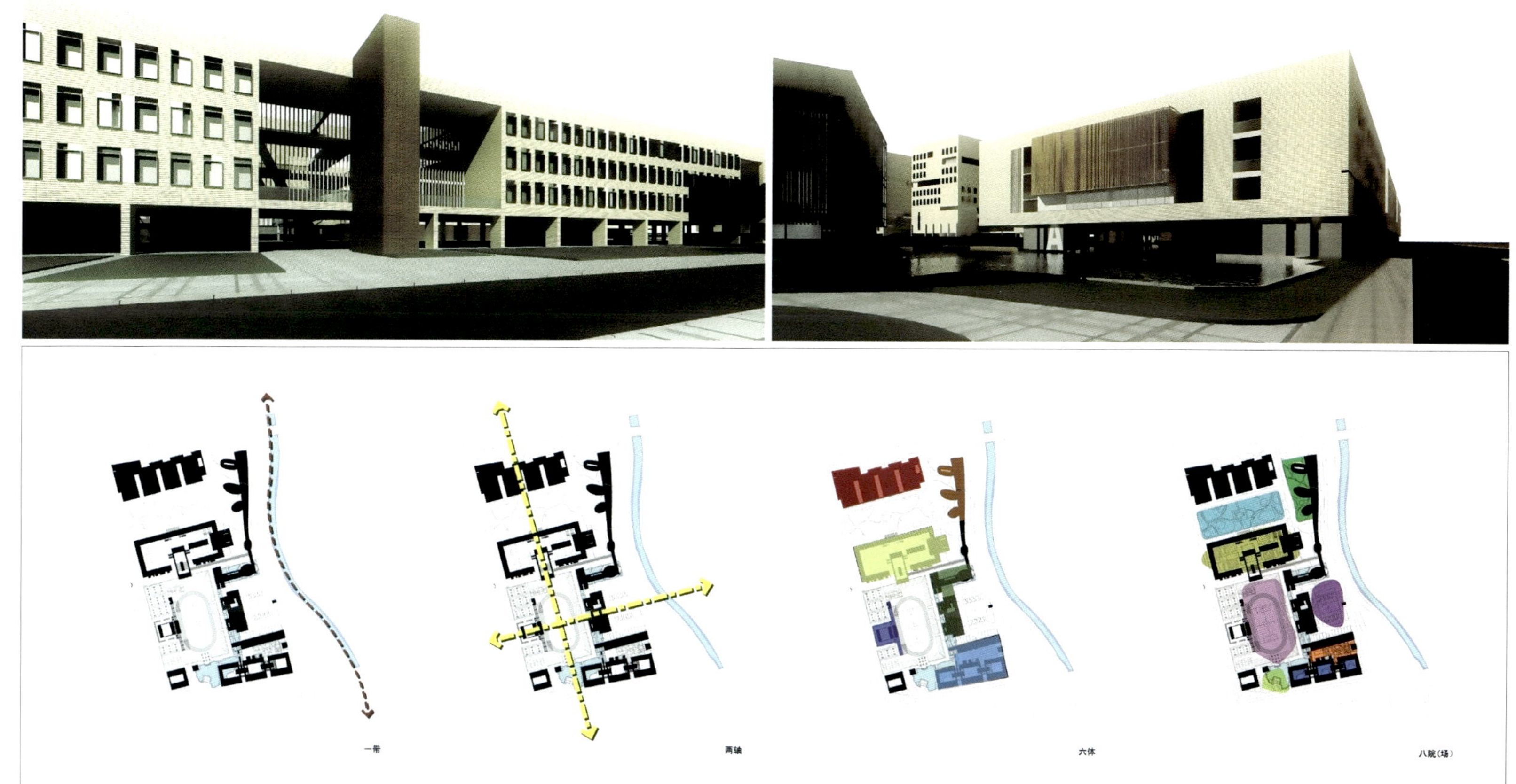

宁波大红鹰学院慈溪分院新校区位于慈溪杭州湾新区，毗邻亚洲最大的跨海大桥——杭州湾大桥，校区占地 300 000 m²，总建筑面积约 180 000 m²。规划校区一期建设用地 67 000 m²，二期建设用地 113 000 m²，计划于 2012 年完成整个校区的总体建设。

校园

"Campus"的字源为拉丁语，意指一个具有连绵不断绿色的场地。"Campus"的字意似乎有三个层面：一是界定了的绿色场地；二是建筑物及其周边的绿地景观结合起来的氛围；最后几乎可以是任何环境中一个相对固定的部分。柯布西埃开放了学校的定义。他认为："每所学院本身就是一个都市单位，无论大小它都是绿色城市。学院自身就是一个世界。"

营造

营造满足多元化的校园空间：当代教育不再是传统意义上的单向的知识讲授与传输，而是从教与学两方面向多元化方向发展。

营造泛生态化校园：网络化的学科生态环境；开放式的知识生态环境；自然优美的景观生态环境；良好的城市社会生态环境营造模拟社会化的交流场所。

空间结构

"一带"：校园东面一湾曲线湖水使城市界面凸凹有致，给人们留下完整明确的学校印象。

"二轴"：两条相互垂直的主轴线清晰有力地界定了校园的空间结构骨架。

"六体"：六块功能复合体形成了若干个场景单元，能够为人们提供一种完整的空间体验，使人们能清晰地辨别自己身在何处，让校园的使用便捷有效。

"八院"：八个院落（广场）根据位置的不同，以内向或外向的形式呈现，与校园建筑空间紧密结合，交融编织出生态、休闲、人景互动的校园景观。

体

在教学综合体区域，由于架空层和空中花园的参与，给人一种既是"内部"也是"外部"的感受，二者之间有着强烈的关联，以至于使人无法分辨是在一座建筑物的内部抑或在建筑之间，空间具有了更强的可进入性，与外部结构得到了更紧密的结合。

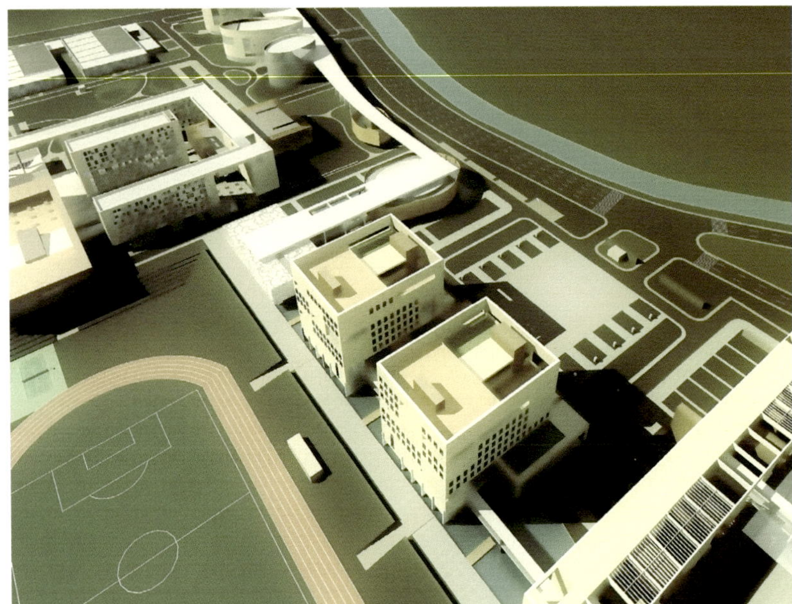

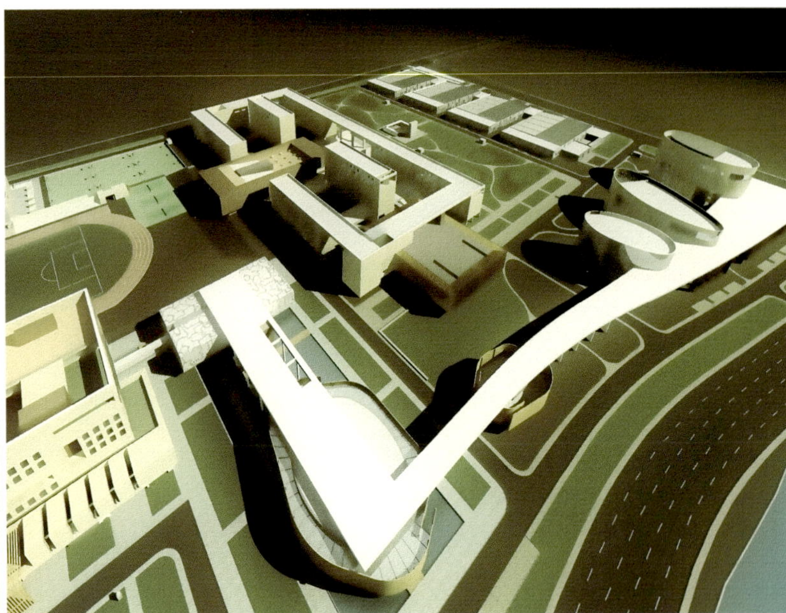

store
PRACTICE FACTORY
SCIENCE EXCHANGE CENTER
RESEARCH CENTER
GREEN PARK
GARDEN
STUDENTS CENTER
STUDENTS HOSTEL
service
service
MEETING CENTER
PLAZA
STADIUM
GYM
COMPOSITE TEACHING
STUDY PLAZA
TEACHER'S HOSTEL

The new campus of Cixi Branch of the Red Hawk College is located in the new zone of Hangzhou Bay in Cixi city, adjacent to Asia's largest cross-sea bridge, Hangzhou Bay Bridge. It covers an area of 300,000 m² with total construction area of about 180,000 m². The first stage of campus planned building site of 67,000 m², and the second 113,000 m². The overall campus construction is planned to be completed in 2012.

Campus

The word "campus" is of Latin origin, meaning a space with continuous green. The word's meaning seems to have three levels: first, a defined green space; second, the atmosphere combining the building and its surrounding green landscape; and the last, a relatively fixed part in almost any environment. Le Corbusier had openned up the definition of the school. He said: "Each college itself is an urban unit, large or small, it is a green city. College itself is an inependent world."

Design

To create a diversified campus space: contemporary education is no longer the traditional sense of the simple knowledge teaching and transmission, but the development of the two aspects to a wide range of directions.

To create a Pan-ecological Campus: the networked environmental disciplines; open knowledge environment; natural and beautiful landscape ecological environment as well as the favorable social environment creates a simulation of urban social exchange places.

Space structure

"One belt" : a curve lake in the east of the campus makes the city interface order with convex and concave, which leaves people a clear impression of the whole campus.

"Two axes": the two main axis perpendicular to each other, unambiguously defined the main spatial structure of the campus.

"Six complexes" : six functional complexes formed a number of scene units to provide people with a complete space experience where they can clearly identify where they are and can conveniently and effectively use the campus.

"Eight yards": eight yards (squares), according to the different locations,are presented outward or inward and closely connected with the campus building space, weaving a blend of ecological and leisure, interactive campus landscape of people and view.

Complex

In the complex of teaching area, due to the involvement of the empty space and the air garden, giving the impressions of both "inside" and "outside", there is a strong correlation between the two, so that you can not tell whether you are inside a building or the space between the buildings. It has more accessibility, and more close connection to the external structure.

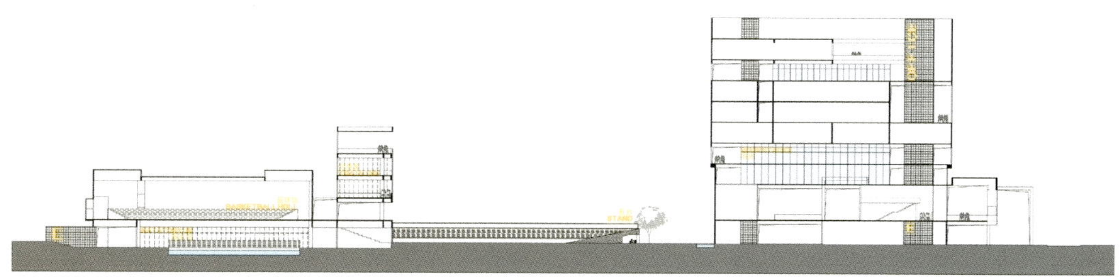

慈溪科技创业中心

CIXI TECHNOLOGY AND INNOVATION CENTER

项目地点：中国 · 宁波　**用地面积**：29 000 m²　**建筑面积**：40 000 m²
建筑设计：DC 国际
建筑师：董屹，揭涌，李光

LOCATION: Ningbo, China　SITE AREA: 29,000 m²　BUILDING AREA: 40,000 m²
DESIGN CORPORATION: DC Alliance
ARCHITECTS: Dong Yi, Jie Yong, Li Guang

慈溪经济开发区（出口加工区）位于浙江省慈溪市北部，北濒杭州湾，是长江三角洲经济圈南翼、环杭州湾地区——上海、杭州、宁波三大中心城市经济金三角的中心。

本项目取名"智城"，其本身具有能自我完成的微型城市结构，区域内空间的层级通道被尽量加长，开放性随着基本动线逐级递减，空间在理性的秩序中呈现一种自组织的状态。在这个项目中，建筑的结构像自给自足的城市，有"街道"、"广场"和独立的建筑单元，"使每一个地方成为一个场所，使每一个建筑和每一个城市成为一系列的场所。

为容纳多样化的功能，并保持可持续发展的能力，建筑在一个模数为 8 m x 8 m x 4 m 的立体格网中构建建筑实体，以一种简单的形式保证较大规模较长时期的发展可能，综合体的这一基本结构就这样等待"填充"的结果，并且从不同表达的综合中获取它的特征。

在追求理性与价值的现代建筑里，诗意总是可遇不可求。但事实上诗意的创造往往来源于一些微小的感受，来自于对建筑细节的体验。建筑的表皮往往带有某种隐喻，引起联想与潜意识里角色的认同，并使人们在若隐若现之间被感动。

Cixi Economic Development Zone (Export Processing Zone) is located in the north of Cixi City, adjacent to the Hangzhou Bay in Zhejiang Province of its north and is the southern wing of Yangtze Delta Economic Zone, and it's also surrounded by Hangzhou Bay Area - in the Golden Triangle Centre of three municipal cities Shanghai, Hangzhou, Ningbo.

The project is called "Smart City", which has self-completed mini-urban structure. The hierarchy access of the space within the area was well lengthened ,and the openness progressively reduces to keep in line with the the basic fixed line. The space presents a self-organized state in a rational order. In this project, the structure is like a self-contained city with "street", "square" and independent construction units, making every place a site and every building and city a series of sites.

To accommodate a variety of functions and maintain the sustainable development capacity we built a construction entity in a dimensional grid of 8 meters by 8 meters by 4 meters module, in order to guarantee a possiblity of large-scale and longer term development in a simple form, the basic structure of the synthesis thus waited for a result of "filling"and got its features from the integration of different expressions.

In the modern architecture that pursues reason and value the poetry mood is always a matter of luck. But in fact the creation of poetic feeling often comes from some minor senses, and the details of construction experience. Building skin often carries a metaphor to arouse association with unconscious recognition of the role, and touch people indistinctly.

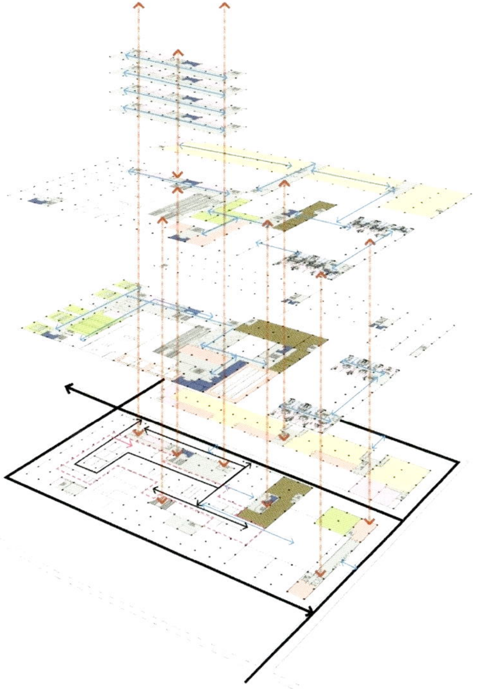

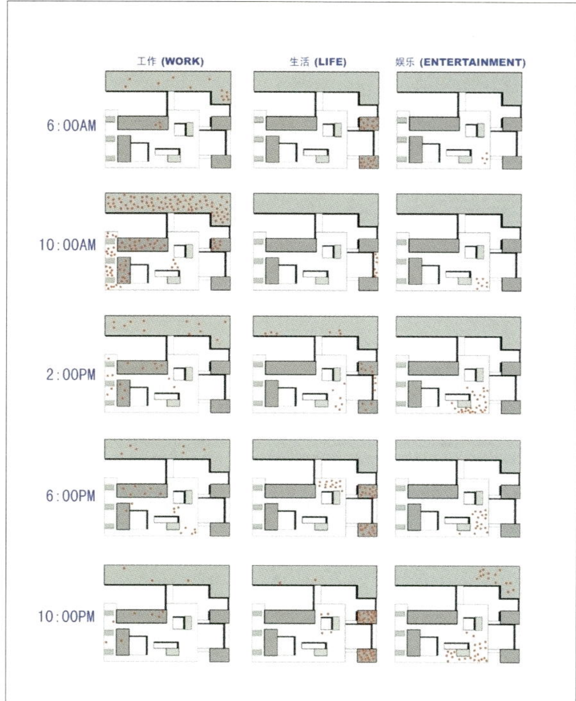

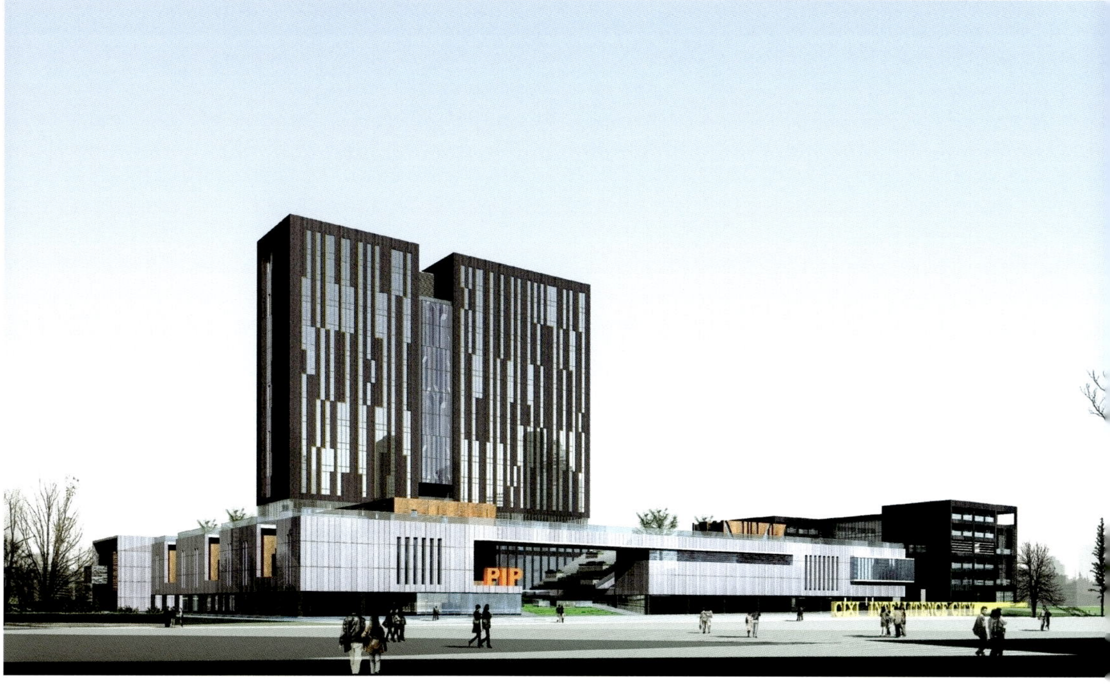

宁波联盛城市商业广场

NINGBO LIANSHENG CITY COMMERCIAL SQUARE

项目地点：中国 · 宁波　**用地面积：**28 000 m^2　**建筑面积：**79 000 m^2
建筑设计：DC 国际
建筑师：崔哲，董屹

LOCATION: Ningbo, China　SITE AREA: 28,000 m^2　BUILDING AREA: 79,000 m^2
DESIGN CORPORATION: DC Alliance
ARCHITECTS: Cui Zhe, Dong Yi

基地位于宁波鄞州区宁南北路西侧，贸城西路北侧，基地为呈狭长不规则形用地。地块的限制使得设计师有了更丰富的想象，多变的形体、流动的街区以及开放的广场构成了良好的购物环境，使顾客在购物的同时享受建筑所带来的乐趣。

后现代的商业空间的最终目标就是传递一种"愉悦"，这一"愉悦"来源于商品和商业空间对于"商品"意义之一的"时尚"的追逐。它声明商业建筑首先应该成为商品，即成为具广告性的标志（商业性），它鼓励人们使用、消费和愉悦。

这一特殊商品将与后现代社会共享着特殊的人文场景，一个庞然大物、功能多样、形态混杂、冲撞、重叠、空间壮观、色彩亮丽迷人、材质可触摸。它着实是一个时尚的商品，一个可以被消费和愉悦的地方，设计就是试图实现建筑从空间向商品的转变，实现消费的自娱自乐即真正的"愉悦"性。

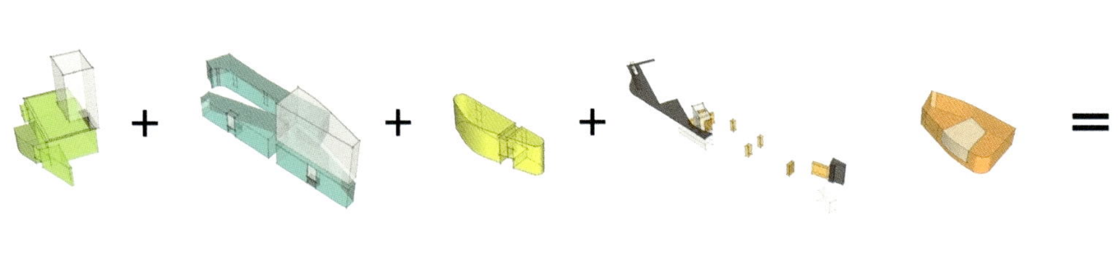

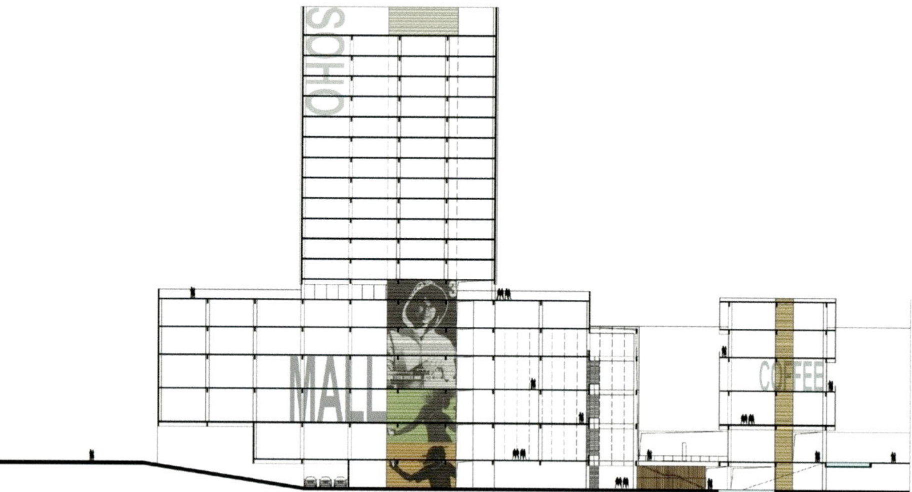

The base is located in the west side of Ningnan North Road in Yinzhou District, north side of Maocheng West Road. The base is a narrow and irregular building site and the site limit allows designers a richness of imagination. The diversity of shapes, flowing blocks and open squares together constitute a fantastic shopping environment for customers to go shopping while enjoying the fun brought by the building.

The ultimate goal of post-modern commercial space is passing a sense of "pleasure" , which comes from the chase of "fashion" in products and commercial space, i.e.one of the meanings of "commodity" It declared that at first commercial building should become a commodity, that is, the advertising logo (commerciality), which encourages the application , consumption and recreation. This special product will share the special human scene with the post-modern society. It's a versatile jumbo with hybrid forms and functions, collisions and overlaps and spectacular space, the tangible materials, which are all colorful and attractive. It's really a fashionable product where consumption is accompanied by pleasure. Design is just the bridge that connects space and goods in architecture to achieve the real "joviality" of self-recreation and happiness in consumption.

DCI DESIGNGR-OUP INTERNATIO-NAL

美国DCI思亚国际设计集团

Company Backgaround

DCI Designgroup International (DCI) is headquartered in New York. Being an international architecture and design firm, DCI has been active in the East Asia and China market for the past two decades, and has set up our Asia pacific headquarters office in Shanghai, and a regional office in Beijing since 2002. DCI integrate the best resources, the latest technology and the local knowledge, to provide focused services tailored to customer's market differentiation and specific stages of their development projects. Our goal is to provide customers high-quality, efficient and comprehensive service based on our global design resources, through our experiences in various regions, information exchange and resource integration.

Scope of Services

Our multidisciplinary practice is built around a number of building types key to the large scale urban mixed-use developments, specialized from retail malls and stores, commercial offices to hotel, hospitality and residential apartments. Within this focused core building types, DCI provide a comprehensive one stop shop, services range from master planning, architecture, interior design to environmental graphics, way-finding design.

Relative Experience

DCI has successfully completed projects in major cities throughout East Asia, in Korea, China and Japan. We are privileged to parter with clients many of the leaders in retail, mixed-use and hospitality operators and developers, such as Shanghai Bailian, Beijing COFCO, Zhejiang Intime, Hong Kong New World Group, Korea Lotte Group, Japan Hankyu Department Store etc.. Our successful projects include: Mix-use projects, such as Lotte C2 World, Lotte main, Osaka City Center, Beijing Joy City, Bailian Jinshan Mall. Department Store projects, such as: Lotte Intime Department Store, Taipei 101 MALL, Wuxi Springland Department Store. Hotels and apartment projects, such as: the Yangzho Slender West Lake Resort & Spa, Taicang InterContinental Holiday Inn hotel, etc.

公司背景

美国DCI 思亚国际设计集团，总部设于美国纽约。近十年来致力于服务中国市场，在上海和北京设有区域性总部。公司拥有国际化的专业规划、建筑和室内设计建筑师及商业策划顾问。将最优的资源、最先进的理念和最新的技术进行整合，针对客户的市场差异化与特定的开发阶段，提供聚焦的服务。目标是以全球性的设计资源，透过各地区的经验累积，资讯的交换与资源的整合，为客户提供更为优质、快捷以及全面性的服务。

服务领域

DCI拥有非常专业的资深设计团队，提供从大型城市综合体购物中心、零售百货项目，到写字楼、五星级酒店和顶级公寓项目，从概念方案到施工图纸的全方位专业设计服务。DCI尤其对大型复合式开发项目，在项目的总体规划、建筑设计、室内空间设计以及室内环境艺术和标志系统设计，各方面都拥有国内外众多丰富的成功实践经验。

相关经验

DCI的设计项目已经遍及韩国和中国境内各大主要城市。拥有跻身亚洲与中国商业地产领袖企业客户群上海百联集团、中粮集团、银泰百货集团、香港新世界地产、韩国乐天集团、日本阪急百货等。其成功案例包含：大型复合式开发项目，如韩国C2乐天世界、韩国乐天百货总店、日本大阪中心、中粮北京西单大悦城、百联金山购物中心；百货商业项目，如北京乐天银泰百货、台北101MALL、华地无锡八佰伴购物中心；酒店和公寓项目，如扬州瘦西湖度假村、太仓洲际假日酒店等。

公司创始人/主创设计师之一

- 张乃尊　Nai-Tsang Chang

张乃尊先生是美国DCI思亚国际设计集团的负责人之一。张先生在美国和亚洲的一些城市及地区从事多功能复合式开发项目与购物中心、零售专卖店、百货公司等规划和设计工作，迄今已有 30 多年的经验。

在入主思亚国际设计集团之前，张先生是美国FRCH 国际设计集团的资深副总裁，为领导一个超过40人的建筑部门的负责人。他主要负责公司业务发展、商业战略和发展、市场行销以及项目设计与监督。

教育背景

1981年哈佛大学设计研究院 建筑与都市设计 硕士

1978年台湾东海大学 建筑专业 学士

美国伊利诺斯州注册建筑师

美国建筑师协会会员

国际购物中心协会会员

乐天 C2 世界

LOTTE C2 WORLD

项目地点：韩国 · 首尔　占地面积：87 770 m^2　建筑面积：607 850 m^2

建筑设计：美国 DCI 思亚国际设计集团

建筑师：张乃尊，Michael Duddy, 金名均，刘灿，施连超，王小燕，沈海燕，曹兢，蒋慧冰，徐建媚，邵云美

LOCATION: Seoul, Korea　PLOT AREA: 87,770 m^2　BUILDING AREA: 607,850 m^2

DESIGN CORPORATION: DCI Designgroup International

ARCHITECTS: Nai-Tsang Chang, Michael Duddy, Jin Mingjun, Liu Can, Shi Lianchao, Wang Xiaoyan, Shen Haiyan, Cao Jing, Jiang Huiping, Xu Jianmei, Shao Yunmei

第一代乐天世界在 1988 年首尔主办奥运会前开幕，为以主题乐园为主，购物、酒店、百货、量贩与运动中心为辅的一个标杆性开发项目，是当时韩国经济起飞的代表性地标建筑。二十年后，邻近同样硕大的基地，乐天正在进行第二代乐天世界的开发。

第二代乐天 C2 世界是一个新世纪的高水平细致的现代复合式开发项目，期望以此项目界定新一代复合式城市核心开发的模式，更进一步创造乐天品牌的再一巅峰及现代韩国的新象征。

乐天 C2 世界中包含了一座由美国 SOM 建筑设计事务所设计的 200 000 m^2、555 m 高 112 层的超高层六星级酒店与超甲级写字楼，建成后将是世界最高建筑之一。另外还有由美国 DCI 设计的包括地下 1 层至地上 11 层购物、餐饮、文化、娱乐的裙楼建筑群。地下 2 ~ 4 层为 2500 个车位的停车场。有别于现代同值性高的超大街廊建筑体的做法，乐天 C2 世界采取了较多元化、多单体的自然形成的城市肌理的取向，具备了丰富多样的使用功能与业种的各单栋建筑主题馆。建筑形式从简洁细腻到凸显、动感，经由整体的基地配置与景观规划设计，有机的、和谐地结合为一体。

各建筑设计以现代当下的设计风格为基调，强调每个组群甚至每栋建筑个别的设计元素与风格多样性的同时，又以时代性的建筑思潮串接相互之间，并与主塔楼的雕塑性及高科技玻璃帷幕设计相协调。如今现代主义步入第二个世纪，已经成熟为一种永恒的、经典的建筑设计样式。乐天 C2 实际的主塔楼与建筑群的设计都蕴涵了这样一个素质，充分运用了简洁有力的造型、体量，高品质材料及结构来表现建筑的现代风格。

乐天 C2 世界地面总共有 13 栋建筑，其配置受超高层塔楼的主导地位的影响，除了超高层塔楼外，以建筑的功能与业种分为四个组群。第一组群位居项目两端的主街交叉口，两栋建筑为高档国际百货名店 AVENUEL 与生活家居馆，以一个玻璃中庭连接；第二组群的四栋建筑以年轻时尚服饰与时尚餐饮为功能，形成一个十字形的玻璃连廊；第三组群位于项目东端，三栋建筑以运动、儿童及娱乐、影院为主题，共 12 层，由一个 80 m 高的玻璃中庭连接；第四组群三栋建筑以文化馆、音乐厅为主题，位居项目南侧，隔着中央广场与东面的超高层塔楼相对应。整个项目面向奥林匹克路，共有 3 个户外广场，以一个十字步行街连接。中央广场面向基地南面的湖面开放，形成一个由城市广场空间连接城市公园空间的开放空间系统。

The first-generation Lotte World was opened before the opening of Seoul Olympics in 1988. It's a series of landmark development projects including the main part of the theme park supplemented by shopping areas, hotels, department stores, hypermarkets and sports center, representing South Korea's economic soaring. Twenty years later, close to the same huge base, the second generation of Lotte worldwide development is just ongoing.

C2, the second generation of Lotte World,is a modern composite development project of high level and delicacy in the new century. It's expected to define the new breed of urban core development model and further create a new peak of Lotte brand and a new symbol of modern Korea.

C2 contains a building of an six-star rate hotel and first rate office building designed by the U. S. architectural design firm SOM, which is 200,000 square meters 555 meters high with 112 storeys expected to be one of the world's tallest buildings. There are also annex spaces for shopping, dining, cultural and entertainment services from the basement floor to the 10th floor.

From 2-4 floors underground are parking spaces accomadating

2,500 cars. Unlike other super high street constructions, C2 Lotte World takes a more pluralistic, multi-monomer natural urban texture orientation , with single theme museums of various functions. From simple and exquisite to outstanding and dynamic,the construction form reveals the harmonious combination of the base configuration and the overall landscape design.

Based on the modern contemporary design style, the architectural designs emphasize multifarious design elements and styles of each block or each building, in the mean time they are connected with each other by Architectural Trends, and coordinated with the main turrical sculpture and the high-tech glass curtains. Today, modernism has been stepped into a new century and become the eternal and classic architectural style. The design of the main tower and buildings of C2 all contain such a quality. It takes full use of the simple but powerful style, body volume, high-quality materials structures to show the modern style.

There are totally 13 buildings in the C2 project. Affected by the leading position of a high-rise turrent, its configuration is divided into four groups of different architectursl functions. The first group at the intersection of the two main streets contains two buildings, the top grade international stores AVENUEL and the living-at-home Shoppes (LIVING HALL), which are connected by a glass atrium ; the second group of four buildings constitue the young and trendy fashion and catering spaces named DOWNTOWN LIVE, looking like a cruciform glass corridor (GALLERIA); the third group at the eastern end of the project contains three buildings for sports, children ammusement and cinema, with 12 storeys connected by a 80-meter-high glass atrium; the fourth group in the south of the project contains three buildings used as cultural centers and concert hall, corresponding to the skyscraper of the southeast across the central square. The entire project is opposed to the Olympic road with three outdoor squares linked by a pedestrian cross. The central Plaza, opening to the the lake in the south of the base, forms a development space system where the urban square is connected to the city park.

DGBK ARCHITECTS INTERNATI-ONAL

张健

Greg Dowling

发展历程

DGBK 1972年成立于加拿大温哥华，目前已在加拿大、美国、中国等国家的多个城市开设分公司，项目遍及美洲、欧洲、澳洲和亚洲的十余个国家，为北美规模最大的综合设计公司之一。2003年—2004年承接工程总额超过了12亿美元，位列加拿大建筑设计公司前茅。

DGBK在中国的业务始于1994年，在10余年的时间内，公司在中国完成了一批标志性建筑。DGBK于2007年在上海成立了亚洲区总部，并在北京、无锡、首尔等城市设立了分公司和代表处。目前，DGBK亚洲公司已在中国、日本、韩国、印度等国家的30余个城市完成了大量重要项目。

DGBK始终将卓越设计、创新技术和精诚合作（DESIGN EXCELLENCE, INNOVATION, COLLABORATION）作为经营宗旨，并通过服务诠释“设计创造价值”的核心理念！

设计团队

DGBK拥有加拿大、美国、中国注册建筑设计师60余人，2位加拿大皇家建筑师协会常务理事（RAY GRIFFIN 和 SEBASTIAN BUTLER），加拿大皇家建筑师协会会员38人，LEED认证建筑师12人，专业建筑师逾300人。国际化的设计团队为公司在世界各地项目的成功提供了技术上和文化上的专业保障。

业务范围

DGBK的项目遍及美洲、欧洲、澳洲和亚洲的10余个国家，业务范围涵盖城市规划、投资策划、大型公共建筑设计、民用建筑设计、室内装饰设计、园林景观设计、建筑节能和智能节能化等领域。

特色专长

DGBK公司一直把特色专长设计作为发展的重要方向，在城市规划、大型公建、木结构建筑、绿色设计等领域取得了突出的成就，形成了业务广泛、特色突出的风格。从1972年起，DGBK就开始了建筑节能设计，并在今后的几十年中把绿色节能建筑标准贯彻到了绝大多数项目中，在国际范围内取得了诸多奖项。公司率先在国际上获得了LEED组织认证，培养了一批LEED认证建筑设计师。

Company History

DGBK Architects entered the design world in 1972 in Vancouver, Canada, and has steadily grown in size and stature to become one of North America's most highly regarded architectural practices. The firm currently has offices in Canada, USA, China and other countries with over 300 employees.Our work has spread throughout the world, including projects in Canada, USA, China, Korea, Japan, Australia, India, and the South Pacific. The 2003-2004 annual project value was over 1.2 billion US dollars, which ranked us as one of the top Canadian firms.

Since 1994, DGBK has been actively involved in the China market for more than 15 years, and has completed a great number of landmark projects. In order to meet the increasing needs of our Asian clients, DGBK established its Asian headquarter in 2007 in Shanghai, and opened several representative offices in Beijing, Wuxi, and Seoul. By 2008, DGBK has accomplished over 100 projects in more than 30 cities in China, Japan, South Korea, and India.

Our success has been a product of our principle of "Design Excellence, Innovation, and Collaboration". DGBK believes in the philosophy of "Design Creates Value" and will consistently address that through our service.

Design Team

DGBK has over 300 associates and professional employees, including over 60 registered American, Canadian, and Chinese architects, 2 RAIC fellows (RAY GRIFFIN and SEBASTIAN BUTLER) , 38 RAIC members, and 12 LEED certificated architects. Our people come from around the world and bring a truly international background of experience and understanding, both professionally and culturally. This international point-of-view contributes to the success of our work by providing a broader insight and a deeper appreciation for the different influences that apply to the design of the built environment in other parts of the world.

Scope of Service

Our project experience is extremely broad, including master planning, investment and programming, large-scale commercial complex, health care, residential, sports and recreational, educational, conventional, hospitality, transportation, interior, landscape, green design, and building-intelligent systems.

Expertise

We are specialists in sustainable design and innovative technologies such as wood framed buildings, and have lectured and written on those subjects throughout the world. Our work has been internationally recognized through awards from both professional associations and clients.

The history of DGBK's green design started from 1972. In the following years, DGBK consistently applied green design into most of our projects, and won numerous international awards. DGBK is regarded as one of the world's leading green design firms, and is accredited by both USA and Canada LEED associations.

银辉集团大厦

YINHUI GROUP SKYSCRAPER

项目地点：中国 · 江苏　规模：150 000 m^2
建筑设计：DGBK 建筑设计咨询（上海）有限公司
建筑师：Greg Dowling, 张健

LOCATION: Jiangsu, China　SCALE: 150,000 m^2
DESIGN CORPORATION: DGBK Architects International (Shanghai) Co., Ltd.
ARCHITECTS: Greg Dowling, Zhang Jian

此超高层为无锡市崇安市二期改造的重点项目。270 m 的高度将成为无锡市的第一高度，也将成为世界最高的 100 栋建筑之一。该项目的裙房部分是一个 4 层的购物中心，5 层到 45 层为商业办公，以上部分为五星级酒店。目前该项目处于前期策划和概念性初步设计阶段。

This project is part of the second phase of the city core commercial area redevelopment. This 270 m skyscraper will be the city's highest building and ranked as top 100 in the world. This complex includes a shopping centre occupying from the ground to the 4th floor, office building from 5th floor to the 45th floor, and the rest for a 5-star hotel.

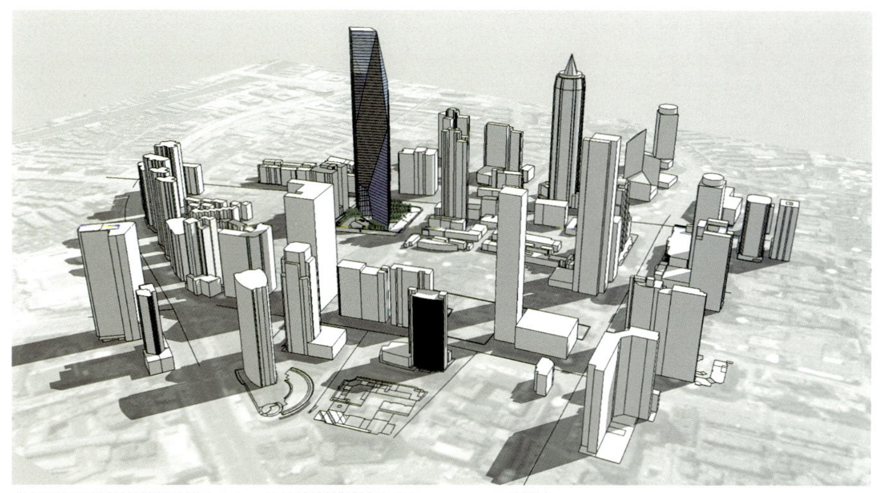

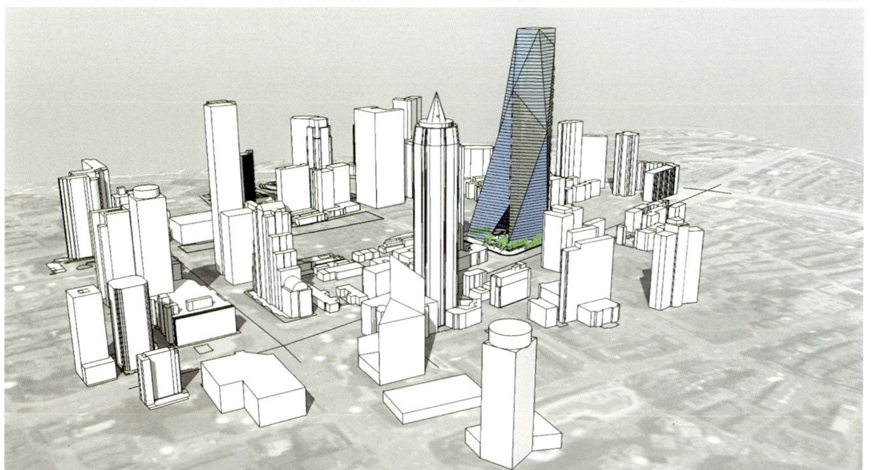

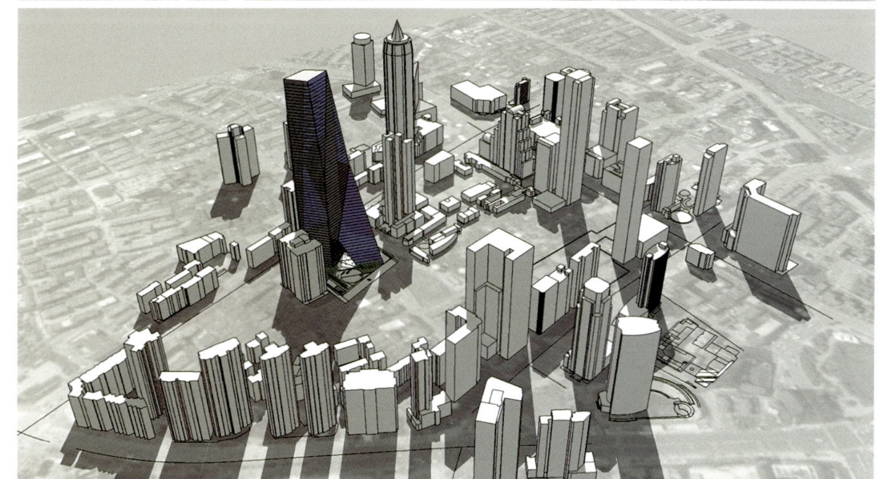

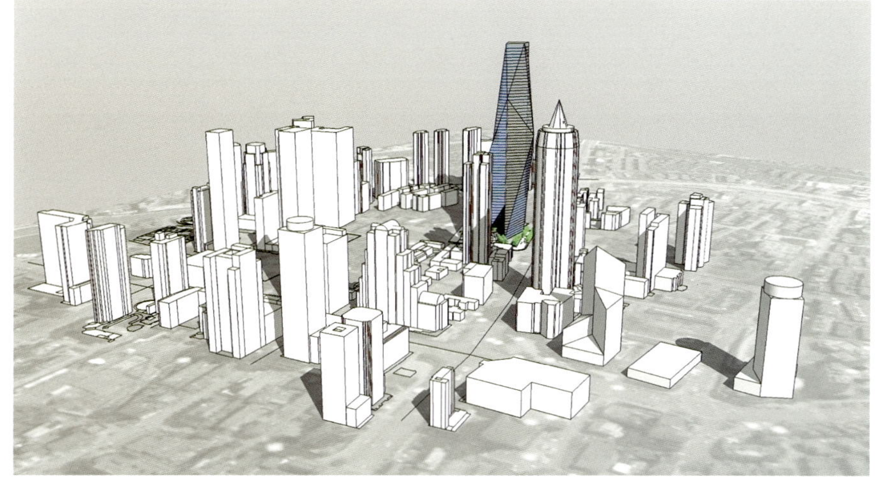

泰州华侨城高尔夫会所

TAIZHOU OCT GOLF CLUB-HOUSE

项目地点：中国·泰州 项目规模：5000 m²
建筑设计：DGBK 建筑设计咨询（上海）有限公司
建筑师：张健，姚嘉，丁盛杰

LOCATION: Taizhou, China PROJECT AREA: 5000 m²
DESIGN CORPORATION: DGBK Architects International (Shanghai) Co., Ltd.
ARCHITECTS: Zhang Jian, Yao Jia, Ding Shenjie

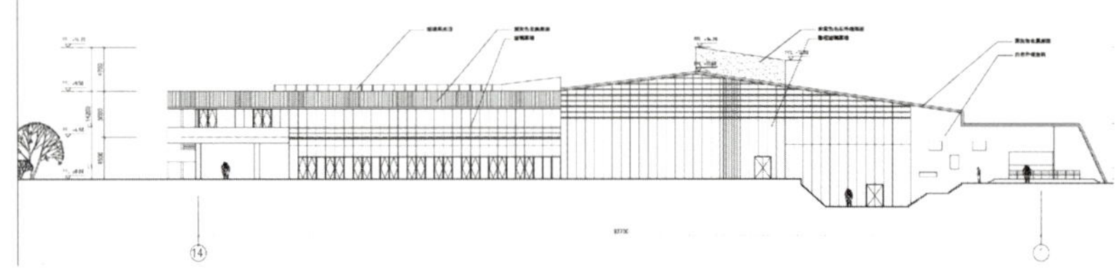

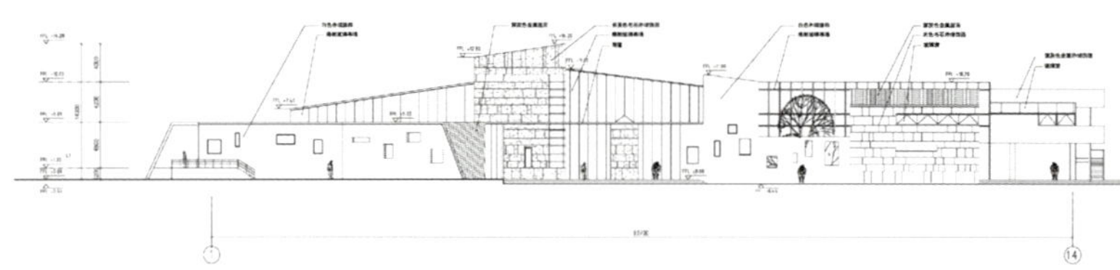

本案位于泰州溱湖风景区西端，建筑面积 5000 m²。项目面临的挑战之一是需要在局促的场地中建设两个流线复杂的建筑单体，同时还要具有相当高的水准。项目拥有极佳的地理条件：地处整个高尔夫球场的中部，享有坡地、草场与树林的绚丽景致。设计结合场地环境，合理布置垂直与水平的交通流线，强调功能及空间节奏的变化，以期通过现代、别致的空间与材质组合创造出具有独特人文气息的休闲空间。

Located in the west Qinhu lake scenic resort, this project has an area of 5000 square meters for the construction. One of the most challenging aspects is that a state of the art architecture should be created to meet the need of building two complicated sole objects within such a small area. This project is in the mid of the golf course which enjoys spectacular views, including sloping lands, lawns and forests. We carefully considered the relationship between the site and surrounding environment, the requirements of horizontal and vertical traffic, so that all the needs could be filled and a well organized design could be provided. Besides, some modern materials have been applied in order to create an unique recreation centre with a literary image.

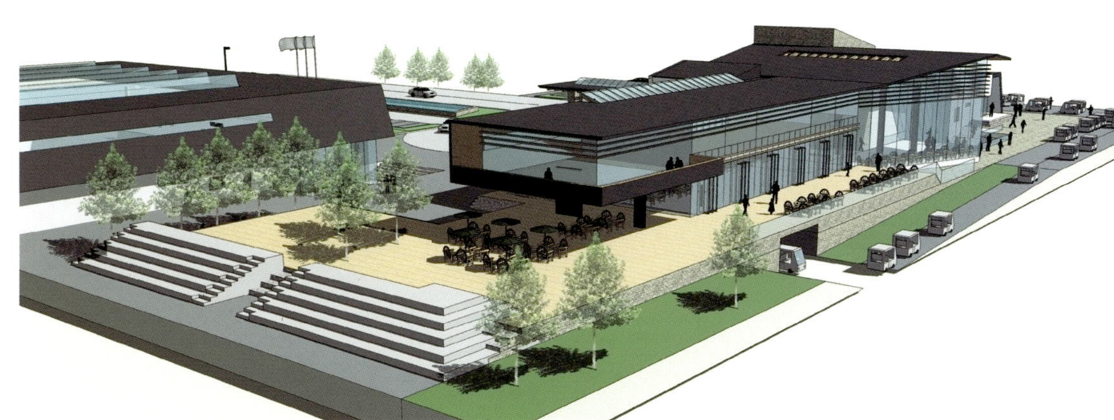

加拿大 BC 省野生动物研究中心

BC WILDLIFE DISCOVERY CENTRE

项目地点：加拿大 BC 省　项目规模：1143 m^2
建筑设计：DGBK 建筑设计咨询（上海）有限公司
建筑师：Greg Dowling, 张健

LOCATION: BC, Canada　SCALE: 1143 m^2
DESIGN CORPORATION: DGBK Architects International (Shanghai) Co., Ltd.
ARCHITECTS: Greg Dowling, Zhang Jian

这个 12 300 ft^2（1143 m^2）的研究中心设于现有的野生动物公园展览中心之中。公园的使命是特别关注集中在不列颠哥伦比亚省的土著物种、它们的栖息地以及诸多影响该省自然环境的问题。为此，新中心的设计是高度可持续的，在一个非常具有挑战性的项目预算的基础上利用许多创新的绿色设计特点获得了 LEED 银级认证。

其中有太阳能和风能为能源系统的示范项目。建筑采用地热式的加热和冷却系统，利用一个废弃的井口作为其来源，大大减少了投资成本以及对建设基地的破坏性。系统中排放的水流入水鸟栖息的湿地，并最终对现有的产卵的鲑鱼和其他野生动物栖息地的保护起到有益作用。

建筑设计形态在采暖季节最大限度地暴露在太阳下进行太阳能采暖，并能限制冷却季节的阳光侵入。创新的木材技术被用于建筑结构，节约利用自然资源。

This 12,300 sq.ft (1143 sq.m.) Interpretive Centre is an addition to an existing Visitor Centre for this wildlife park. The Park's mandate is in the process of changing to focus exclusively on indigenous species in British Columbia, their habitat and the many environmental issues that effect the natural environment of the Province. In keeping with this mandate, the design of the new centre is highly sustainable, utilizing many innovative green design features, within a very challenging project budget, to achieve a LEED Silver Certification.

Among these innovations are demonstration projects for solar and wind energy systems. The heating and cooling for the building is from a geothermal system that utilizes an abandoned well for its source – greatly reducing capital costs and disruption of the site. Discharge water from the system charges a wetland for waterfowl and ultimately returns to an existing stream for spawning salmon and other wildlife habitat restoration. The architectural design is shaped by the sun to maximize solar exposure in the heating season and limit soar intrusion in the cooling season. Innovative wood technologies were used for the building structure to economize on the use of natural resources.

DNA_ DESIGN AND ARCHITECT-URE

DnA _Design and Architecture的建筑实践着眼于当代社会，关注各个科学领域，跨越不同的尺度。我们认为建筑并非孤立的一个学科，而是涉及当代社会的各个层面，各个领域的多维度的立体构架。我们认为文脉（context）、功能（function）以及这二者之间的相互作用是决定设计（design）、诠释建筑（architecture）的基本元素，亦建筑的基因"DNA"。由此展开的研究和讨论，不仅激发每一个项目独特创新的理念，也使得我们的设计充分适应并融合到当代多样与复杂的社会，最大程度地参与社会变革。

文脉、功能以及二者之间潜在的内在关联将建筑发展成为一种多维度的表达，尤其是为建筑的使用者带来新的体验、发现和变革。我们相信建筑会持续地影响、拓展、激发我们生活的灵感。

DnA _Design and Architecture is an interdisciplinary practice addressing our contemporary living environment, both physically and socially, from scales small or large. Our approach to projects starts with research and discussion on context, program, and their interaction, which we believe are the fundamental elements, or the "DnA"of building, that will define design and architecture, to adapt, engage, and contribute to our society of multiplicity and complexity.

Context, program, and their potential relationship, will cultivate architecture into a multi-dimensional expression and generate new experiment and exploration for users. Architecture will continue to influence and inspire our contemporary life.

公司合伙人及主要设计师

- 徐甜甜　Xu Tiantian

中国清华大学建筑设计专业，学士学位

美国哈佛大学城市设计专业，硕士学位

2000—2004年实践于美国、荷兰等国际性建筑事务所

包括鹿特丹OMA（RemKoolhaas）和波士顿LWA Achitects

2004年至今在北京主持DnA（Design and Architecture）工作室

DnA 建筑事务所及主创建筑师获奖情况

DnA被美国城市设计协会Urban Land Institute 评选为2008年十家新锐建筑事务所

主创建筑师徐甜甜获得2008年英国《建筑评论(Architectural Review)》ar+d 国际青年建筑师奖 AR Award

主创建筑师徐甜甜获得2008 纽约建筑联盟(Architecture League)青年建筑奖(Young Architects Award)

主创建筑师徐甜甜入选纽约Rizzoli出版社2006年出版的《中国十位建筑师》

2007－2008年清华大学《世界建筑（World Architecture）》WA中国建筑奖

鄂尔多斯美术馆

2005－2006年清华大学《世界建筑（World Architecture）》WA中国建筑奖

宋庄美术馆

水岸会所

VILLA BY WATER

项目地点：中国 · 鄂尔多斯 **用地面积**：20 000 m² **建筑面积**：1200~1500 m²
建筑设计：DnA_Design and Architecture
建筑师：徐甜甜

LOCATION: Ordos, China SITE AREA: 20,000 m²
BUILDING AREA: 1200-1500 m²
DESIGN CORPORATION: DnA_Design and Architecture
ARCHITECT: Xu Tiantian

水岸会所位于鄂尔多斯市考考什那文化旅游区南端，面向考考什那水库，沿岸的 30 亩（约 20 000 m²）规划为会所区域用地，保留原生态的沙丘和绿色植被；会所选址在临水的沙丘上，功能要求包含 5 套私人套间，其余为公共空间，总建筑面积 1200~1500 m²。这个唯一明确的功能要求，即 5 套私人套间，给了我们最初的也是贯穿设计始末的灵感：5 个单元像花蕊般向上向光和空气生长；其余的公共功能则如同花瓣在地面延伸交织。居住是私密的，脱离地面，朝向水景，不互相干扰；公共部分则是贯通的，花瓣的交接形式产生不同的空间和光影。

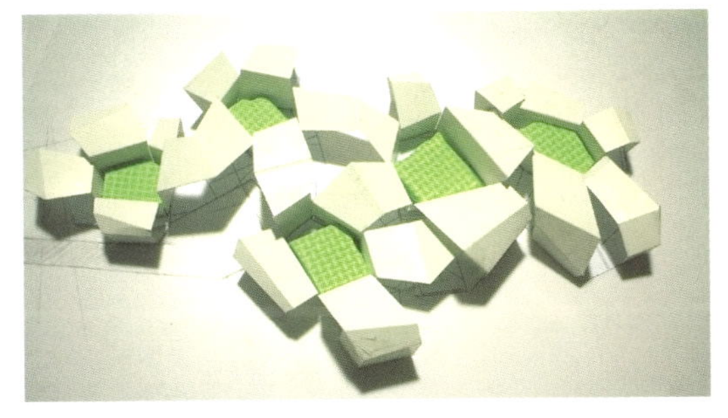

Villa by water is a private residence located on a vast mongolian land facing KaoKaoShiNa Lake in Ordos, Inner Mongolia. This will be a villa or private clubhouse for the owner to invite and host guests' families. In the program of 1200 square meters, there're 5 private suites which will accommodate 5 families, public functions including living room, dinning hall, swimming pool, library, winter garden, etc.

The natural Mongolian landscape, sky, sand dune, lake, the crispy air and breeze became key elements that define the atmosphere and inspire constantly, building became flowers blooming in Mongolian sky. The 5 private suites are flower cores striving up for light, air and view; each suites is a tower on second level starts with a double height living space, a stair leading to a bed mezzanine, and finally coming to a balcony looking over to the landscape beyond. The rest public programs are covered by petal-formed roof and walls merging and stretching on ground. The articulation of flower petals not only allows view on all directions and defines light and shadow, but also create an organization of interior public spaces and a series of courtyards. Supporting functions (such as kitchen, wine cellar, bathroom, sauna, etc.) are embedded between petals under the bedrooms with setbacks allow corridors and stairs becoming circulation for public spaces and bedrooms, and also create vertical shafts interfering with the continuous horizontal flow on ground level.

Materialization on each program further allows the differentiation between private or public: all the petals are painted white while bedroom suites are covered by ceramic tiles that shines under sunlight.

宋庄艺术公社

SONGZHUANG ARTIST RESIDENCE

项目地点：中国 · 北京 **基地面积**：1400 m^2 **建筑面积**：5300 m^2
建筑设计：DnA_Design and Architecture
建筑师：徐甜甜 **摄影**：周若谷

LOCATION: Beijing, China SITE AREA: 1400 m^2 BUILDING AREA: 5300 m^2
DESIGN CORPORATION: DnA_Design and Architecture
ARCHITECT: Xu Tiantian PHOTOGRAPHER: Zhou Ruogu

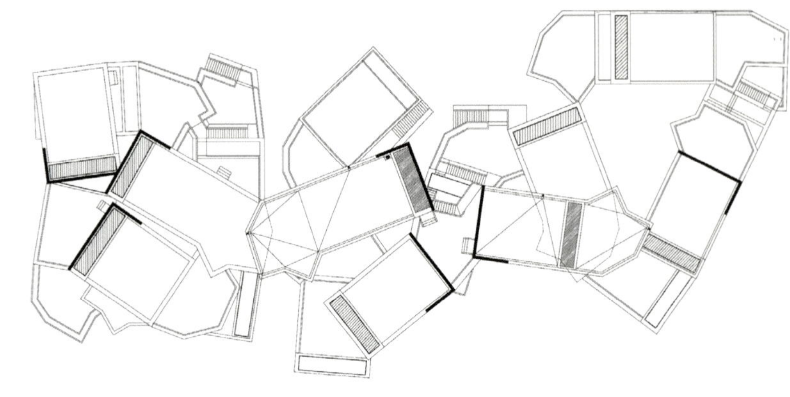

随着中国当代艺术家的激增，艺术家对居住和工作空间的需求也随之扩大。坐落在北京六环以东的宋庄艺术家村就正经历着由于这种增长所带来的变化，20 个面向鱼塘的艺术家工作室应运而生。

按照工作室的居住和工作两种主要功能，每一个单元都包含两种不同的尺度与组合：工作部分为 6 m 高的简洁的方盒子，居住部分高 3 m，由一系列几何形空间容纳起居厨厕功能。体量高差决定了每个单元的工作和居住功能或在同层联系或通过楼梯联系。整个建筑立面采用压型钢板，水平活动平面则是当地典型的红砖。

这 20 个工作室就像叠加在一起的集装箱，不仅和场地原有的工业气息相呼应，而且创造出富有张力的外形和独特的建筑空间。体量的虚与实，光与影，创造变化的室外交流空间，为艺术展示提供了更多的可能性。当地艺术节开放工作室时，这片区域可以为参观者带来多样的空间体验，同时成为丰富的创作展览现场。

这个带有艺术创作和居住功能的建筑何尝不是位于艺术村的一种另类美术馆？

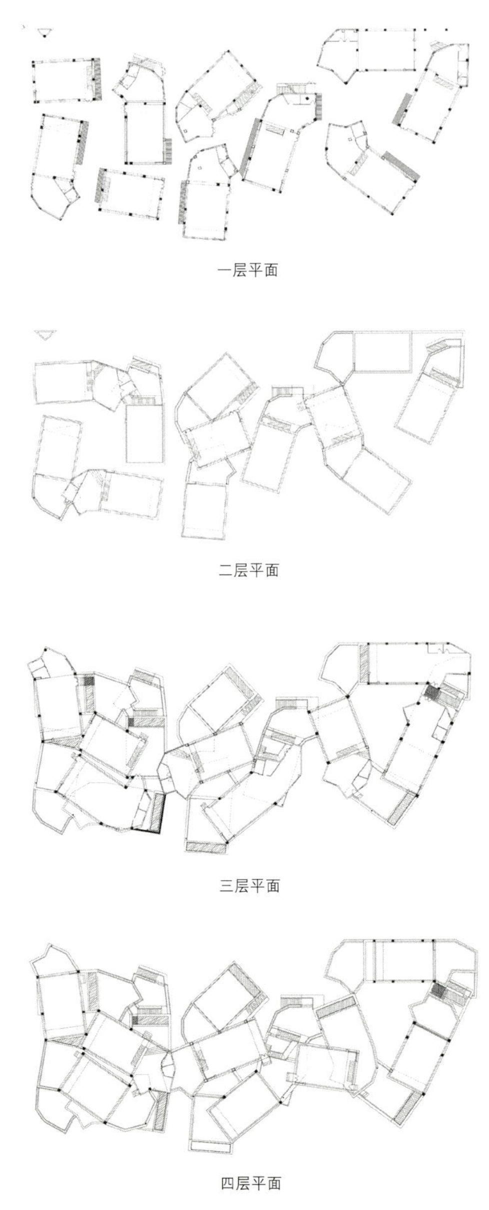

一层平面

二层平面

三层平面

四层平面

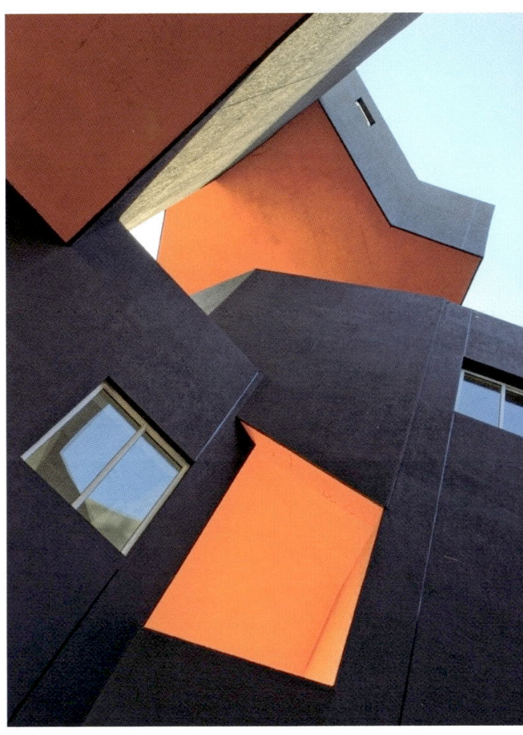

Located right next to east sixth ring road of Beijing city. Songzhuang Artist Village is undergoing a dramatic expansion of artist population and increasing demand of artist's working and living space, and this 20-units artist residence facing a fishpond at a former outdoor storage lot is one of the local development targeting such demand.

The programmatic requirement of working and living defines the height and geometry of both volumes: 6 m height for working and 3 m for living; a simple rectangular box for studio and a complex geometry for living indicating bedroom, kitchen and toilet. Living volume is plugged into working volume either on the same level or led by stair to upper level.

These 20 units are regarded as containers stacking up on this former industrial outdoor storage lot, creating an expressive configuration and spatial quality. The interplay of volume and void, light and shadow allows artists and visitors to constantly explore and experiment the outdoor community space, which could be the extension of art production and presentation as well as linking these 20 units as 20 individual showrooms on open studio days.

In other word, this complex becomes an alternative museum for living art creation and exhibitions.

西溪休闲中心

XIXI LEISURE CENTER

项目地点：中国 · 杭州 **建筑面积**：6300 m^2
建筑设计：DnA_Design and Architecture
建筑师：徐甜甜

LOCATION: Hangzhou, China BUILDING AREA: 6300 m^2
DESIGN CORPORATION: DnA_Design and Architecture
ARCHITECT: Xu Tiantian

西溪休闲中心是坐落在西溪国家湿地公园的 12 个艺术和文化建筑群的单体建筑之一。这个建筑群是由杭州市政府委托给 12 名中国建筑师的集群项目。杭州以其悠久的历史、丰富的文化和西湖景观而众所周知，而西溪艺术和文化建筑群则恰恰促进了其文化及旅游业的发展。

休闲中心延续了原有湿地的路径和形态，营造出由地面上的休闲功能空间和下沉的活动池塘组成的连续、开敞的的空间形态。在二层，特色水疗室则成为相对独立的空间，就像漂浮在水上的睡莲一样，娱乐空间从它下方经过。

这种荷叶的组织和连接方式使人们从户外的步行道通过中间的各层荷叶到达屋顶上的小池塘，从而最终在屋顶揭开了这片湿地面纱。

在荷叶间的逗留会勾起人们的似曾相识的感觉，好像我们祖先千百年前的一个幻想，经过这么多年的紧张的心理和生理上的变化，我们，作为人，仍然可以轻如一片鸿毛，敏捷如一条祥龙，在这充满微风和熏香的自然环境中逗留？

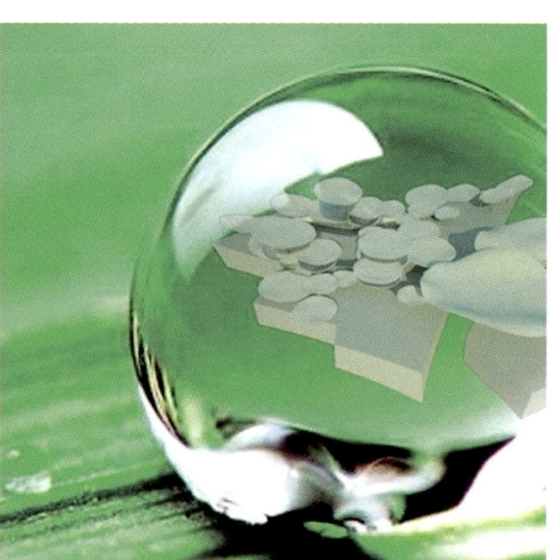

Xixi Leisure Center is one of the 12 buildings located in an art and culture compound in Xixi National Wetland Park, a cluster architecture commissioned to 12 Chinese architects by Hangzhou city government. Hangzhou has been well known for its long history, rich culture and the West Lake landscape. Xixi art and culture compound is part of development to stimulate cultural tourism.

The leisure center will house leisure function as open and continuous circulation on ground level and sunken activity pools imitating paths and ponds of wetland topography, while on second level, specialized SPA rooms become rather individual spaces, like the thumbnail floating in water lily leaves, carried by leisure circulation underneath.

In such an atmosphere, tender and bulk columns are applied like a bamboo forest supporting the volumes and leaves.

The format and organization of leaves leads an outdoor promenade from the paths up to roof terraces discovering small water lily ponds, eventually unveiling the wetland to a panoramic view from above.

A pause on these leaves might evoke a déjà vu, a fantasy traced back thousands years ago from our ancestors' belief, that after years and years of intense mental and physical training, we, human beings, are able to pause upon water, light as a fur, alert as a dragonfly, meditating the breeze and aroma in our nature?

AUBE DESIGN

欧博设计

AUBE（欧博设计）团队由极具设计创意和专业技能、熟知中国市场规律的法国欧博建筑与城市规划设计公司、深圳市博艺建筑工程设计有限公司、深圳市欧博设计有限公司及中外设计师共同组成。

AUBE（欧博设计）的设计经验引导了规划、建筑、景观、工程领域。近年来，AUBE（欧博设计）中国区的业务范围立足于深圳，扩展到国内大中城市，并多次在国际竞标中胜出，代表作品有贵阳国际会议展览中心规划建筑景观工程设计、深圳长富金茂超高层建筑设计、深圳南油购物公园、深圳康佳研发大厦、深圳半岛城邦一期住宅区规划建筑景观工程设计、深圳北站交通枢纽中心景观工程设计、合肥澜溪镇建筑设计、华润置地成都翡翠城、深圳市华侨城生态广场、深圳华侨城中央教科所附属学校等。

AUBE（欧博设计）品牌的树立和发展，建立在设计师资源和设计平台协作共享的基础上，同时也建立在初始法国合伙人之间和陆续加入的不同国籍合伙人之间所表达的个人独立设计风格上。

AUBE（欧博设计）的发展基于参与重要的国际竞赛和接受经过特别选择的委托设计。

AUBE（欧博设计）的多国设计师自1993年开始介入中国的重大建设工程项目。截止到目前，AUBE（欧博设计）参与和完成的中国重大设计项目300余个。设计师的工作经历普遍具有国际化特征。

AUBE（欧博设计）始终坚持"国际化经验、地域化实践"的设计理念，与每一位关注它的同仁共创、共享、共同发展。

AUBE Design Group is composed of French Aube-architecture and Urban Designing and Planning Co., Ltd., Shenzhen Boyi Construction Design Co., Ltd. and Shenzhen AUBE Design Co., Ltd. and foreign and domestic designers, who are initiative and professional in design and know well about market decipline.

The design experience of AUBE has led the areas of planning, architecture, landscape design and engineering. In recent years, the business AUBE in China is based in Shenzhen and extends to domestic metropolis and cities, and has won many international competitions. Its representative works are Building Planning and Landscape Engineering design of Guiyang International Convention and Exhibition Center, Shenzhen Chang Fujinmao high-rise building design, Shenzhen South Oil Shopping Park, Shenzhen Konka R & D Building, Shenzhen Peninsula Primary Urban Residential planning and landscape design, Shenzhen north Station transportation hub landscape design, Hefei Lanxi architectural design , China Resources Chengdu Emerald City, Shenzhen City Ecology Square and Shenzhen Overseas Chinese Town Central Education and Scientific Affiliated School, etc.

The establishment and development of AUBE (AUBE design) brand are based on the shared design resources and design platform as well as the individual design styles expressed by the initial French copartners and the continuously added partners of different nationalities. AUBE (AUBE Design)'s development is based on participation in important international competitions, and the accept of specially selected commissioned design.

AUBE (AUBE Design)'s multi-national designers have involved in major construction projects in China since 1993. Up to now, the major projects designed or participated by AUBE in China has reached the number of more than 300. Designers' work experience generally assume internationalized characteristics.

AUBE (AUBE design) has always insisted the "international experience, geographic-based practice" design concept to co-create, share and develope together with every colleague that focuses on it.

贵阳中天世纪新城三号地块三组团

GUIYANG MERIDIAN TRANSIT NEW CENTURY CITY, THE 3RD BLOCK OF LOT NO.3

项目地点：中国·贵阳　用地面积：175 428.5 m^2　建筑面积：236 819.3 m^2

建筑设计：欧博设计——法国欧博建筑与城市规划设计公司　深圳市博艺建筑工程设计有限公司

LOCATION: Guiyang, China　SITE AREA: 175,428.5 m^2　BUILDING AREA: 236,819.3 m^2

DESIGN CORPORATION: Aube Design — Aube Conception D'Architecture

Shenzhen Boyi Architectural Engineering Design Co., Ltd.

毗邻贵阳市中心的中天世纪新城，将成为未来中国真正花园城市之一。在整个新城中，经典成熟的规划、精巧构思的建筑和景观让人们享受到私家住宅和花园带来的清幽、舒适、雅致的生活。新城中拥有商业区、步行街、学校、体育中心、办公空间、酒店等其齐全的配套设施，让人们如同生活在市中心一样便利。

倚山而居的生活处处与自然、景观紧密相连。即便是位于新城核心地带的商业中心也是如此，联系新城东、西两区的景观步行系统，沿途散落的空中花园、休闲广场与商业相融合，带来了别具特色的体验式商业消费模式，集购物、休闲娱乐为一体，正是未来新城高品质生活的重要组成。这不仅仅是新城的规划特色之一，也将成为贵阳城市中的亮点。

3 组团工程：低密度叠拼住宅区。

作为新城先期开发的 3 组团，该组团住宅依山傍水，布局呈环状。优美的环境、较低的容积率足以彰显社区居民的高贵身份。开阔通透的视野、徜徉驻足的花园平台，无与伦比的自然生态景观，却不失充满朝气与活力的都市氛围，界定了该组团独有的居住品质。

顺着沿河道路精心布置了该组团的商业立面。露天咖啡座、各具特色的餐厅，让人们在此与友人相聚，或独自饮啜沉思都别有一番风味。这一商业立面为整个社区营造出节日的氛围，同时也为新城居民激情四溢的都市生活拉开序幕。

在沿河商业立面的中心是两幢 SOHO 建筑，它们界定了住宅区的主要入口。这一入口同时也是与河流垂直的主要中轴线起点。沿着中轴线拾阶而上，房屋顺着山势，在不同的高程上错落有致地排列着，逐渐延伸至高处。轴线的尽头是学校和体育中心。

3 组团中的大部分产品都为联排叠拼住宅和花园洋房。联排叠拼住宅坐落在山腰之上，花园洋房则依偎在南明河畔。

这一布局借鉴了传统经典的模式，营造出尊贵、气派的氛围，但建筑却处处流露着时尚、现代的气息。这并不是一个简单的风格与形式的选择，而是从新城整体发展的角度，基于对山地居住模式充分的研究和构思，辅以丰富的住宅设计经验所得出的结论。为了打造完美家园，我们注重了各方面的细节：遵守节能规范，提倡环保；保证住宅及其花园私密性；合理利用、塑造地形以创造开阔的景观视野；注重夏日遮阳处理，确保住宅的自然通风；在材质的选择上，我们挑选了持久耐用、易于保养，并与景观相协调的建筑材料，诸如用于立面装饰的劈开砖、用于墙基的本地石材，为取得通透的视野而选用的玻璃栏杆，以及易于保养的仿木质铝型材窗框、百叶等。

Meridian Transit New Century City adjacent to the center of Guiyang will be the real garden city in China. In the whole new city , the mature calssic programing, ingeniously conceived architecture and landscape are reminiscent of the tranquil, comfortable and elegant life in the private residence and garden. There's the commercial area, pedestrain mall, school, sports center, official space, hotel and other well equipped facilities, which are as convinient as in the center of the city.

Life beside the mountains is intimately conncted with the natural landscape, and so is the commercial area in the core of the new city. The scenic walking system connecting the east and west areas of the new city, the dispersed air garden and casual square integrated with the commercial area bring chic experience of consuming style that combines shopping and racreation, which is just the essential part of the future high-quality life of the new city. This is not only one of the characteristics of the new city but a shining point of Guiyang City.

The 3rd block engineering: low-density puzzled residential area.

The earlier developed 3rd block of residential houses is surrounded by mountains and water, and the layout is ring-shaped. The beautiful environment and the relatively low plot ratio totally show the noble identity of the owners. The open field of view and the good ventilation condition, the spacious garden platform, and the unparalleled natural ecological landscape are enhanced by the vigrous and lively city atmoshpere, which defines the unique living

quality of the bock.

We elaborately arranged the commercial facade along the path by the river. The open-air coffee seats and the uniquely characterized resturant offer a fancy experience to meet with friends or drink or sip in meditation alone. It creates a festive atmosphere for the whole block and in the meantime it enlights the passionate urban life for the residents in the city.

In the center of the commercial facade are two SOHO buildings, which identify the main entrance of the residential area. The entrance is also the start point of the main axis vertical to the river. Up along the stairs along the central axis, the houses are well-proportioned in arrangement complying with the mountain shape in different elevations and gradually extend to the higher positions. At the end of the axis are the school and the sports center.

Most houses of the 3rd block are joined houses or garden houses. The joined houses are posted on the mountain side while the garden houses lean close to the Nanming River.

This arrangement borroewed the traditional classic style to create the dignified and grand atmosphere, however the construction reveals stylish and modern flavor erverywhere. It's not simply an option of style and form, but a result based on the full research and conception of the mountain residence added with the rich residential design experience from the perspective the whole development of the new city. To build a perfect homestead we pay close attention to every detail of the construction: to comply with energy saving norms, to promote environmental protection; to ensure privacy of residence and the garden; to rationally use and create the landform in order to create a broad landscape perspective; to focus on the form of the summer sunshade and ensure natural ventilation; in the material selection, we selected durable building materials with easy maintenance and coordination to the landscape,such as split brick for the facade decoration, local stone materials for the wall base, glass railings to achieve transparent vision, as well as wood-emulating aluminum window frames and shutters with easy maintenance and so on.

成都麓山国际超高层住宅

CHENGDU LUSHAN INTERNATIONAL SUPER HIGH-RISE RESIDENCES

项目地点：中国 · 成都　用地面积：3.61 ha　建筑面积：13.47 ha

建筑设计：欧博设计——法国欧博建筑与城市规划设计公司　深圳市博艺建筑工程设计有限公司

LOCATION: Chengdu, China　SITE AREA: 3.61 ha　BUILDING AREA: 13.47 ha

DESIGN CORPORATION: Aube Design — Aube Conception D'Architecture
Shenzhen Boyi Architectural Engineering Design Co., Ltd.

主题思想：户户尽享景观系统。

基于片区居住质量高、居住人群高层次的状态，设计中采用了独具特色的户型结构；朝向景观是单元布置的出发点，尽量采用了“户户朝南、家家取景”的处理手法，分别布置不同面积指标的户型，使得户户都有良好的朝向和景观系统。

每户均高度保证私密性，保证每一户的客厅和餐厅及部分房间可以直接观看高尔夫球场的美景，室内外环境做到自然交融，希望住户可以体验空中的别墅生活。每类户型均设计了与之相称的房型；平面尺度舒适，动静分区明确，功能完备；面积分配、房型比例、交通组织合理，结构规整清晰，便于组合分隔。

住宅立面构想：出水芙蓉的意念赋予了每一栋建筑与生俱来、与众有别的特殊气质，让建筑与场景彼此和谐融入，而非平面的简单生成，是我们在整个项目中所坚持的设计原点。本设计中的每一单体，其总的建筑架构都依循“自由、流畅、轻盈、通透”的设计思路。“自由”指在建筑造型上，立面打破常规的几何化的外立面设计，大量运用多变的平面曲线及空间曲线来塑造多角度的造型变化；“流畅”则是一个自然的设计，也是建筑平面功能的外在表达，通过强调水平向弧线的多重组合而达到视觉的连续移动；“轻盈”意在表现花朵的柔美，通过体量的限定与拆分弱化了建筑自身巨大的体量，减少对周边住宅的强烈压迫感；“通透”则是一方面进一步强化“轻盈”之感，另一方面则最大化地加强室内外景观的互动交融。

错动的空中花园除了具有动态特征外，也表现了生活的本生形态——随意而生动。空中花园赋予了建筑立面另外一个重要特征，此花园的存在主要有两种意义。其一，为整个社区构成了三维的立体景观系统；其二，常规的楼盘设计中只会出现半私密庭院，而在本设计中，各组建筑各自独立的空中花园则为住户带来了此小区内部各住宅之间的半私密庭院更接近全私密。在这里，住户可无忧无虑、不受干扰地享受与自然的交流。

Main idea: every household to enjoy the landscape system.

As for the high-quality housing and the high-level inhabitants, we designed the unique interior structure. The starting point of the layout is the direct view to the landscape, so strived to achieve the goal of " every household faces the south and every family finds a view". We arranged structures of different sizes so that every house has a good orientation and view system. Each household have a high degree of privacy, and in the living room and dining room of every household one can watch some of the beautiful golf course. The indoor and outdoor environments reach natural blending, so that people can experience the life of the air villa. Each type of the house is matched with the commensurate chambers of comfortable dimensions, clear movement areas, full functions and equippment. The reasonable area size distribution, ratio of the chamber and the traffic orgnization as well as the neatly defined structures make it easy to combine and separate.

The residential facade idea: the waterlily idea endows each building with an inborn and distinguished temperament,so that a harmonious integration of architecture and scenes come into being ,while the simple non-planar formation is the original point we adhere to throughout the project.The total building structure of each monomer follows the design concept of "free, smooth, light and transparent"."Free" refers to the construction style, i.e. the unconventional and geometriacal facade design extensively uses changing space plane curve and curve to form the multi-angle shape diversity; "smooth" is a natural design, which is also the external expression of functional architecture. by emphasizing the the multiple combinations from the flat to the curve to achieve the vision of continuous movement; "light" is intended to show the tenderness and beauty of the flower; through limit and separation of the body amount to achieve the reduction of the huge volume of the building ,and then to relieve the strong sense of pressure to the surrounding residence; "transparent" is, on the one hand to further strengthen the "light" feeling, and on the other hand to maximize the interaction and blending of the indoor and outdoor landscape. Air Gardens of dislocation not only show the dynamic characteristics, but also reveal the natural form of life - casual and lively. The air arden gives the facade another important feature. The existence of the garden mainly shows two kinds of significance. First is to constitute a three-dimensional landscape system for the whole community, and second, the conventional designed premises just exposes semi-private courtyard, while in this design, the independent gardens of separate households brought almost full-private courtyard within the community, so that the residents can enjoy uninterrupted communication with nature.

IMAGINE ARCHITECTS

畅想建筑设计事务所

畅想建筑设计事务所专业从事于城市设计、建筑设计、景观设计及由此衍生的泛设计。对畅想建筑而言，设计是一系列动态平衡区间或瞬间，在不定中寻找确定，在无界中寻找边界，在限制中寻找突破……

畅想设计目前正专注于一系列城市综合体、校园文化建筑和创意产业园区的规划与建筑设计，并积累了丰富经验。

畅想设计近年来获得了包括迪拜高层标志物（magic mirror）在内的一系列国际国内设计大奖和重要规划建筑设计投标的第一名，设计作品多次参加国内外设计展，广泛刊登在国内外期刊书籍上。

Imagine Architects specializes in urban design, architectural design, landscape design and the derived pan-design from the three. As for Imagine Architects, design is a series of dynamic equilibrium intervals or moments to find certainty in uncertainty, boundary in boundlessness and the breakthrough in confinement...

Imagine Architects is currently focusing on planning and design of city complexes, campus culture constructions and creative industry park constructions, and has accumulated rich experience.

In recent years Imagine Architects has won domestic and international design awards, including Dubai high-level landmark (magic mirror), and the first place of important architectural planning and design bid. Its design works has participated in design exhibitions at home and abroad, which has been widely published in international journals and books.

公司合伙人及主要设计师

- 罗四维　Luo Siwei

罗四维1963年7月生于南京，1985年毕业于东南大学建筑系，中国首批一级注册建筑师，厦门大学兼职教授。1998—2001年任厦门新区建筑设计院副院长，2001—2003年任福建省建筑设计研究院厦门分院副院长兼总建筑师，2004—2006年任美国WWCOT建筑规划事务所（中国区）总建筑师，2006年至今任上海畅想建筑事务所设计总监。2009年获第十一届ThyssenKrupp国际执业建筑师设计竞赛第二名。2008年入选《made in china》。2006年11月入选“荷兰建筑协会（NIA）中国建筑展”等，2006年11月入选《DOMUS与中国78位建筑师作品集》及《中国环境艺术设计》。2005年12月入选首届“深圳国际城市建筑双年展”，《中国房地产报》2005年12月人物专访。2005年入选英国《New Chinese Architecture》。2004年度入选《新地产》杂志100位中国最具影响力建筑大师，2004年底荣获“WA中国建筑设计奖”。2002年《时代建筑》杂志第三期建筑师专访。2001年入选《中国青年建筑师33》等。工程实例及文章广泛刊登在国内外重要学术刊物和书籍上。

杭州加利利科技集团科技综合体

SCIENCE AND TECHNOLOGY COMPLEX OF HANGZHOU GALILEE TECHNOLOGY GROUP

项目地点：中国·杭州　用地面积：56 755 m^2　建筑面积：181 000 m^2
建筑设计：上海畅想建筑设计事务所
建筑师：陈思源，徐文才，郑万敏，罗四维，陈晓南，卢珊，莫少军

LOCATION: Hangzhou, China　SITE AREA: 56,755 m^2
BUILDING AREA: 181,000 m^2　DESIGN CORPORATION: Imagine Architects
ARCHITECTS: Chen Siyuan, Xu Wencai, Zheng Wanmin, Luo Siwei, Chen Xiaonan, Lu Shan, Muo Shaojun

本方案中以具有灵气的水晶为主要灵感，使建筑间相互串联，形成水晶之链，单个的建筑体则代表水晶钻、水晶柱、水晶塔，围绕主入口水晶广场的四栋建筑相连像朵莲花，使整体布局既有联系又彰显灵气。

水晶数码塔：计算机、手机等数码产品是由电子元器件构成的，二进制在电气、电子元器件中最易实现，它只有两个数字，用两种稳定的物理状态即可表达，而且稳定可靠，比如磁化与未磁化，晶体管的截止与导通（表现为电平的高与低）等。0、1构成另一个虚拟世界，在建筑中，我们抽象0、1为基本设计元素，水晶数码塔就是由不断变化的0、1构成。

条形码：条形码也是采用二进制，高层办公楼采取水滴形状平面，立面为抽象化的条形码，造型圆润细腻，富有现代感。

The main inspiration of the program is the aura of the crystal which makes the buildings in series to form a crystal chain. The independent building structures stand for the crystal diamond,crystal pillar and crystal tower, connectedly surrounding four buildings at the main entrance of the Crystal Plaza like lotus flowers, making the overall layout appear attached and smart.

Crystal Digital Tower: computer, mobile phone and other digital products are composed of electronic components, in which binary signal are the easiest to implement, because it has only two digits,and can be expressed with two physical conditions, both stable and reliable,such as magnetic and non magnetic, the transistor cut-off and conduction (showed by high or low electric levels) and so on. 0 and 1 can constitute another virtual world. In the construction, we abstract 0 and 1 as the basic design elements, which constantly change to cnstitute the Crystal Digita Tower;

Bar code: bar code also uses the binary signals, high-rise office building takes the drop-shaped flat, and the vertical surfaces are abstract bar codes, sculpted smooth and delicate with a very modern sense.

长宁银河绿地小游园

GALAXY GREENLAND CHANGNING

项目地点：中国 · 上海　用地面积：25 000 m^2　建筑面积：300 m^2
建筑设计：上海畅想建筑设计事务所
建筑师：罗四维，朱渊，卢珊

LOCATION: Shanghai, China　SITE AREA: 25,000 m^2　BUILDING AREA: 300 m^2
DESIGN CORPORATION: Imagine Architects
ARCHITECTS: Luo Siwei, Zhu Yuan, Lu Shan

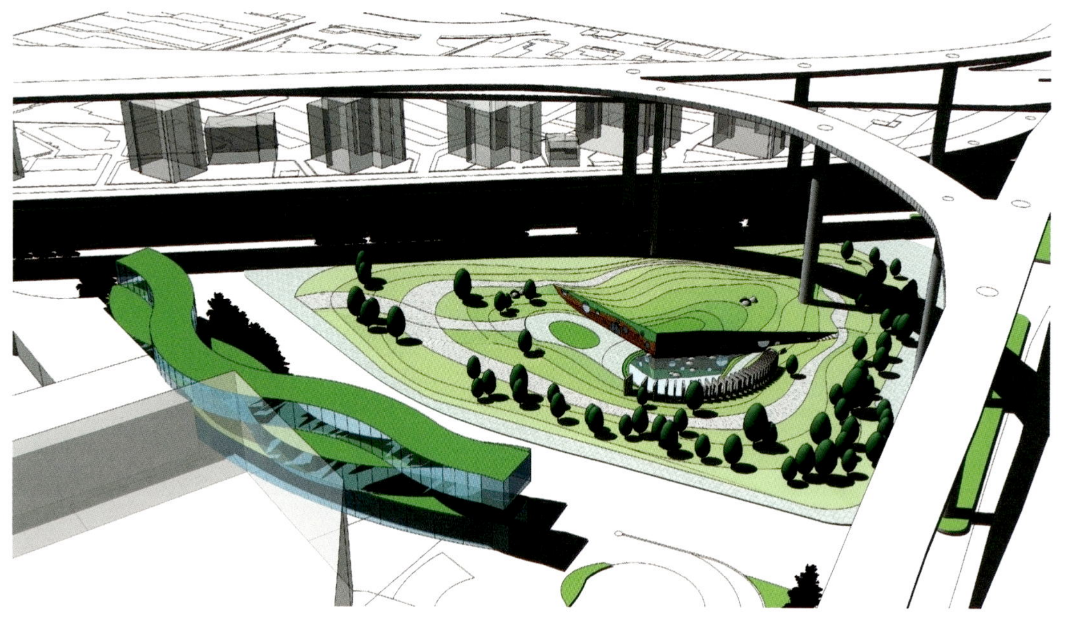

本项目为城市公共绿地中的茶室小品设计，构思来源于对断层地貌的理解与再创作，通过将建筑与基地地形巧妙结合，把建筑体量隐藏于地貌之中，最大限度地将绿色还给城市。

设计从三类问题入手。

(1) 小游园为封闭式布置，有围墙，有特定的门，出入不便。

(2) 由于银河宾馆、虹桥宾馆沿新华路一侧安排的是辅助用房，小游园无法与宾馆共享，反而在两者交界处形成城市消极空间。

(3) 小游园的衰退，政府投入不足是一大原因。此轮改造完成后如何在维护、管理上进入良性循环，在人气、内容、地段潜能、城市节点价值等方面形成良性互动是本次设计的重点。

通过两个方面，用两种手法来解决上述问题。

(1) 把虹桥宾馆、银河宾馆沿新华路一侧的功能置换纳入到小游园改造的内容中去，变消极空间为积极空间，为小游园边界带来切实的人气和消费使用场所，通过小游园环境的改观来提高该场所的品质。

(2) 拆除小游园围墙，变封闭小游园为全开放绿地。保留绝大部分现有树木，通过适当的地形改造，把任务书要求的功能房间埋入在地下，使建筑密度为零。

为此提出凹凸两种设计手法，通过垂直方向的地形凹凸和水平方面的地形凹凸形成相应的绿地景观场所、室外活动场所、兼有雨水收集功能的水池、室内空间等。由垂直凹凸形成的切面保持土壤断面原形，经加固处理后直接作为分隔空间的内外界面。

结果获得一个概念扩大化了的景观场所。在这里处于各种行为中的人的景观和物的景观同样重要。

The project is the garden ornament design of a teahouse in urban public greenland.The idea comes from the understanding and re-creation of fault topography.The project maximumly returns green to the city by seating the building in the landform through ingenious combination of the two.

Our design start from three problems:

(1) The inconvinient transportation was resulted from the closed-end layout of the small garden,the enclosure and the specific door.

(2) The small garden can not be shared by the hotel because the back-up rooms are arranged in the Galaxy Hotel and Rainbow Hotel along the Xinhua Road but form negative city space at the junction of the two.

(3) The recession of the small garden was largely due to lack of government investment.The focus of this design is how to make positive cycle in maintenance and management, and positive interactions between popularity, content, site potential and the value of urban nodes after this reform.

We use two methods to resolve these issues in two aspects:

(1) To make the function of the Rainbow Hotel and Galaxy Hotel along one side of Xinhua Road a part of the transformation and change the negative space into a positive one;to bring the small garden border tangible popularity and consumption functional area,and to improve the quality of the place through environment changing of the garden.

(2) To remove the enclosure to change the closed garden into an open green;to retain most of the trees and, through appropriate landshape transformation, build the functional rooms required in the task book underground to create a zero building density.

Thus we propose two design methods of concave and convex. The vertical and horizontal concave and convex form corresponding green landscape and outdoor areas, the rainwater pool and indoor space, etc. The section formed by vertical concave and convex retains the prototype of the fault topography, which, after the strengthening , is used as the boundary between the internal and external spaces.

The result is we got an enlargement of the concept of landscape areas, where the scenery of the people with various acts is as important as that of the materials.

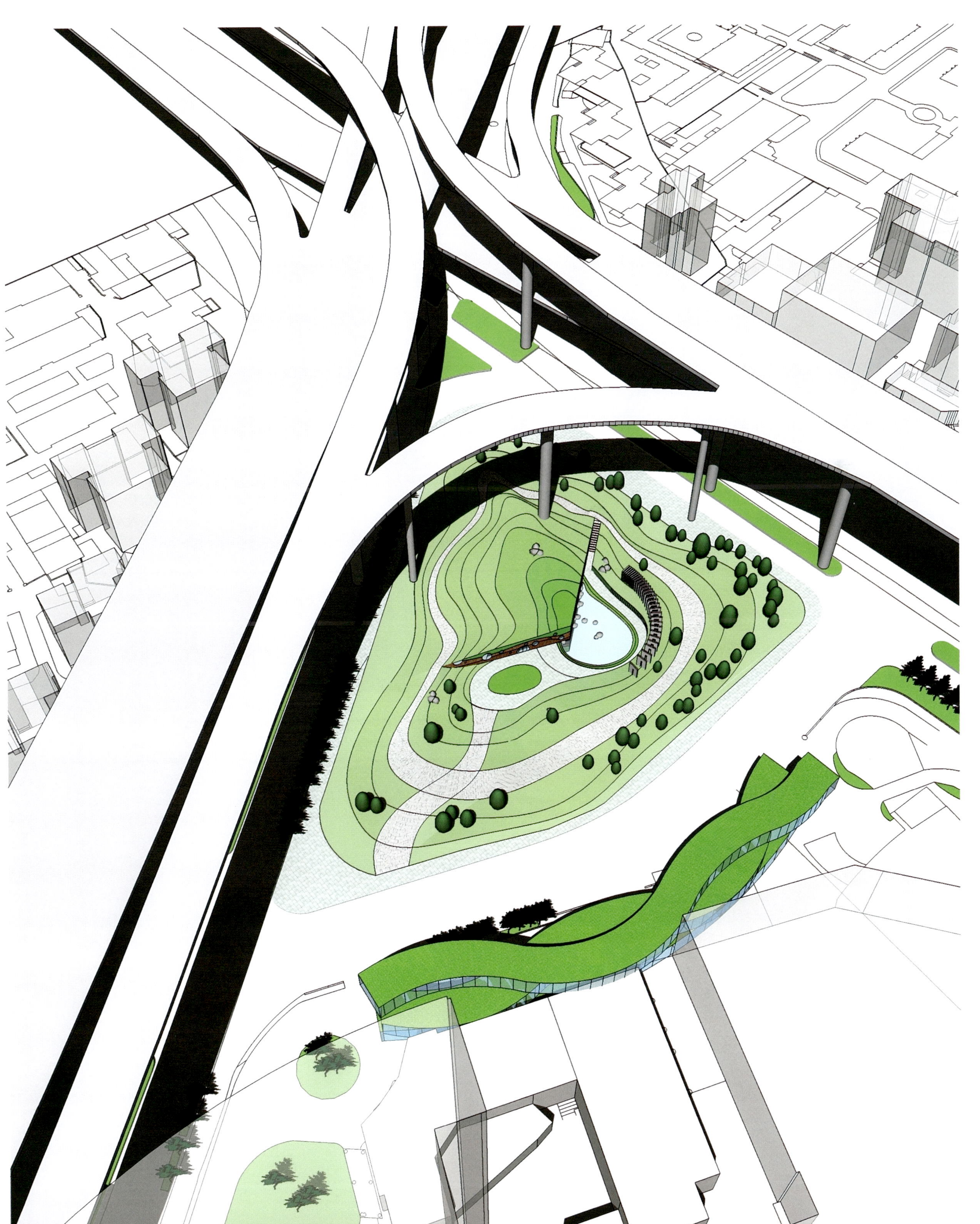

复旦大学国权路商业街

COMMERCIAL STREET OF FUDAN UNIVERSITY, GUOCHUAN ROAD

项目地点：中国·上海　用地面积：4900 m^2　建筑面积：7592 m^2
建筑设计：上海畅想建筑设计事务所
建筑师：罗四维，周伟，周凡，卢珊

LOCATION: Shanghai, China　SITE AREA: 4900 m^2　BUILDING AREA: 7592 m^2
DESIGN CORPORATION: Imagine Architects
ARCHITECTS: Luo Siwei, Zhou Wei, Zhou Fan, Lu Shan

方案一主要表达了方与圆的对话，一层沿街的不规则弧形体量排列以获得商业界面最大，商业流线最便捷；二至四层则根据功能需要采用方正的建筑体量，且与商业区形成对比。

方案二构思来源于堆积木的游戏，各个建筑体块之间使用堆积木的方式相互搭接组合，并形成丰富的空间。

方案三来自于对 DNA 双螺旋的理解，两条建筑相互弧形连接，使空间以及建筑造型都别具特色。

方案一

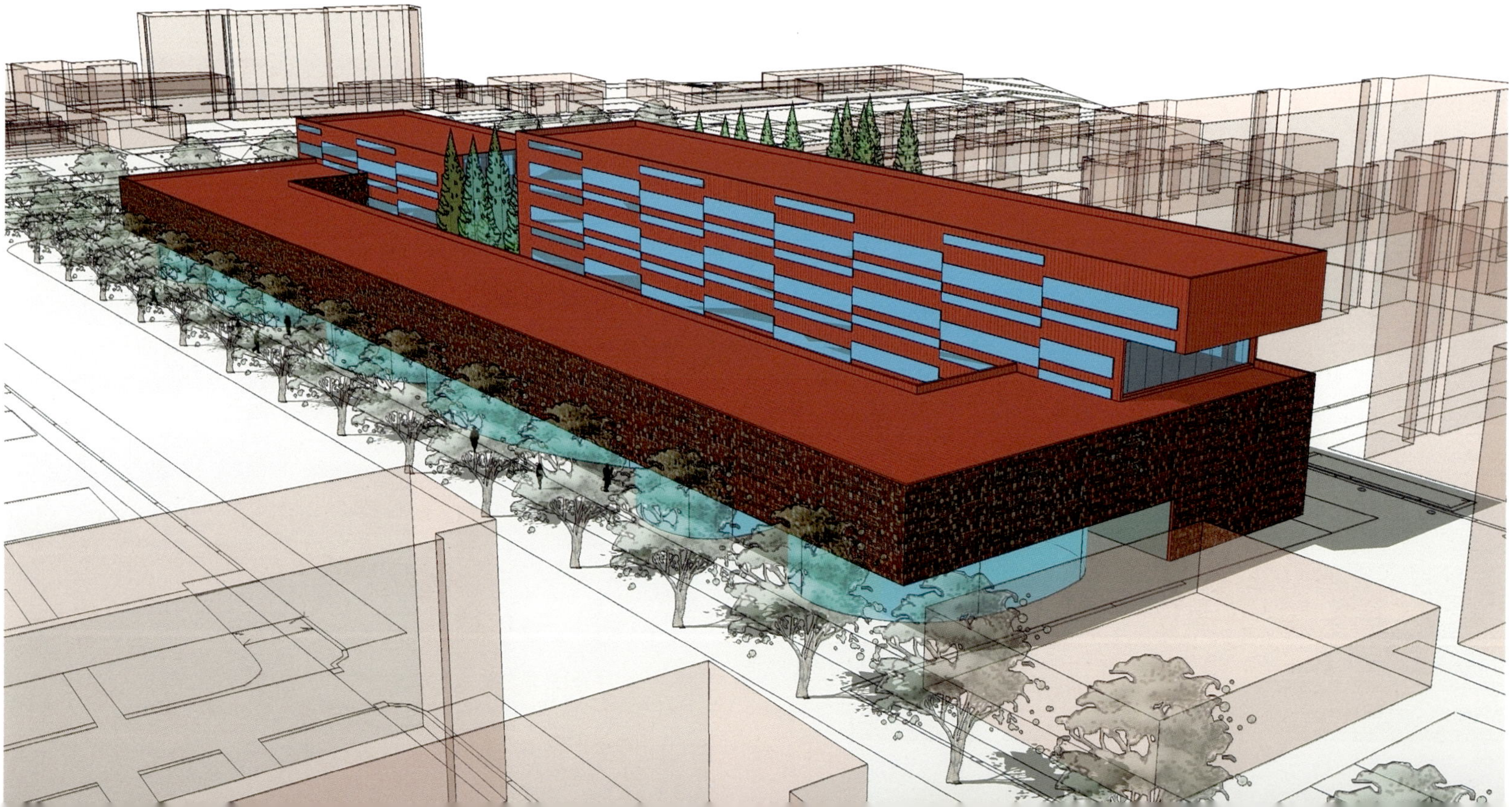

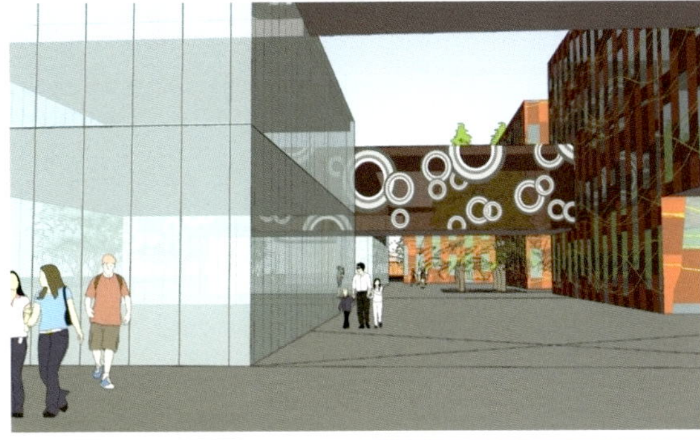

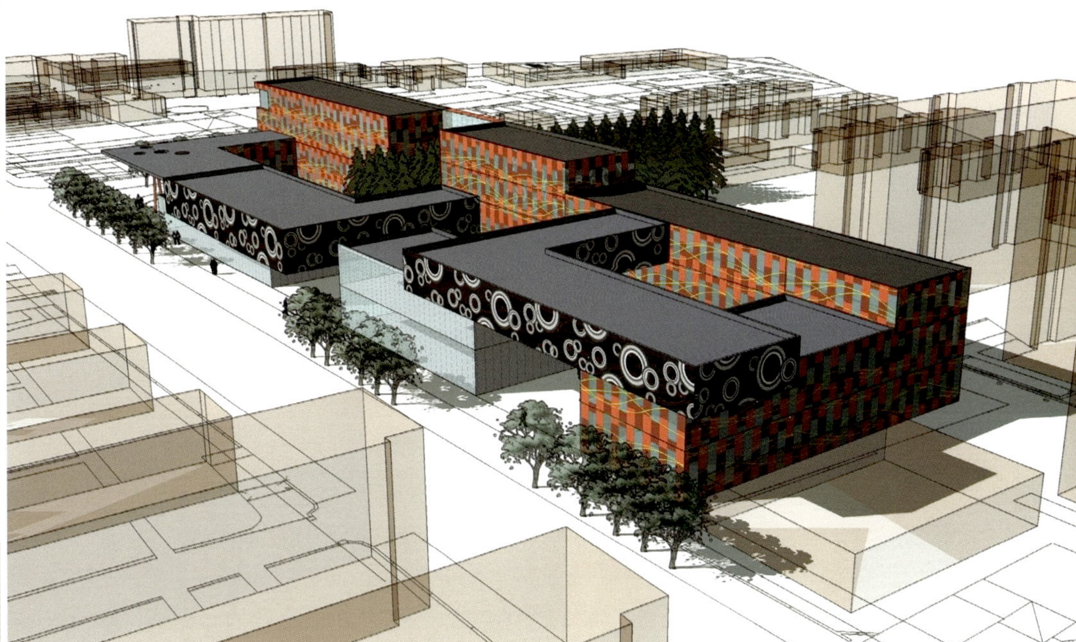

方案二

Program I mainly shows the dialogue btween square and round. The irregular arc body masses along the street in the first floor are arranged to achieve ultimate business interface and commercial streamline convenience;the 2-4 floors take the regular square building body masses in contrast to the commercial street.

The idea of Program II comes from the toy brick game. The building blocks are connected and combined by the way of putting together the toy bricks to form abundant space.

Program III comes from the understanding of double helix of DNA. Two buildings are connected by the arc bodies,making the space and the whole architecture unique.

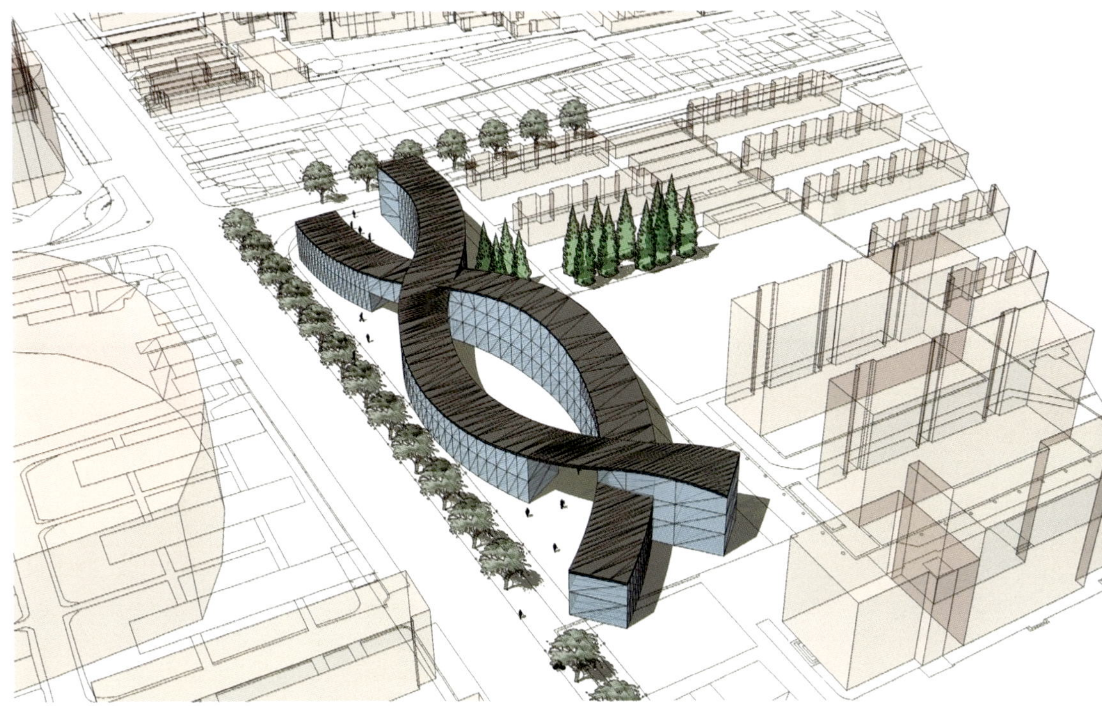

方案三

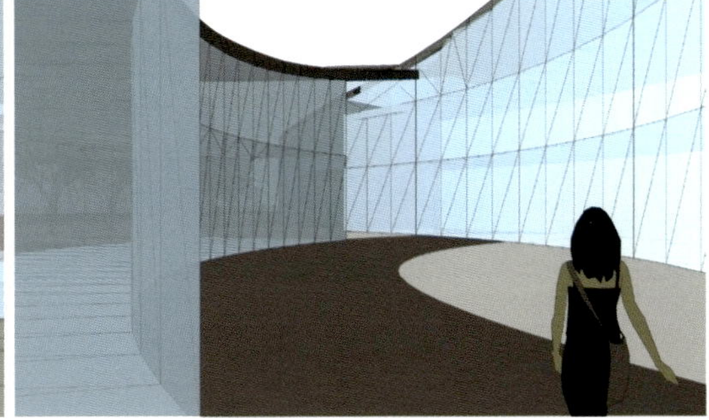

嘉定新城东云街 2–4 地块建筑设计

ARCHITECTURAL DESIGN OF 2-4 BLOCKS OF JIADING NEW HAVEN STREET

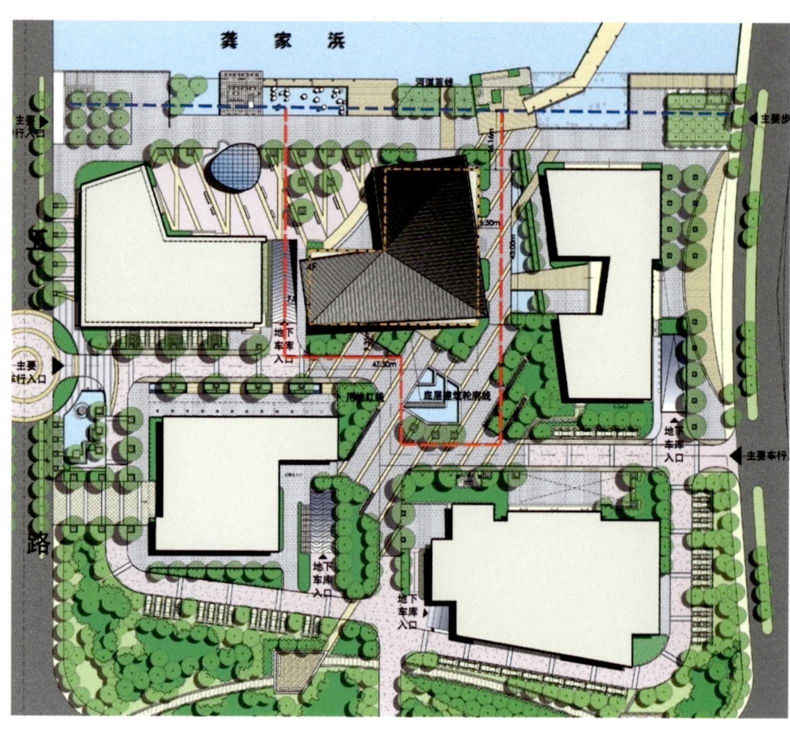

项目地点：中国 · 上海　基地面积：4267 m^2　建筑面积：9379 m^2
建筑设计：上海畅想建筑设计事务所
建筑师：罗四维，卢珊，卢浩星

LOCATION: Shanghai, China　SITE AREA: 4267 m^2　BUILDING AREA: 9379 m^2
DESIGN CORPORATION: Imagine Architects
ARCHITECTS: Luo Siwei, Lu Shan, Lu Haoxing

由于本项目基地位置处于龚家浜河道景观带辐射区，用地以江南特色餐饮及商业休闲为开发重点，强调生态的消费环境和江南文化氛围，故建筑借鉴江南民居的特征，采用 L 形的院落布局，强调聚合感，同时与周边的建筑相对应。为了实现项目投资的合理化和利益的最大化，综合基地的各种条件，商业用房高度控制在 24 m 以内，按五层进行设计，空间可自由划分。并设有地下一层以供停车。

在造型设计上遵循现代承载传统的原则。形体从古代楼阁中获取灵感，层层叠加，运用每层与每层之间体块的错位穿插来丰富整体空间，充满节奏感。屋面则沿袭了江南一带的坡屋顶，分成几个不规则的折面并与周边建筑的坡屋顶相呼应，以现代设计手法为载体传达传统文化意境。立面造型借鉴传统文化神韵，依托现代建筑手法，古典与现代相结合，在统一中寻求变化，以丰富的建筑语言来表达嘉定优秀的地域文化，形成了层次丰富、生动活泼、与周边建筑不失协调的立面肌理。

在建筑色彩部分以灰色、青色为主，运用磨砂玻璃、彩釉玻璃来表达江南建筑的神韵。在玻璃细部采用带有文化气息的漏窗疏密相间排列，表达其独特的中国传统美感建筑形式，达到江南风情和现代生活、人与建筑、人与景观的和谐共融。

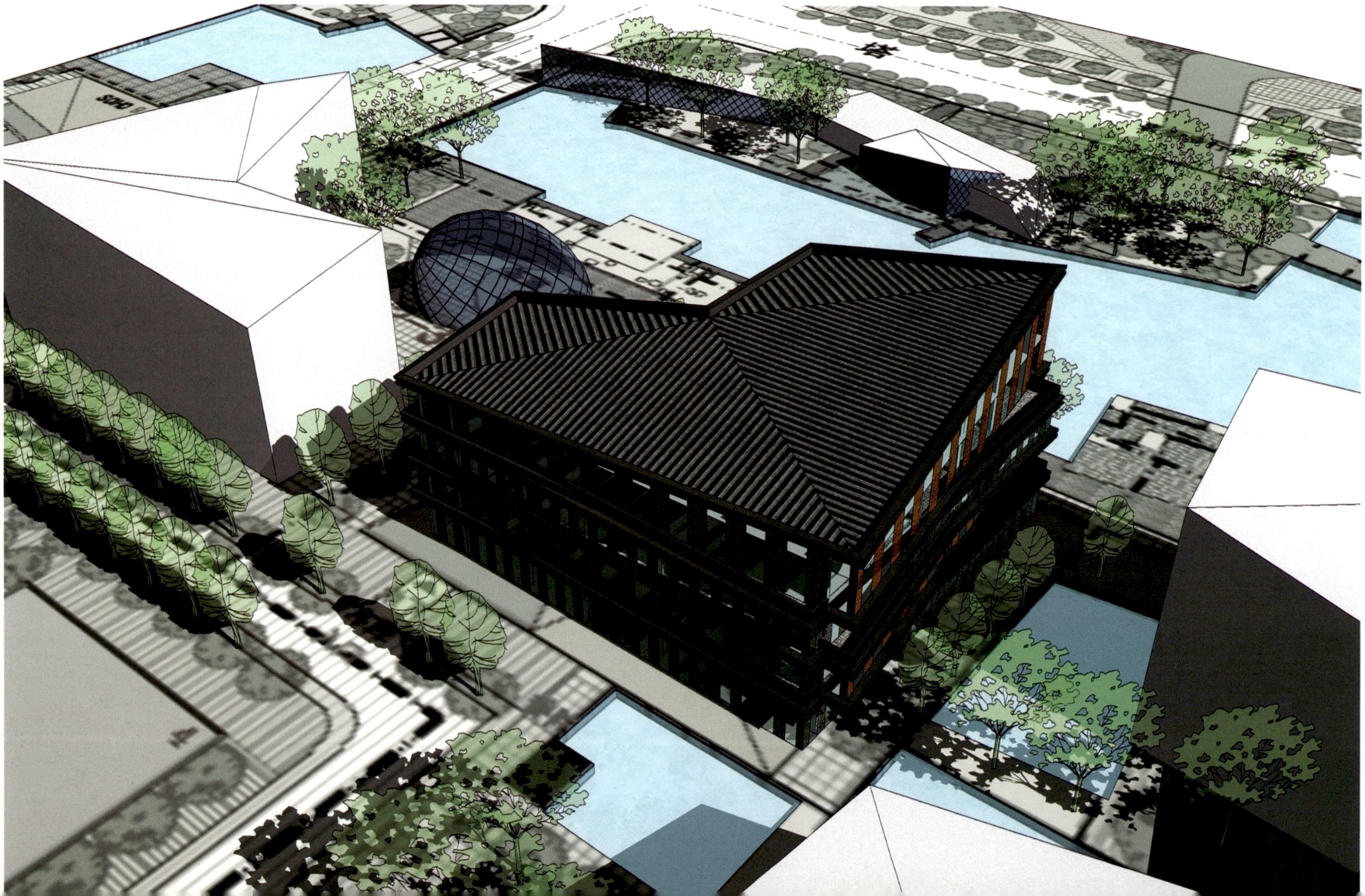

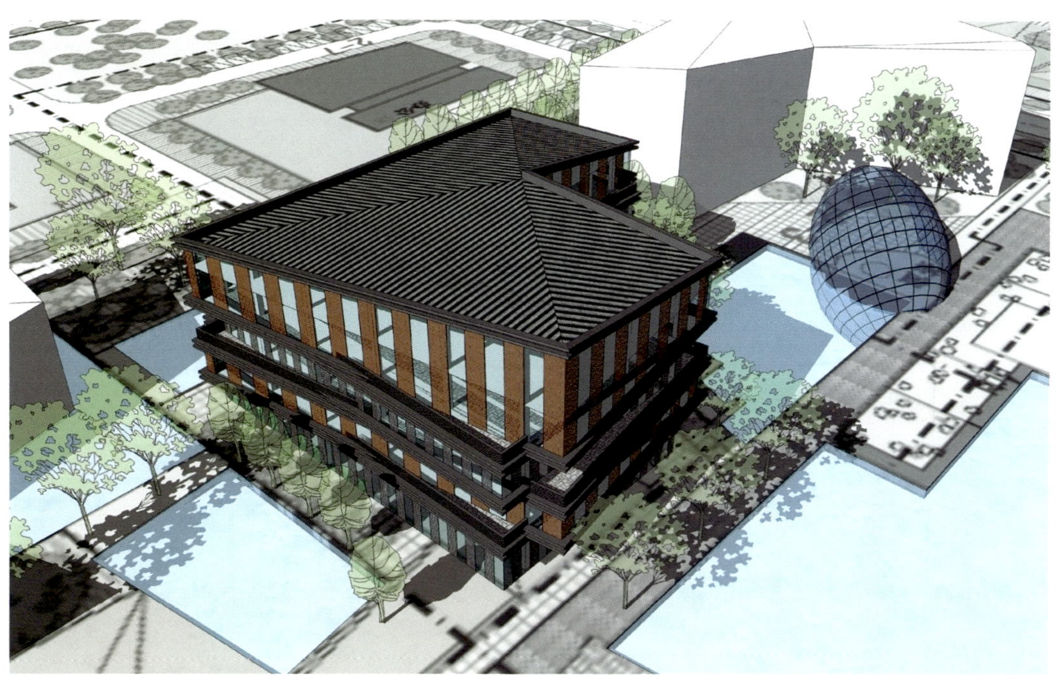

Since the project is based in Gong Jia Bing river landscape radiation zone, focusing on the development of southern catering and business leisure, that is the development of southern China culture and ecological consumption environment, therefore the construction draws the features of south residential houses to use the L-shaped compound layout, emphasizing aggregate sense and corresponding with the surrounding buildings. In order to achieve the rationalization of investment and maximum benefits based on the comprehensive conditions of the building,the height of the commercial buildings should be within 24 meters for five storeys ,with the space freely divided. And there is a basement for parking.

In the shape design it follows the principle of modern bearing the tradition. The shape gets inspiration from ancient castles,with layer upon layer, and the blocks interspersed between them rhythmically enrich the overall space. The roof shape follows the sloping roof of the southern area, divided into several irregular folding surfaces echoing with the pitched roofs of surrounding buildings, which conveys traditional culture conception of the carrier with modern design techniques. Based on modern architectural techniques the surface design takes elements of traditional culture charm to reach the combination of classical and modern it seeks diversity in unity and expresses the excellent local culture with rich architectural language,forming rich sense of gradation and liveliness and surface texture coordinated with the surrounding buildings.

The building colors mainly use gray and blue, combined with the frosted glass and glazed glass to express the romantic charm of Southern architecture. In the details of the glass it uses decorative openwork windows densely or sparsely arranged to express its unique beauty of Chinese traditional architectural form, to reach the coexistence of the Southern style and modern life, as well as harmony among man, landscape and nature.

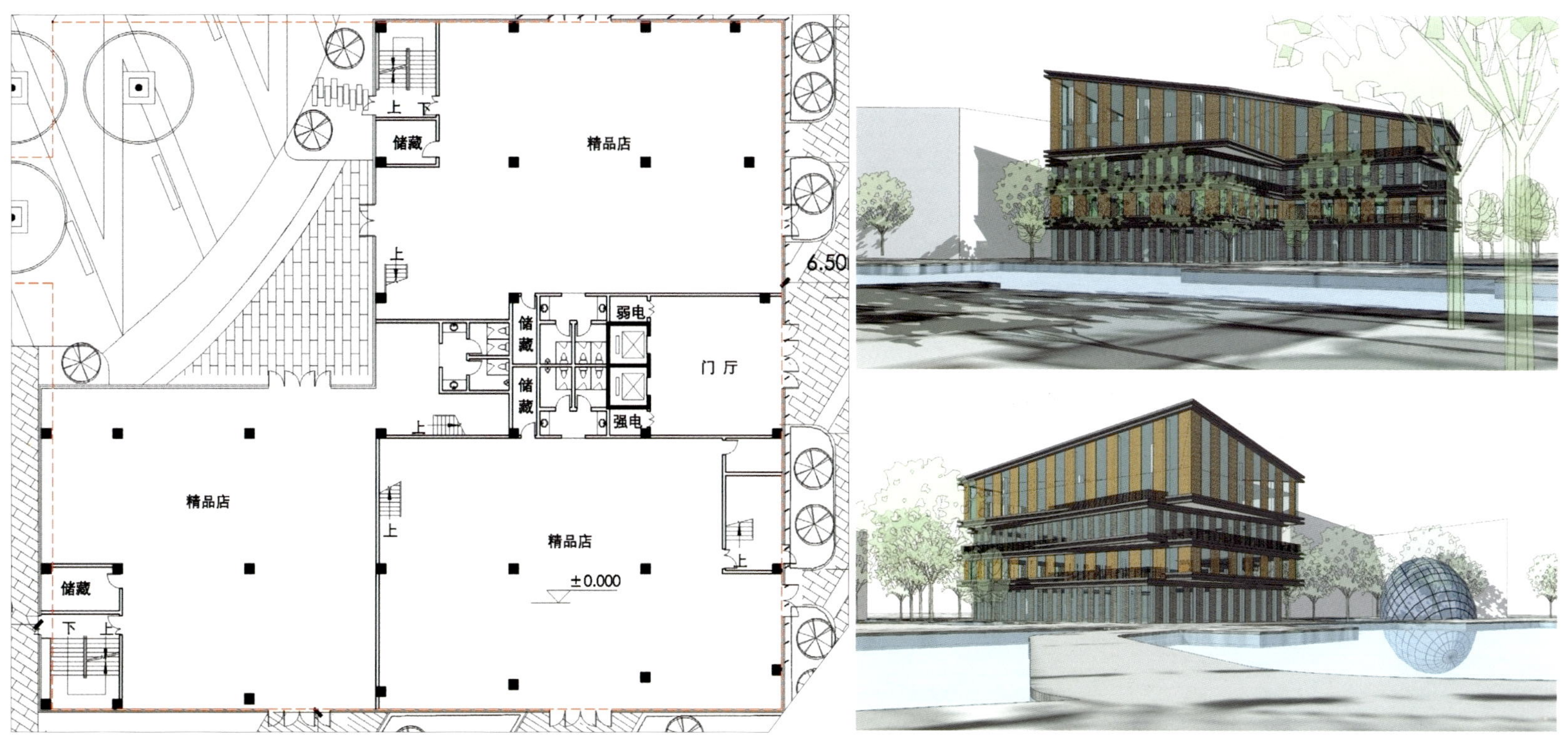

波兰历史博物馆

POLISH HISTORY MUSEUM

项目地点：波兰华沙　用地面积：33.8 ha　建筑面积：20 000 m^2
建筑设计：上海畅想建筑设计事务所
建筑师：罗四维，周伟，王晨军，任斯诚

LOCATION: Warsaw, Poland　SITE AREA: 33.8 ha　BUILDING AREA: 20,000 m^2
DESIGN CORPORATION: Imagine Architects
ARCHITECTS: Luo Siwei, Zhou Wei, Wang Chenjun, Ren Sicheng

立意：历史是一面镜子。

陈述：任何历史都是主观史和客物史的混合物，事实与谎言的混合物，客观与遮掩的混合物……

设计：项目处于华沙市中心区域森林公园，基地满是古树名木、古堡雕塑。基地中间有城市道路穿过，形成半峡谷。为了保持基地的优美环境，把建筑物紧贴道路两侧峡谷布置，并在道路下方和道路上方相互连通，建筑上部连接体既是建筑，又是桥梁、广场、景观……把一分为二的公园连成片。满铺镜面玻璃，映射天空、古木、古堡、四季景观，而自身消失在这种映射之中。人可在镜上行走，看风景、看自己……

诗意：让 MHP 成为一面镜子
让镜子反映天空
让天空中承托行人
让人思考历史

悖论："历史是一面镜子"陈述的是一个事实，每一个民族或当事人都自觉地不自觉地用它来照别人的缺点和自己的优点，否则历史不会惊人地相似了。但历史不应该是一面镜子，它就是历史自身，一旦成为镜子，变成了权力者掌握的准信仰，你必须表明你的态度。多数时候、多数情况下、对多数人而言你必须附合他，自觉地不自觉地，历史就这样被扭曲了。MHP 设计，陈述了事实，却扭曲了历史。

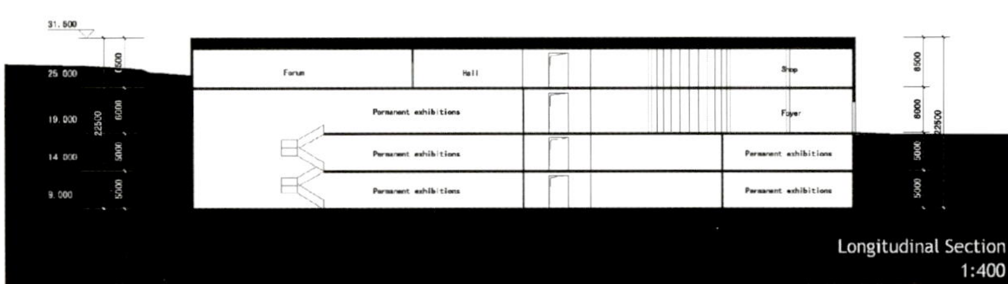

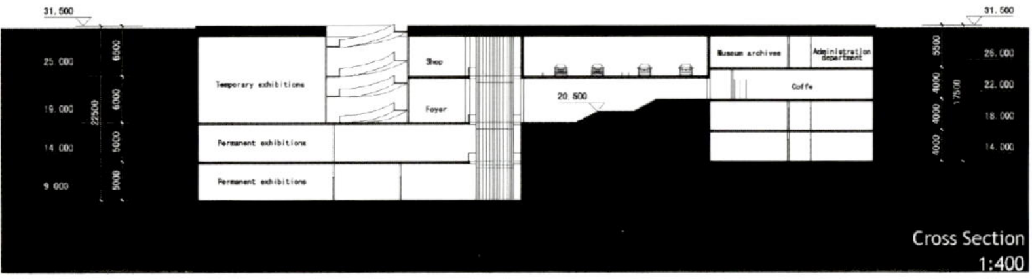

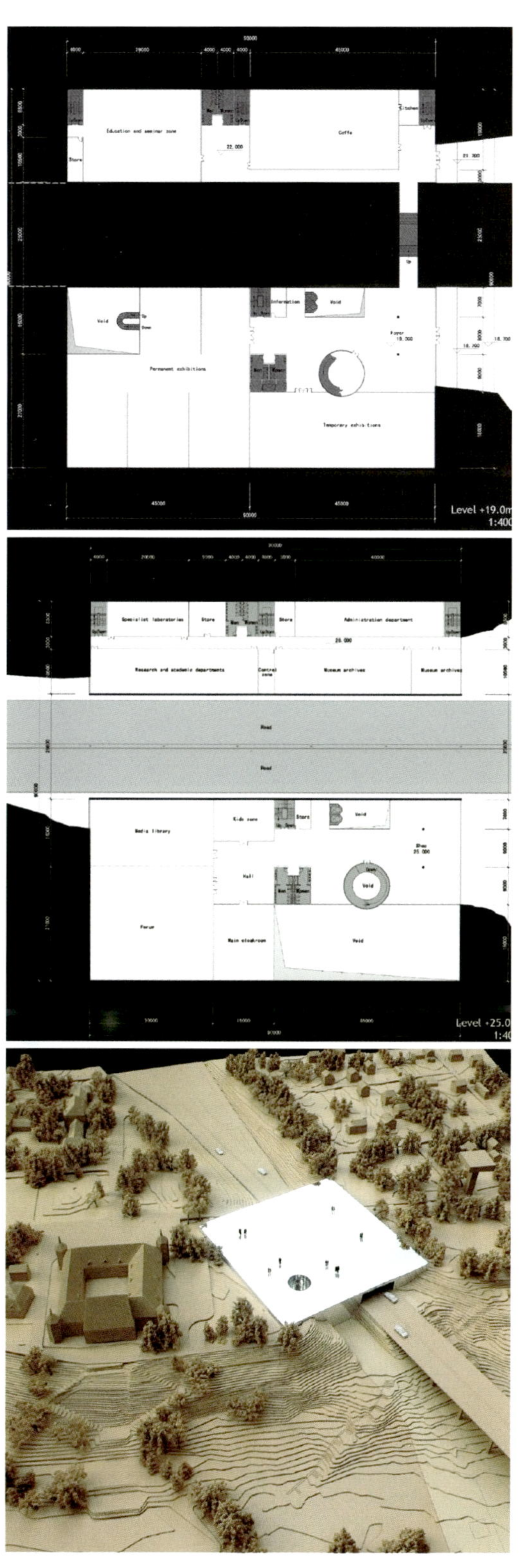

Conception: History is a mirror.

Statement: Any history is a mixture of subjectivity and objectivity, of truths and lies as well as objectivity and concealment.

Design: The project is in the forest park, the central area of Warsaw. The base is full of old trees castles and sculptures. The city road running through the base makes it a semi-valley. In order to maintain the beautiful environment, we set the buildings close to the valleys on both sides of the road,and made to link up between the bottom and top of the road. The the upper connecting body of the building also functions as a bridge, a square as well as the landscape ... The splitting park is made into an intact part. The mirror glasss, mapping the sky, old trees, castles as well as the four season landscape, but lost itself among them. People can walk on the mirror enjoying the scenery, and themselves.

Poetry: Let MHP be a mirror

Which reflects the sky

Where pedestrians are borne

To make people ponder over the history

Paradox: "History is a mirror" is a statement of fact, where every nation or people, consciously or unconsciously, uses it to find the defects of others and the merits of themselves, otherwise histories would not be strikingly similar. However history should not be a mirror, but itself. Because once it becomes a mirror,it will play the role as a quasi-religion of the powerful, and what you have to do is to show your attitude. Most of the time, in most cases and for most people, you must assent to him, consciously or unconsciously, when history would be distorted. MHP design presents facts but distorts the history.

深圳能源大厦

SHENZHEN ENERGY BUILDING

项目地点：中国 · 深圳　用地面积：6427.7 ha　建筑面积：96 420 m^2
建筑设计：上海畅想建筑设计事务所
建筑师：罗四维，周伟，王晨军，任斯诚

LOCATION: Shenzhen, China　SITE AREA: 6427.7 ha　BUILDING AREA: 96,420 m^2
DESIGN CORPORATION: Imagine Architects
ARCHITECTS: Luo Siwei, Zhou Wei, Wang Chenjun, Ren Sicheng

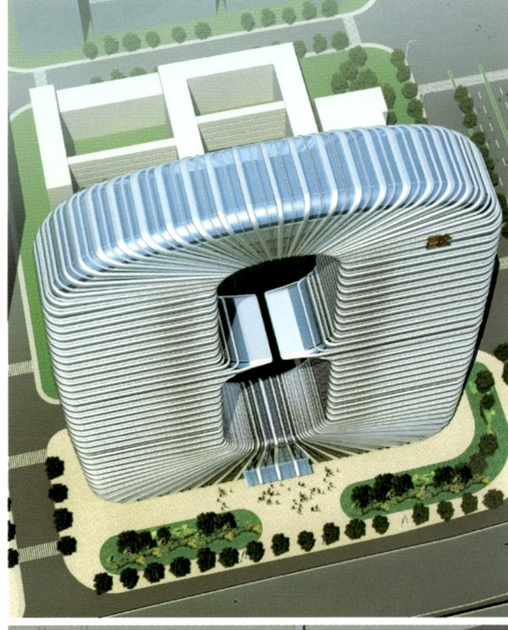

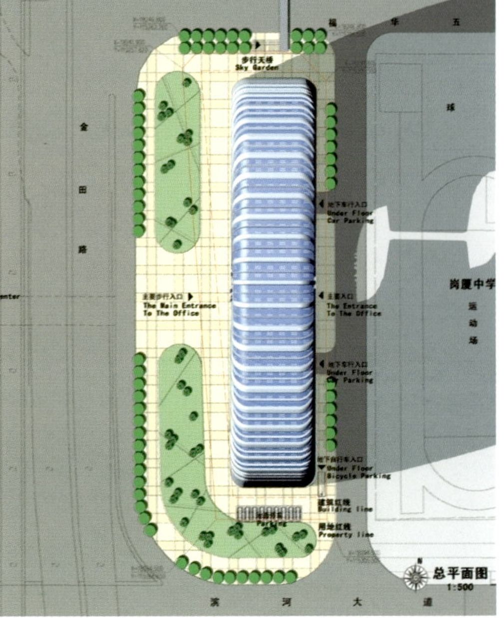

深圳能源集团股份有限公司以发电为主业，深圳国际能源大厦的基本形态设计以发电机的重要核心——定子为出发点，结合基地规划现状抽象组合而来，把她命名为“能量环”。能量环的具体形象是“0”和“∞”的组合体，“0”代表二氧化碳的零排放，“∞”代表可再生能源的持续使用。能量环也可看做是两个“E”——ENERGY 和 ENVIRONMENT 的组合。能量环的主要功效表现在以下四个方面。

（1）公司象征——明确地表达了深圳能源公司的基本性质、未来发展方向，并巧妙地将两栋高层建筑组合成具有多种意蕴的整体。

（2）城市地标——简约、抽象、独特、实用的现代建筑造型，城市中心地带东侧高层天际线醒目的起点。

（3）科普装置——高压放电和磁力线具象化，既有科普教育作用又成为奇特的城市景观，它们定时开放，具有报时功能。

（4）时间坐标——作为景框和节气时点，在春分、秋分等特定节气之季展示阳光与大地的关系。

Shenzhen Energy Group Co., Ltd. mainly works on power generation. The starting point of the basic design form is the important core of the generator - the stator,combined with an abstract composition of the current base planning status. We named her "energy ring." The specific image of the energy ring is the combination of "0" and "∞". "0" represents zero carbon dioxide emissions, and "∞", the sustainable use of renewable energy. Energy ring can also be viewed as two "E"s - ENERGY and ENVIRONMENT.The energy ring maily functions in the following four aspects:

(1) The company symbol - a clear expression of the fundamental nature, the future direction of development of the company, and skillfully combined the two high-rise buildings into an entirety with a variety of meanings.

(2) City landmark - simple, abstract, unique and practical style of modern architecture. It's the visible starting point of the east skyline at the city center.

(3) Science popularization device -- the high-pressure electric discharge and the magnetic field lines visible are both good for science education and to create a unique urban landscape. They are regularly open with the timekeeping function.

(4) Time coordinate -- as a scenery frame it reflects the intimacy between the sun and the earth at the particular solar terms such as spring equinox and autumnal equinox.

罗昂是德国独资的国际设计公司，致力于城市规划设计、建筑设计和景观设计，根据项目位置、社会经济环境量身定制完美的整体解决方案，充分满足客户的具体需求。罗昂创意团队汇聚了来自国内外的建筑师、景观设计师、室内设计师、城市规划师、工程师、3D 设计师、社会学家以及市场专家。他们紧密合作，协力完成所有项目，从而保证罗昂的设计紧跟国际潮流，同时符合地方建筑法规。此外，我们相信，通过各学科之间的互动形成的理念能够在创作卓越设计方案的同时全面兼顾建筑技术、建筑用途以及场地的历史背景。罗昂的众多专业人才一直在进行精诚合作。同时，公司拥有深厚的国际专业技术背景，并且在中国积累了丰富的行业经验。凭借以上优势，罗昂得以提供从可行性研究到优质设计方案、从现场服务到质量控制等各类服务。为了确保高质量的建筑、景观和设计整体设计方案，所有的项目均由四个主要部门通力合作完成。研究发展部（R&D）全面分析场地、原有建筑用途、交通流量以及影响当地社会形态的消费者行为和社会经济状况。商务市场部以研究发展部提供的调研结果为基础，提炼出商务概念并向客户提供利益最大化的解决方案。设计部把商务概念转化为具有很高美学价值的作品。技术部则保证设计兼顾结构标准，并确保在建造时采用最佳的施工方法。

logon is a Germany-based international design office that integrates urban planning, architecture and landscape design to respond to our client's challenges by partnering with them to find extraordinary and comprehensive solutions, perfectly tailored to situation and context. From our Shanghai office, local and international architects, landscape designers, interior designers, urban designers, engineers, 3D artists, sociologists and marketers partner together on all projects. This guarantees that world-class design ideas respect local building code and regulations. Also, we believe that the knowledge produced by the interaction of multiple disciplines creates not only beautiful design, but holistic design that responds to vehicle & human flows, business concept and the site's history. Thanks to extensive experience in China and international expertise along with a number of industry professionals challenging one another daily, logon offers a wide range of services from feasibility study, to quality design, right through to site service and quality control. In order to ensure a quality full package approach in urban architecture, landscape and design, projects are carried to completion by four main departments. The Research & Development (R&D) department analyses the site, former building uses, traffic flows or local consumer behavior and socio-economic factors that influence local society. Business Marketing Services refines the business concept based on research provided by R&D, to provide the client with the most profitable solution possible. The Design Department ensures the concept is translated into an aesthetically compelling design. And the Technical department guarantees structural quality during the design stage and top construction implementation in the building phase.

柯复南

王芳

公司合伙人及主要设计师

- 柯复南 Frank Krueger

设计总监 建筑学硕士 城市规划设计博士

柯复南先生是德国改建项目的优秀建筑师，2001年在德国柏林合伙创建了罗昂建筑设计公司。初期与中国建立了合作关系，并和柏林科技大学合作，为共享设计技术资源奠定了良好的基础。2005年在中国上海开设分部，现为罗昂建筑设计公司设计总监及合伙人，管理城市规划设计、建筑设计及景观设计部门。在他的带领下，其团队在中国设计了许多精品项目，如位于上海市中心地段静安区的800秀项目，位于济宁市的4万人文体中心项目及丽江束河88栋高档别墅等。

- 王芳 Wang Fang

执行总裁 建筑师

王芳2004年获得同济大学建筑学硕士学位，其间参加了同济大学和柏林工大学的联合培养硕士项目。取得同济大学硕士学位同时，获得与德国柏林工业大学联合培养硕士学位。留德学业结束后，被高薪聘请为技术顾问任职于德国技术合作公司(GTZ)，完成了大量的城市可持续规划和管理方面的重要课题。期间王芳的工作重点为中德政府长期共同关注的社会课题“可持续性的城市更新”。2006年扬州市市政府以该研究成果为主题成功申请并获得了联合国人居署颁发的最高人居环境奖项“联合国人居奖”。

王芳对中欧文化的深入了解，对城市规划、建筑设计和社会经济学等跨学科知识的掌握以及大量跨学科领域的的工作经验，同时熟悉中德政府相关事宜，帮助她和她的合伙人成功创立了上海罗昂并担任执行合伙人职务。期间，她对项目运作和对市场的深刻洞察对上海罗昂的迅速成长起到了决定性的作用。在罗昂王芳主要负责市场拓展、人事、财务管理以及与德中政府或公司合作的相关事宜。她充分了解城市发展问题，在如何建立高标准的城市发展模式和公司的经营运作方面颇有建树。

创意产业园 800秀

800SHOW CREATIVE PARK

项目地点：中国·上海 用地面积：15 000 m^2 建筑面积：30 000 m^2
建筑设计：罗昂建筑设计咨询有限公司
建筑师：柯复南

LOCATION: Shanghai, China SITE AREA: 15,000 m^2 BUILDING AREA: 30,000 m^2
DESIGN CORPORATION: logon
ARCHITECT: Frank Krueger

德国罗昂建筑在“800秀”的国际招标中异军突起，成为最终的赢家。“800秀”是上海中心城区的最大改建工程，是2010年世博会的重要组成部分和2009年国际创业产业周的主会场。罗昂的获奖设计团队将旧钢铁制造厂改建成20 000 m²的创意产业园区，专注地为每个不同时代且具有不同历史的建筑创造其独特的个性语言，成功地实现了国际化的使用功能与改造后建筑的完美结合。在整个工作过程中，罗昂设计有意识地突出了历史建筑材料，开发了新的原创建筑材料，通过成功地使用历史和原创材料创造了超现代的表达方式。设计融合了国际化的生活方式，国际社区将诞生在昔日老旧的工业厂区内，各种国际化的生活、办公和消费方式产生，包括画廊、创意办公室、餐馆、酒吧和艺术品商店等。令人印象深刻的100 m长的旧生产车间，无论是空间形态上还是照明效果上都堪称绝品，也是核心城区内可举办各类创意活动、聚会和时装发布的最佳创意场所。罗昂的设计使“800秀”成为上海和新城区静安区的新一代的地标建筑。当原始建筑物在顶尖创意和技术水准的建筑中融合之际，连接过往的桥梁终于待建完成。

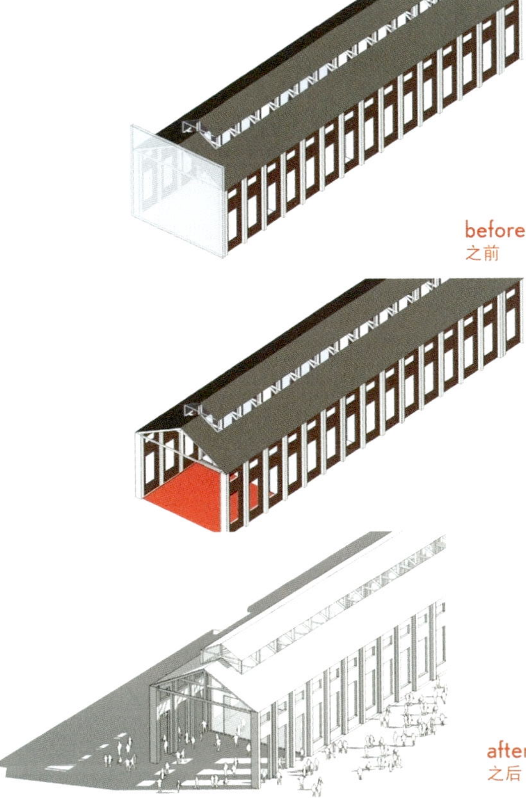

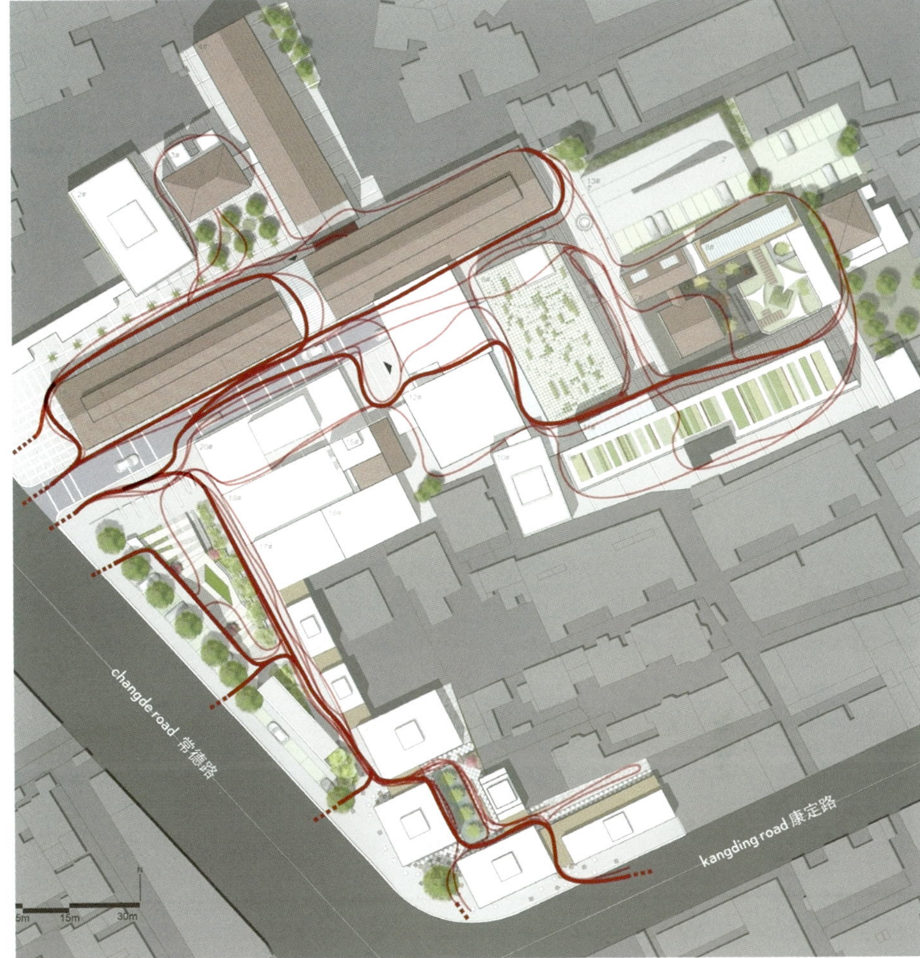

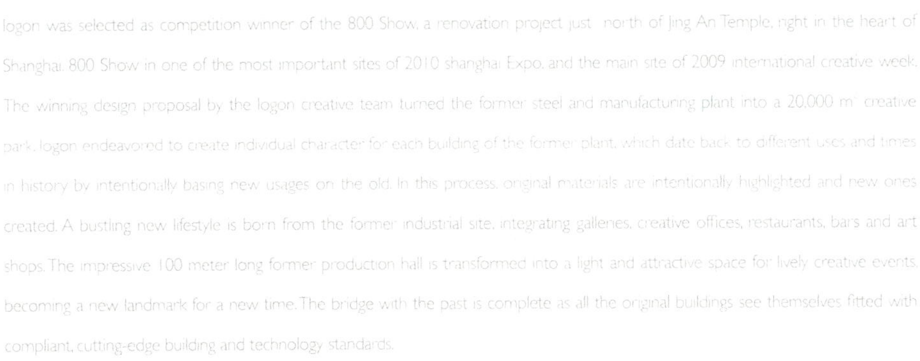

logon was selected as competition winner of the 800 Show, a renovation project just north of Jing An Temple, right in the heart of Shanghai. 800 Show in one of the most important sites of 2010 shanghai Expo, and the main site of 2009 international creative week. The winning design proposal by the logon creative team turned the former steel and manufacturing plant into a 20,000 m² creative park. logon endeavored to create individual character for each building of the former plant, which date back to different uses and times in history by intentionally basing new usages on the old. In this process, original materials are intentionally highlighted and new ones created. A bustling new lifestyle is born from the former industrial site, integrating galleries, creative offices, restaurants, bars and art shops. The impressive 100 meter long former production hall is transformed into a light and attractive space for lively creative events, becoming a new landmark for a new time. The bridge with the past is complete as all the original buildings see themselves fitted with compliant, cutting-edge building and technology standards.

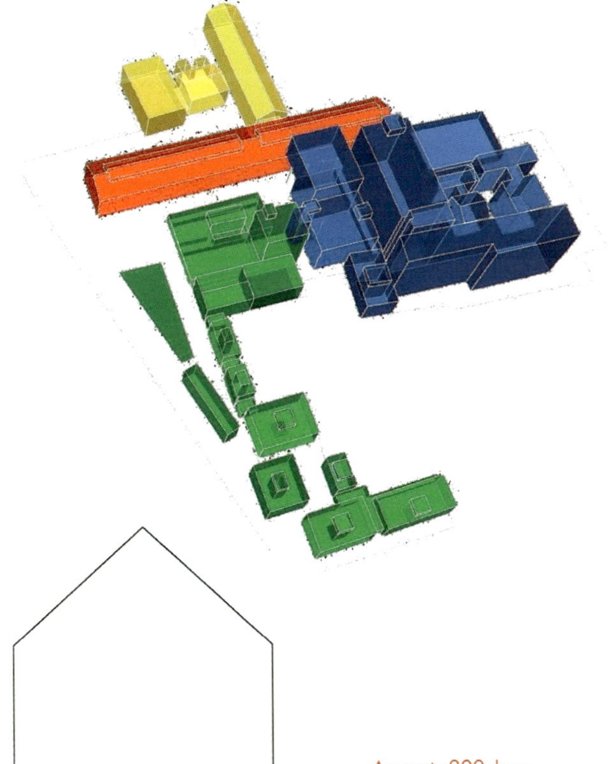

上海映巷视觉创意产业园

IIINSHANGHAI CREATIVE PARK LANDSCAPE

项目地点：中国·上海 建筑面积：9000 m^2
建筑设计：罗昂建筑设计咨询有限公司
建筑师：柯复南

LOCATION: Shanghai, China BUILDING AREA: 9000 m^2
DESIGN CORPORATION: logon
ARCHITECT: Frank Krueger

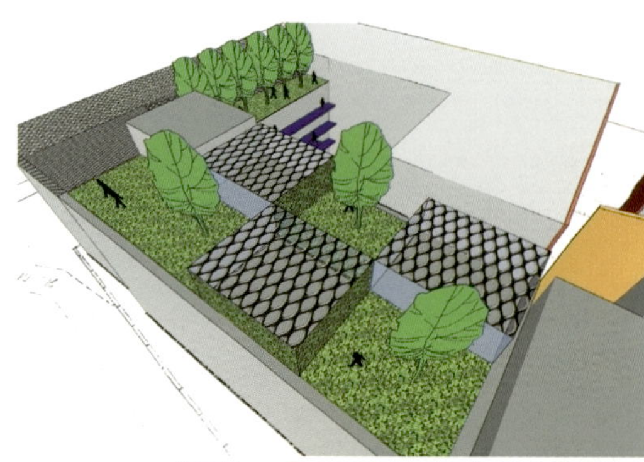

罗昂把原上海金属塑料制品模具厂改造为上海市长宁区的第一个创意产业园。厂区内有四栋厂房，建筑面积总计9000 m²，它们共同组成了一个狭窄但是具有吸引力的内庭院，提供了一个可以回溯历史的独特视角。厂区内的建筑大多建于20世纪80年代中期，室内空间大气简洁，建筑立面却缺少自身特征。与大多数改造项目不同，罗昂并非重新建造一个全新（但是虚假）的立面，而是从厂房原始的功能——金属塑料制造出发，试图找到属于厂区自身特质的独特语言并将其运用到建筑改造的设计中去。因此，建筑最终的立面上较多地选用了金属元素，其中相当的一部分甚至就直接来源于原厂房工业制造中的成品。金属杆件被重新整合成为了庭院颇具艺术气质的院墙；废弃的金属片被用在了地块入口的拱形连桥和建筑立面的金属窗框中；两根被用做结构连接的工字钢与新的玻璃砖结合后成为了一个透明的空中桥梁。新旧两种工业元素的结合延续了工业建筑的记忆，又为这里的工作和娱乐休闲提供了一种全新的体验。

logon reconverted the former Shanghai Metal and Plastic Module Factory into the first creative centre of Changning District. With a total floor area of 9000 sq.m, spread over four buildings, the factory forms a narrow but exciting yard, providing insight into the history of the site. The original buildings date only from the 1980s, with great interior spaces but lacking character on the exterior. Instead of creating a new (but fake) style, logon has picked up the former function, metal and plastic processing, and turned it into the new character of the creative centre: the main elements of the new facades are made of metal, crafted by man just like the factory had produced its ware: leftover metal pieces were turned into an entrance bridge and window frames, steel pipes on a facade provide the basis for vertical gardens, and a crane was turned into a transparent bridge.

SUPERBEE

青浦新湿地体验

THE NEW QINGPU WETLANDS

项目地点：中国·上海　用地面积：68.6 ha
建筑设计：罗昂建筑设计咨询有限公司
建筑师：柯复南，陈一萍，汤锦燕

LOCATION: Shanghai, China　SITE AREA: 68.6 ha
DESIGN CORPORATION: logon
ARCHITECTS: Frank Krueger, Chen Yiping, Tang Jinyan

the structure

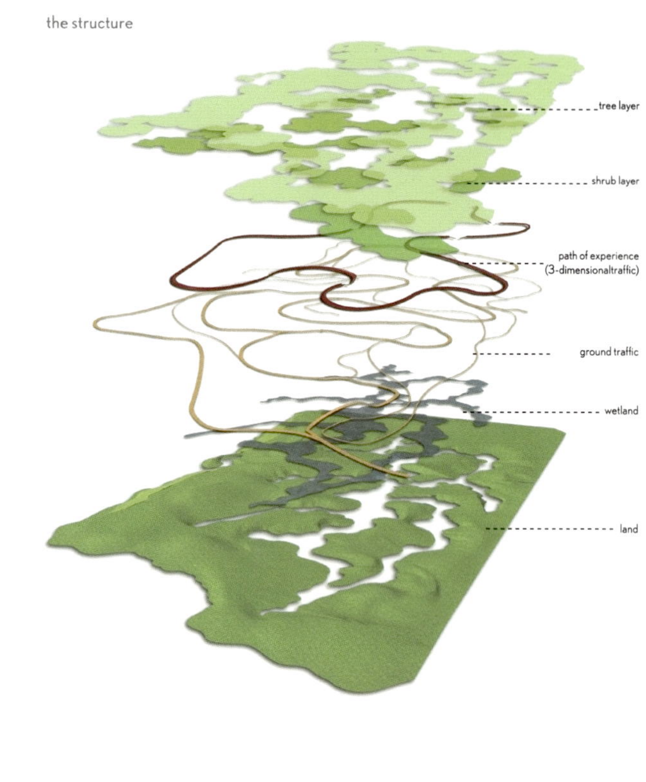

the three-dimensional path

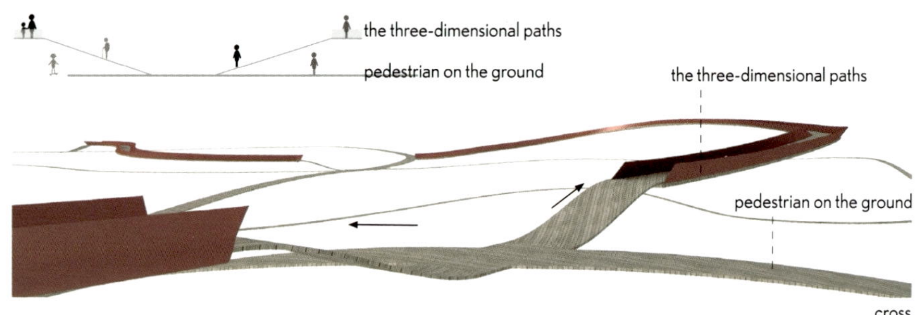

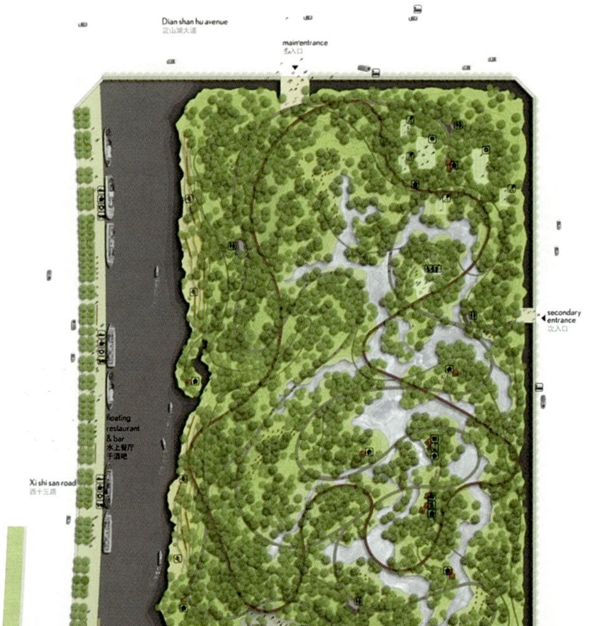

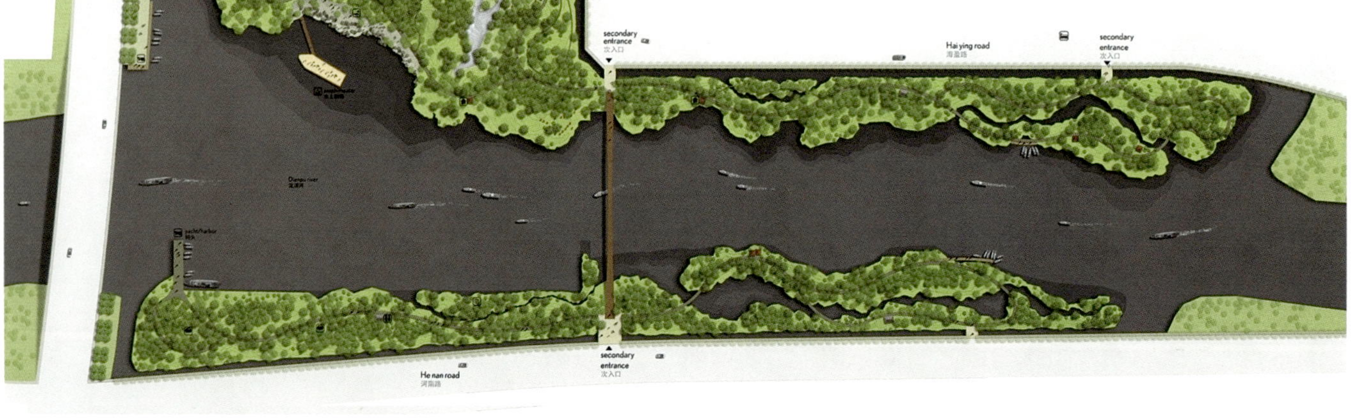

new view of unknown nature

activities in the park

the wetlands

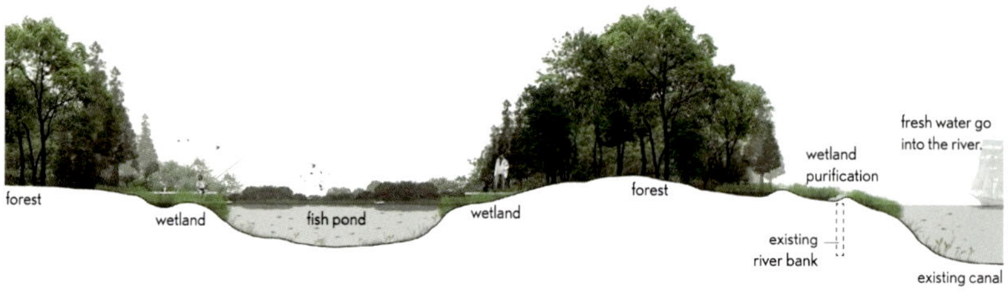

forest
wetland
fish pond
wetland
forest
wetland
purification
fresh water go
into the river
existing
river bank
existing canal

在现代化和城市化的进程中，中国的一些地区，特别是大城市的周边城市，正经历着翻天覆地的变化。青浦距离上海市中心约30 km，是上海周边的几个工业区之一。罗昂提出的“青浦新湿地体验”城市景观方案在城市中营造了一处自然的生态湿地，致力于创造一个生态可持续系统，具有经济投入低、维护成本低等特点。考虑到基地本身的潜质，植物、水和土地成为设计的最原始素材。介于这些元素之间，在距离地面4 m高处，一条蜿蜒曲折的“体验之路”横空出世，游人和当地居民将能在不同的高度体验自然。设计旨在还自然于其最初之面貌，在城市中创造一处宁静的、全新的生态体验空间。

In the process of modernization and urbanization, various regions in China, especially those around megacities, are undergoing dramatic transformation. Located around 30 km south-east of downtown, Qingpu is one of several industrial districts of the urban megalopolis Shanghai. logon's urban design proposal of the New Qingpu Wetlands is to build a natural ecological wetland to create a sustainable system with smallest financial investment and smallest maintenance input. By utilizing the site's potential, the park first develops plants, water, and land. Then, respectfully around and between these elements, the Path of Experience, weaving up to 4 m above ground, allows visitors and local residents an extraordinary experience among the nature. Great effort is made to help nature return to its original beauty, to devise a wholly unique, quiet space in the city center.

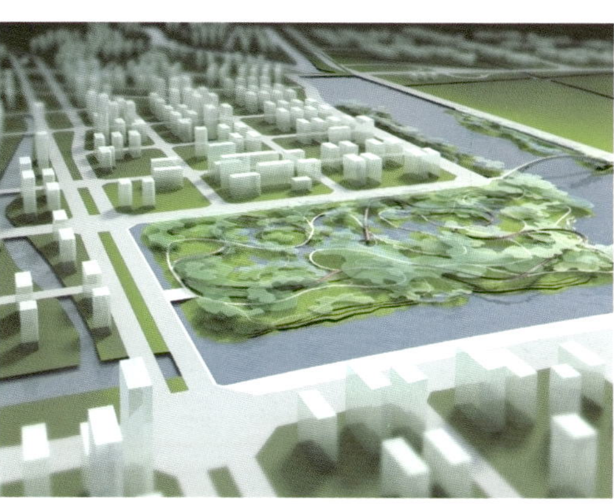

天津“星耀五洲”德国别墅

GERMAN VILLAS “THE WORLD”, TIANJIN

项目地点：中国·天津　用地面积：7820 m^2　建筑面积：6194 m^2
建筑设计：罗昂建筑设计咨询有限公司
建筑师 ：柯复南

LOCATION: Tanjin, China　SITE AREA: 7820 m^2　BUILDING AREA: 6194 m^2
DESIGN CORPORATION: logon
ARCHITECT: Frank Krueger

星耀五洲位于中国第四大城市天津的西南部，项目将世界版图浓缩于7.3 km^2的土地上。罗昂创意总监柯复南先生及团队受邀作为“十大建筑师”之一主笔“星耀五洲”德国别墅的设计，包括一系列湖岸高级定制别墅以及一座集商务和娱乐一体的岛上俱乐部别墅。别墅以德国技术和设计质量为标准，强调生活品质、舒适性和可持续性，致力于为用户提供最佳的生活体验和健康的生活方式，成为中国低能耗家用别墅的典范。

"The World" is emerging in the south-east of Tianjin, the 4th largest city in China, to create a replica of the globe scaled down to the size of 7.3 sq.km. logon creative director Frank Krueger and his team were selected amongst ten foreign"Master Designers" to design the German villas of "The World". Frank and his team delivered the architectural design and landscape design for a series of prestigious, tailor-made luxurious villas on the lake's front as well as a business and recreation clubhouse on a private island. The villas feature German design and technology to provide the best possible living comfort and health lifestyle. The villas set a landmark for energy-efficient single family housing in China.

13
14
15
16
17
18
19
20
-3.60
D10
D11

重庆新世界购物中心

CHONGQING NEW WORLD SHOPPING CENTER

项目地点：中国·重庆　用地面积：9.42 ha　建筑面积：701 100 m^2

建筑设计：罗昂建筑设计咨询有限公司

建筑师：柯复南

LOCATION: Chongqing, China　SITE AREA: 9.42 ha　BUILDING AREA: 701,100 m^2

DESIGN CORPORATION: logon

ARCHITECT: Frank Krueger

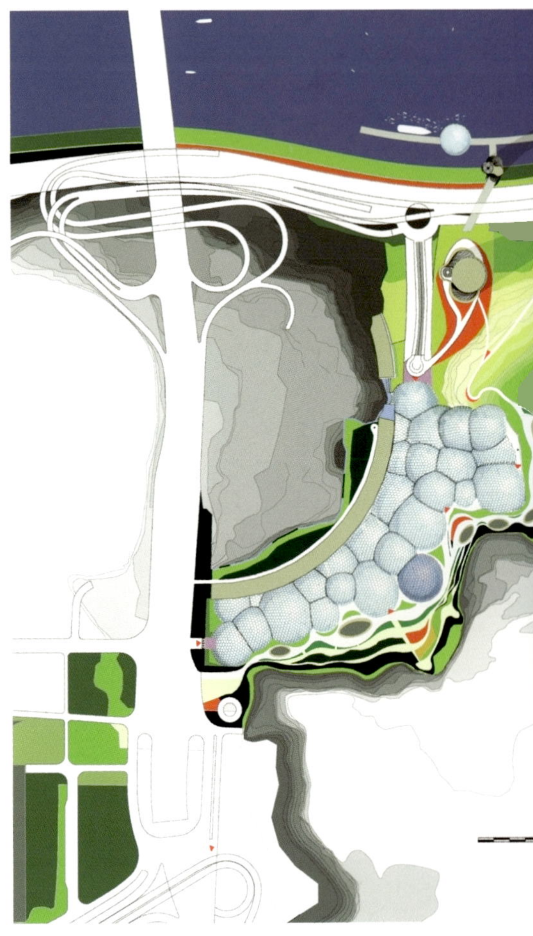

重庆新世界购物中心位于嘉陵江和长江的交汇处，与重庆旧CBD隔江相望。整个地块落差达到40 m，设计采用将购物部分划分成4级退台式的模式，通过封闭的构筑模式隔绝来自城市的污染和灰尘，充分利用风、光、水、景观等自然资源，创造舒适的购物环境，提供多功能的休闲娱乐空间。该购物中心与场地中的豪华住宅楼及一幢拥有5星级酒店的超高层塔楼相连接，摩天大楼将成为重庆新CBD的标志性建筑物。罗昂从商业概念角度入手，针对性地进行规划和建筑设计，项目建成后将成为全世界第十三大的休闲娱乐购物中心。

Chongqing New World Shopping Center is located at the intersection of the Jialing and Yangtze River and just opposite the old CBD area in Chongqing. Despite the considerable plot fall of 40m, the shopping paradise is spread across 4 terraced levels, encapsulated in a protective glass mold to keep out the pollution and dirt of the city, fully use natural resources available such as wind, light, water and landscape, creating a comfortable shopping atmosphere and providing a multi-function leisure and entertainment space. The shops are connected to a strip of luxury residential housing as well as a super highrise tower serving luxury guests with its 5-star hotel. The super highrise will become the landmark of Chongqing new CBD area. The planning and architectural design was well devised based on the business model, upon completion, New World Shopping Center will become the 13th largest leisure, entertainment and shopping center in the world.

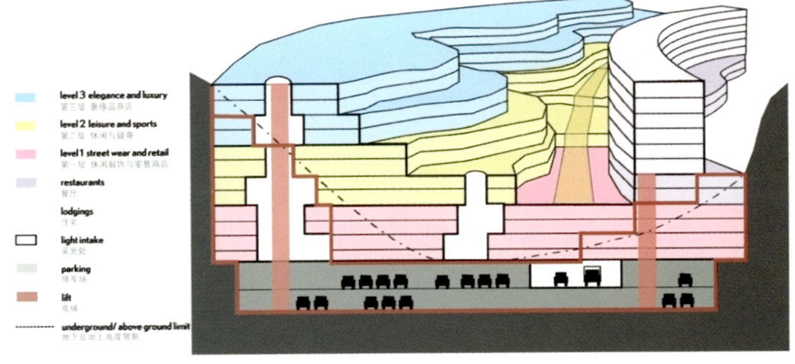

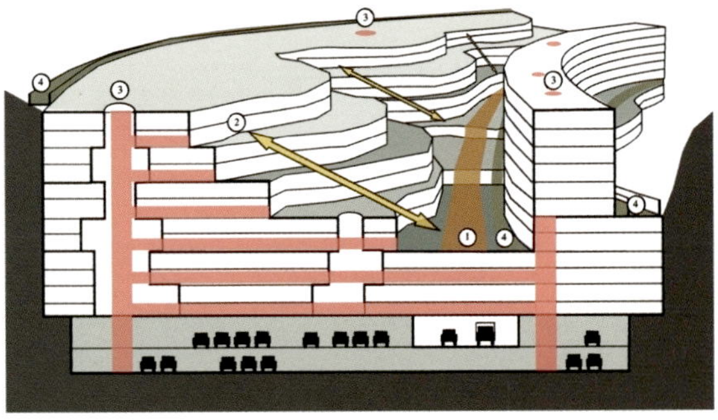

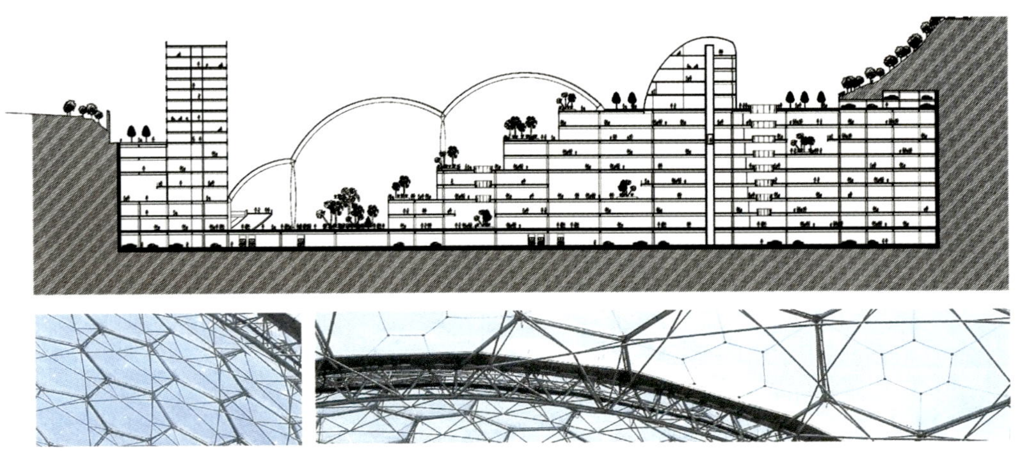

Coca-Cola
NEW WORLD
MEXX

NEW WORLD
NEW WORLD

NEW WORLD

MAD事务所是一家具有创新精神的建筑师事务所，关注当代的建筑、景观设计及相关的城市、文化问题，并希望通过超前的理念为未来的城市提出独特、有效的设计方案。

MAD目前在世界范围内有各种规模和类型的建筑项目，其中包括正在施工中的多伦多梦露大厦（2006年MAD赢得该项目设计竞赛第一名），位于天津滨海新区的358 m超高层——中钢国际广场，三亚凤凰岛的整体规划设和建筑设计项目，北海滨海居住社区，以及在丹麦、中国香港、迪拜、新加坡、马来西亚、日本、哥斯达黎加等地的大型公共建筑和住宅项目。

MAD的设计作品得到世界范围的关注与报道。MAD获2006年度纽约建筑联盟青年建筑师奖。MAD还举办了一系列个展。其中包括2006年MAD在意大利威尼斯举办的“MAD in China”展览，与威尼斯双年展同步展出，在北京Tokyo Gallery举行了名为“MAD Under Construction”建筑设计展览。2007年MAD在丹麦哥本哈根的丹麦建筑中心举办名为“MAD in China”的漂浮城市系列个展。2008年9月，受第11届威尼斯双年展邀请，MAD在军火库展出“Super Star-A mobile China Town”。

MAD is a Beijing-based architectural design studio dedicated to creating innovative projects. It combines a sophisticated design philosophy with advanced technology in exploring contemporary architecture, social, and cultural issue in today's China. We examine and develop our unique concept of futurism through current theoretical practices in architectural design, landscape design, and urban planning. In 2006, MAD was awarded the Architectural League Young Architects Forum Award.

MAD's ongoing projects include: the Absolute Tower in Toronto, Canada, an international competition MAD won in 2006; the Sinosteel International Plaza, a 358 m high-rise building in Tianjin, China; and some large-scale public complex and residential housing in Denmark, Costa Rica, Dubai, Singapore, Malaysia, Japan and Hong Kong China.

MAD's works has been published worldwide. In 2006, MAD was awarded the Architectural League Young Architects Forum Award. The office has also presented its designs in a series of exhibitions, including the "MAD in China" exhibition at the Venice Architecture Biennial, and the "MAD Under Construction" exhibition at the Beijing Tokyo Art Projects Gallery in Beijing. In 2007, "MAD in China," a floating city of MAD's work, was shown at the Danish Architecture Centre in Copenhagen, Denmark. MAD's concept proposal, Super Star-A Mobile China Town was on exhibition in the Uneternal City of the 11th Architecture Biennale in Venice.

MAD创立人

- 马岩松　Ma Yansong

1975年出生于北京，毕业于美国耶鲁大学（Yale University），获建筑学硕士以及Samuel J. Fogelson优秀设计毕业生奖。曾经在伦敦的扎哈·哈迪德建筑事务所和纽约埃森曼建筑事务所工作。

马岩松2004年回到中国并成立了北京MAD建筑事务所，同时任教于中央美术学院。

马岩松获2001年美国建筑师学会（AIA）建筑研究奖金，2006年度纽约建筑联盟青年建筑师奖。2008年他的建成作品红螺会所被英国伦敦设计博物馆入选2007年度设计奖并被ICON杂志提名为当今最具影响力的20名设计师之一。2008年马岩松被香港的南华早报评为年度建筑师。马岩松的作品包括曾在2002年引起国内外建筑界广泛关注和讨论的“浮游之岛”——重建纽约世界贸易中心方案，这件作品后来被中国国家美术馆馆藏。2006年MAD建筑事务所在加拿大多伦多ABSOLUTE超高层国际竞赛里中标的“梦露大厦”设计(2009年建成)，成为历史上首位在国外赢得重大标志性建筑项目的中国建筑师。目前正在建设中的项目包括天津滨海新区358 m高层——中钢国际广场，内蒙古鄂尔多斯博物馆以及迪拜世界岛东京岛的规划建筑设计。

马岩松的艺术装置作品“鱼缸”、“墨冰”曾分别在中国国家美术馆和中华世纪坛展出。2006年，MAD在意大利威尼斯举办了“MAD in China”展览，与威尼斯双年展同步展出；同年在北京Tokyo Gallery举行了名为”MAD Under Construction” 的建筑设计个展。2007年MAD在丹麦哥本哈根的丹麦建筑中心展出一个名为MAD in China” 的漂浮城市系列个展。2008年9月，受第11届威尼斯双年展邀请，MAD将在军火库展出“Super Star-A Mobile China Town”。

台中会展中心

TAICHUNG CONVENTION CENTER

项目地点：中国 · 台湾 基地面积：70 318 m^2 建筑面积：216 161 m^2

建筑设计：MAD 建筑事务所

设计师：马岩松，党群，Jordan Kanter, Jtravis Russett, Irmi Reiter, Diego Perez, 戴璞 , Rasmus Palmquist, Art Terry, Chie Fuyuki

LOCATION: Taiwan, China SITE AREA: 70,318 m^2 BUILDING AREA: 216,161 m^2

DESIGN CORPORATION: MAD

ARCHITECTS: Ma Yansong, Dang Qun, Jordan Kanter, Jtravis Russett, Irmi Reiter, Diego Perez, Dai Pu, Rasmus Palmquist, Art Terry, Chie Fuyuki

这是一组连绵起伏的建筑群落，褶皱状的“山体”模糊了建筑、景观和城市公共空间的界限，构成一幅展现东方自然精神的未来世界。方案传承了中国对建筑群体和空间序列追求的传统，并把东方文化中与自然和谐相处的精神气质贯穿于其中。在这个规模庞大的建筑群中，重要的不再是某个建筑单体本身，建筑物的形象是一体化的，而它们所围合的空间则成为主体，那是一种在空气、风、光线之间形成的自然秩序，以及由此建立起来的人与自然之间的情感共鸣。

台中所需要的是一件超越地域，重新定义城市文化景观和社会生活的大型都市艺术品，以独特的建筑观和全新的建筑宣言，使台中跃升成为世界文化先锋。

当今地标式建筑的特征已经由对高度的原始追求转为面向未来和自然的文化诉求。地标不仅仅是视觉上的冲击，更应该是一个聚集城市活力，激发交流和想象力的戏剧性生活场景。

这块基地本身就有着它丰富的生命力——平静的表皮下蕴藏的能量渴望被表现出来，地形本身就是一件具有潜力的自然艺术品。建筑物好像当地的“环形山”，相互之间牵动，围合，转化为连绵起伏，具有中心汇聚点的建筑形体，形成建筑与自然地景的对话。

包裹“群山”的建筑表皮是由一系列的绿色技术构成的复合生态皮肤。褶皱状的连续外表皮为建筑提供自然的空气流动，收集太阳能并维持最低的能耗。

由“群山”围成的院落相互连接，构成了室外空间的自然序列。正如传统的紫禁城和中国园林中对于人与自然和谐共生的追求一样，这个建筑群落的意义更多地表现在其非物质的属性，即围合空间极其自然的精神——一棵树，一片竹林，一潭池水成为了空间的主体。这是基于传统哲学和美学的可持续发展观，而不是基于技术的。

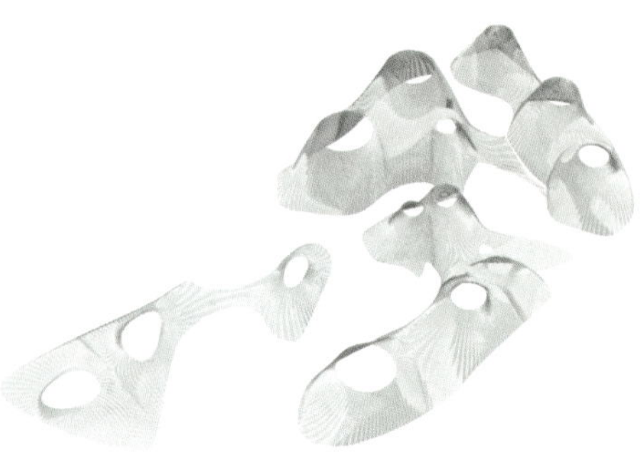

ECO-FRIENDLY
PLEATED SKIN SYSTEM
遮阳节能表皮

STANDARD
CONSTRUCTION
基本土建结构

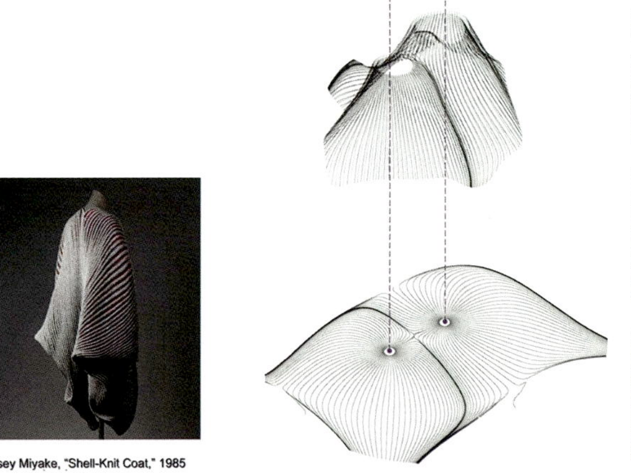

Issey Miyake, "Shell-Knit Coat," 1985

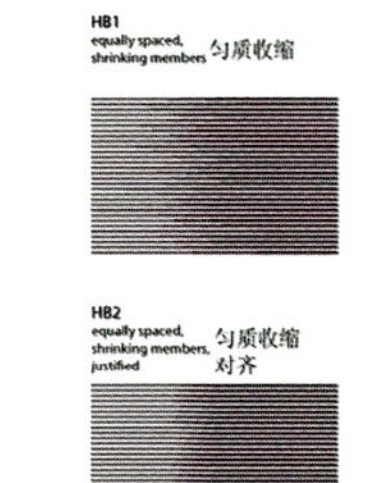

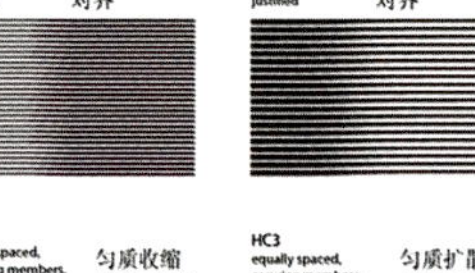

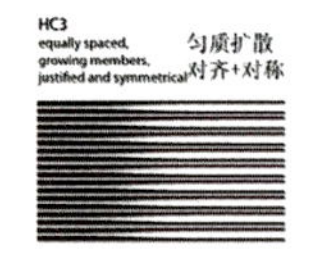

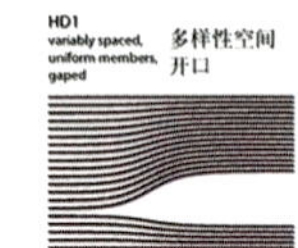

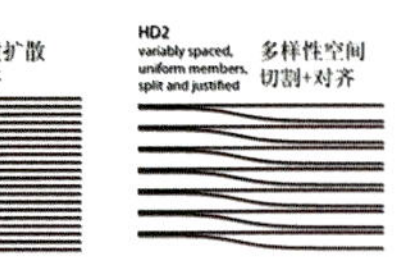

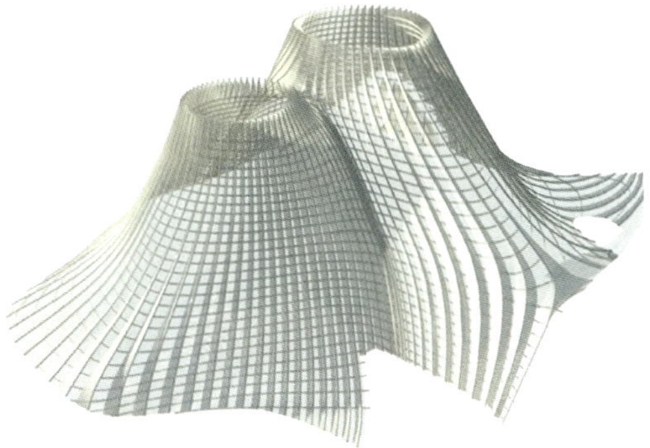

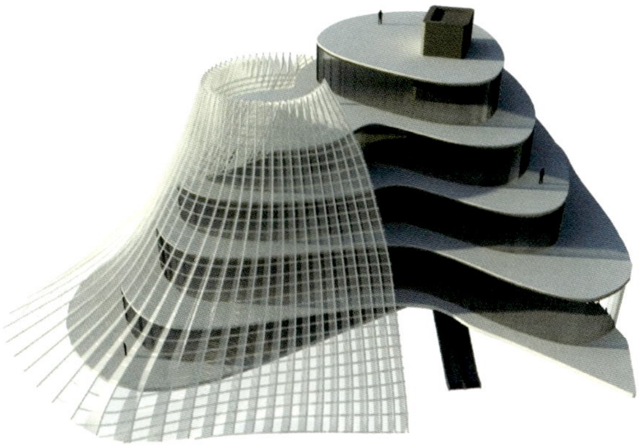

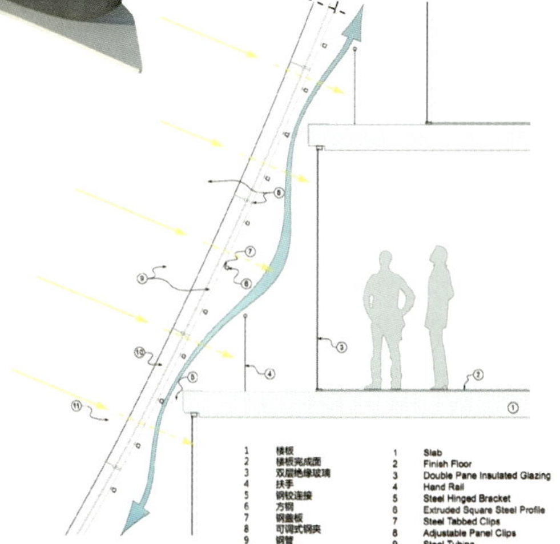

1 楼板
2 楼板完成面
3 双层绝缘玻璃
4 扶手
5 钢铰连接
6 方钢
7 钢盘板
8 可调式钢夹
9 钢管
10 中空层
11 半透明板

1 Slab
2 Finish Floor
3 Double Pane Insulated Glazing
4 Hand Rail
5 Steel Hinged Bracket
6 Extruded Square Steel Profile
7 Steel Tabbed Clips
8 Adjustable Panel Clips
9 Steel Tubing
10 Semi-Solid Panel
11 Transparent Panel

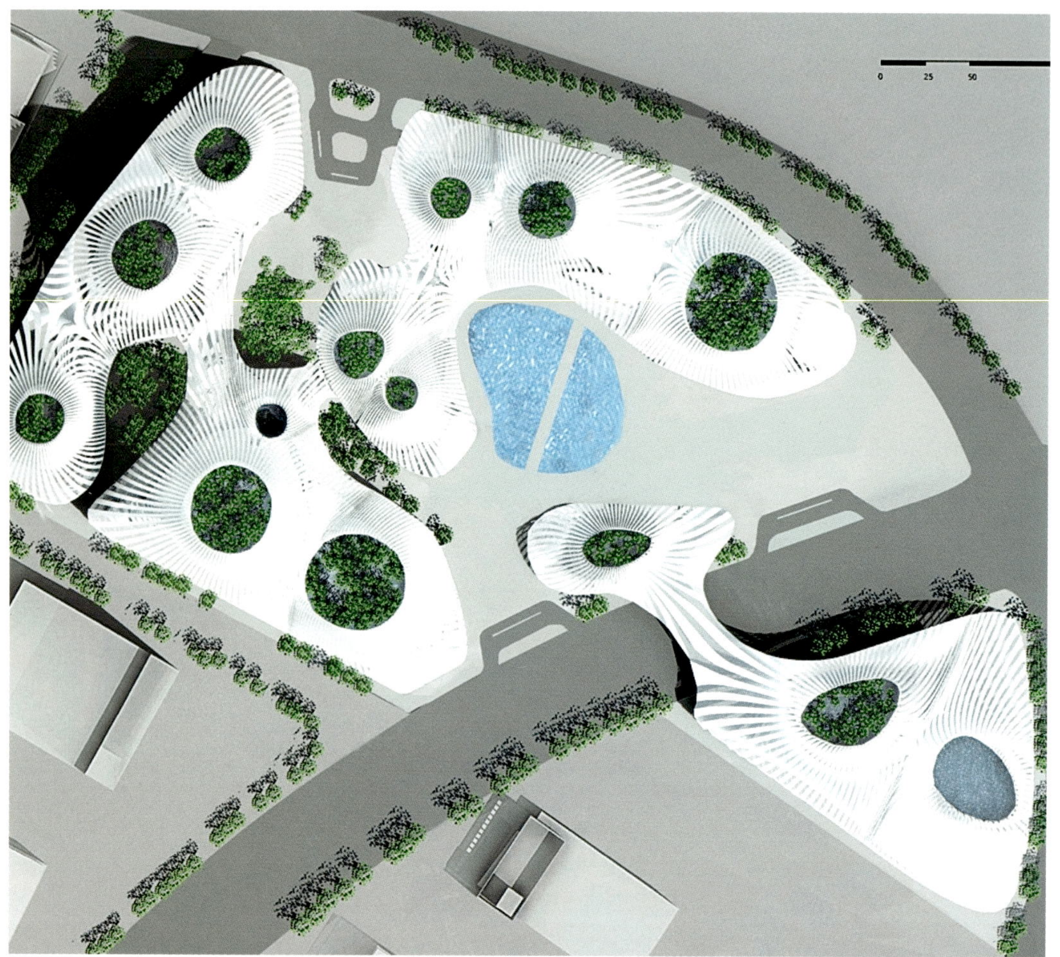

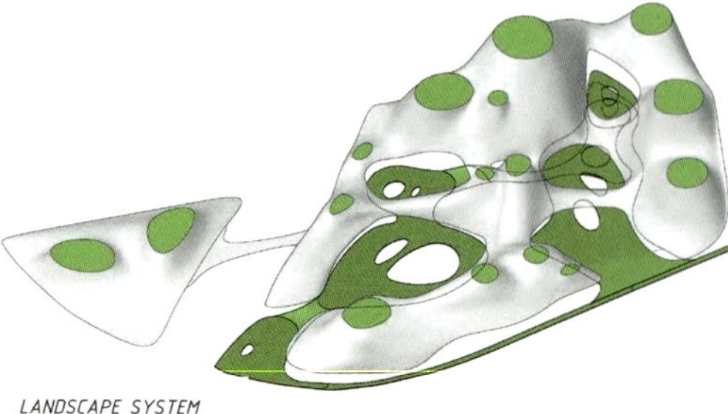

LANDSCAPE SYSTEM
景观系统

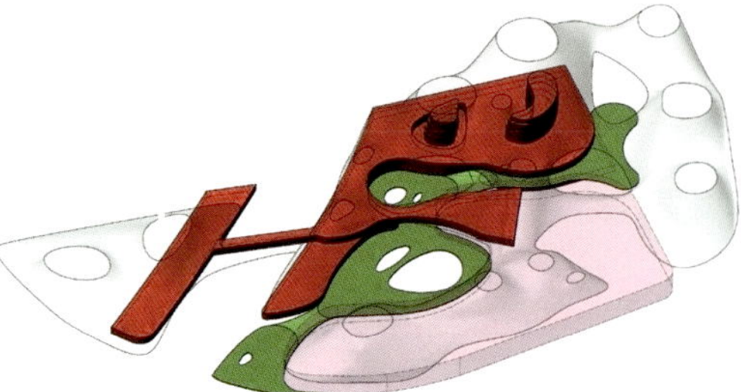

RETAIL CORRIDOR
商业连廊

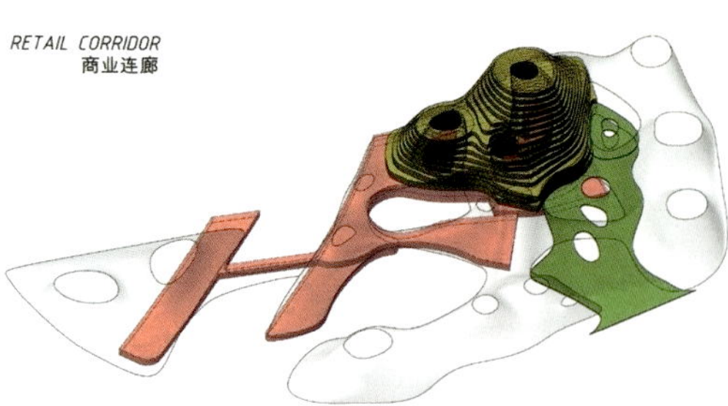

HOTEL & RETAIL RELATIONSHIP
酒店&商业关系

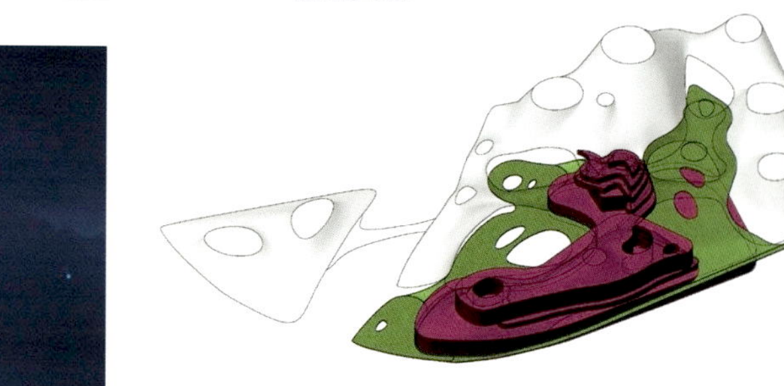

CONVENTION & LANDSCAPE RELATIONSHIP
会议&景观关系

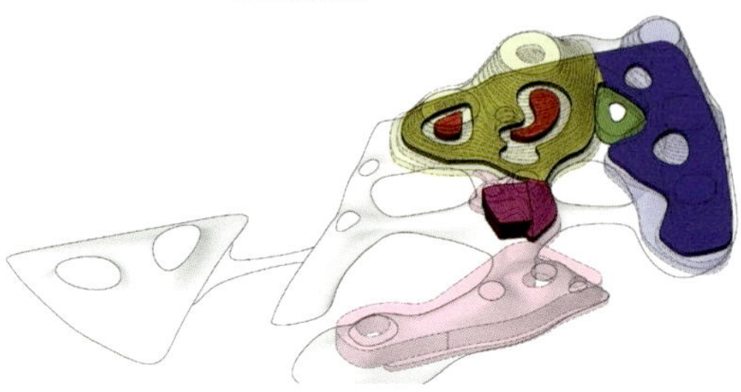

PROGRAMME CONNECTIVITY BAND
功能联系图

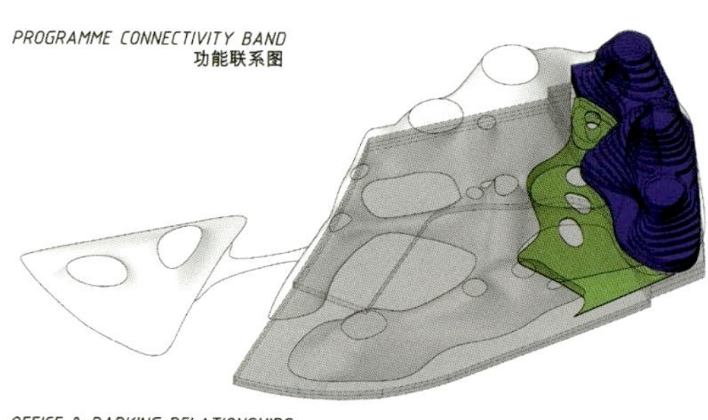

OFFICE & PARKING RELATIONSHIPS
办公&停车关系

The design is conceived as a continuous weaving of architecture and landscape that blurs the boundary between folding architecture, public space and urban landscape, proposing a futuristic vision based on the East's naturalistic philosophy. This project inherits Chinese architecture's long-standing attitude towards holistic integration and order of space. It employs the Eastern philosophy of a harmonized synthesis between human and nature. In the face of the project's enormous scale, the architecture no longer exists as a series of individual blocks, but instead is unified as a collective form. The resultant space enclosed within comes into focus, in a natural order emerging from air, wind and light, fostering a resonance between human and nature.

The city of Taichung requires a metropolitan landmark that goes beyond the local to renew urban life and redefine the cultural landscape of the city, that, through unique architectural concepts and proposing a new kind of architectural philosophy, launch Taichung into the arena of world class cultural cites. Today's landmark buildings are no longer characterized by mere considerations for height, but have turned to cultural inquiries into the future and nature. More than making visual impacts, landmark buildings should foster public recreation, and encourage communication and imagination.

The site for this project is inherently characterized by an energy-rich landscape. Under its calm surface, topological potentials await to

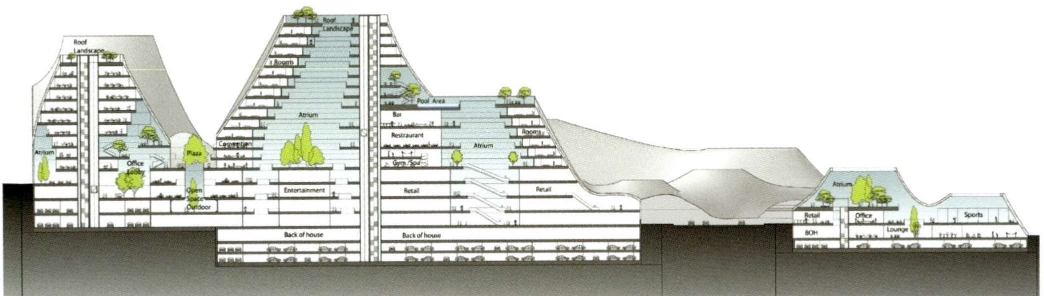

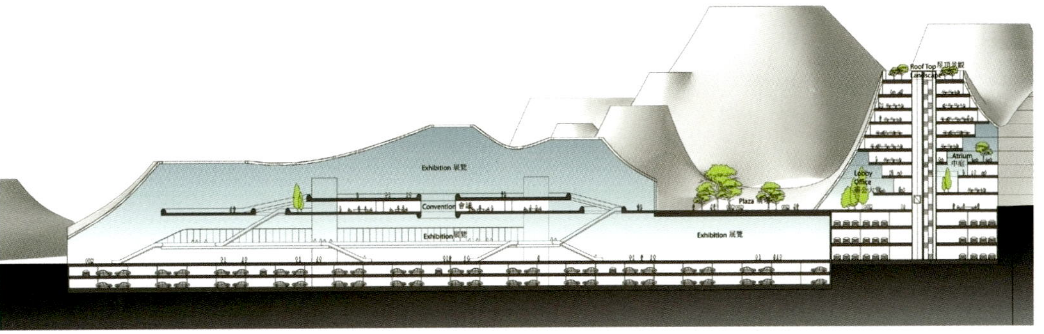

Solid Material 实体材料
Transparent Material 透明材料

be discovered and expressed as urban landmarks. On one hand, the architecture's crater-shaped formation and resulting rotundas are the outcome of found site conditions. On the other hand, it simultaneously shapes and influences the surrounding environment, opening up a dialogue between architecture and landscape. The surface of the 'mountains' is a high-tech, eco-friendly pleated skin system. The smocking-like envelope provides air flow to the building while keeping energy consumption at a minimum by utilizing solar energy.

The open courtyards that connect the individual 'mountains' are integrated into a natural sequence of outdoor spaces. Like the quest for a harmonic coexistence between people and nature exemplified by Forbidden City and ancient Chinese gardens, this project seeks greater meaning in its non-material qualities, spaces encircled with the upmost naturalistic spirit. A single tree, a patch of bamboo, or a pond become central figures of the space. This approach to sustainable development is based not on technology, but on traditional philosophy and aesthetics.

NINGBO URBAN CONSTR-UCTION DESIGN CO.,LTD.

宁波都市营造建筑设计有限公司

宁波都市营造建筑设计有限公司创立于2008年，公司设计着眼于对设计的系统性研究，立足建筑、室内、景观空间的系统设计和研究。公司强调设计的创造性和探索性，自成立以来完成了宁波长丰滨江"舟宿夜江"、宁波江北北门户区启动区城市设计、宁波市第五医院、宁波市镇海建设与交通局办公楼改造、镇海公路段管理房、永康宏盛公司办公楼等项目。

Ningbo Urban Construction Design Co., Ltd. was founded in 2008. The design of the company focuses on the exploration of design systematicness, based on the study of the architecture, interior and landscape space designs. The company emphasizes the originality of the design. Since the establishment it has completed projects including "Zhou Su Ye Jiang" at Changfeng, Ningbo, Urban Design of North Gateway Starting Area along Yangtze River, the Fifth Hospital Ningbo, Zhenhai Construction and Transportation Department Office Renovation in Ningbo, Zhenhai Highway Administration Office, Yong Kang Hongsheng Office Building, etc.

宁波江北门户区中策创意广场方案

CSI INNOVATIVE SQUARE IN JIANGBEI PORTAL DISTRICT, NINGBO

项目地点：中国 · 宁波　用地面积：33 333.4 m^2　建筑面积：69 329 m^2
建筑设计：宁波都市营造建筑设计有限公司
建筑师：翁建祥，陈超，裴涛，童美云

LOCATION: Ningbo, China　SITE AREA: 33,333.4 m^2　BUILDING AREA: 69,329 m^2
DESIGN CORPORATION: Ningbo Urban Construction Design Co., Ltd.
ARCHITECTS: Weng Jianxiang, Chen Chao, Pei Tao, Tong Meiyun

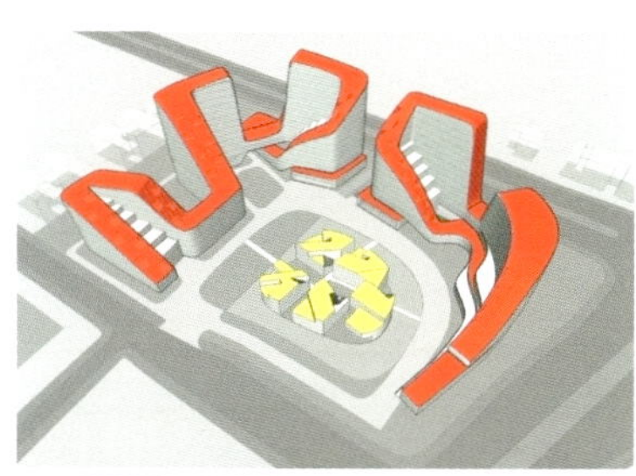

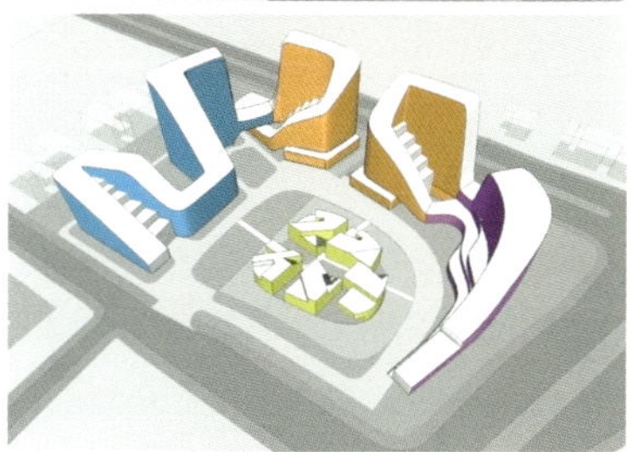

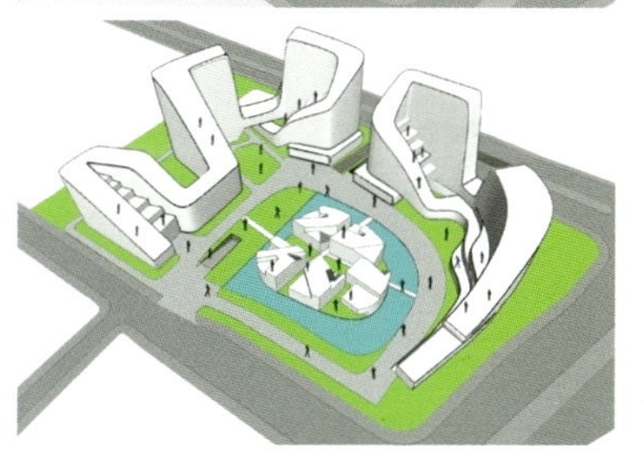

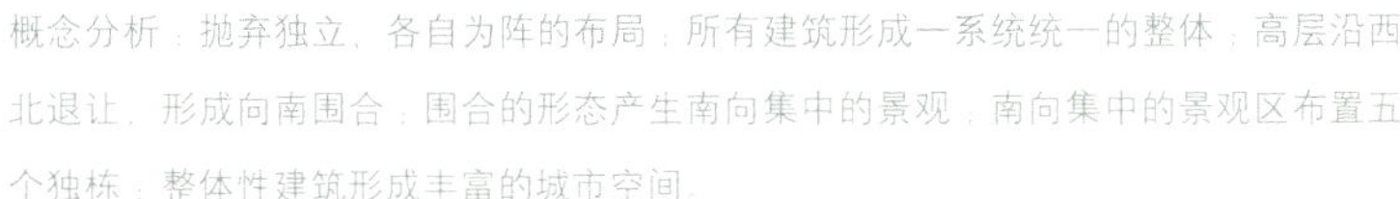

概念分析：抛弃独立、各自为阵的布局；所有建筑形成一系统统一的整体；高层沿西北退让，形成向南围合；围合的形态产生南向集中的景观；南向集中的景观区布置五个独栋；整体性建筑形成丰富的城市空间。

建筑并不是独立的，而是整体设计考虑，以高低错落、连续起伏的建筑形态，创造出丰富的城市空间形象，并围合出相对静谧的场地内部空间环境。整个带形建筑的凹凸创造出新的景观空间。中央的独栋办公区被水域和森林所包围，创造出曲径通幽的办公环境。

景观：连续整体的建筑屋面层层错位内凹形成空中露台，层层退台绿化，共同构成高层区空中景观带。

整体性建筑创造出丰富的城市空中景观。

梦想岛：中央独栋办公区被水域和森林包围，层层露台，并通过台阶连通形成互动的交流空间。办公区形成水中岛景观，三条木栈道从外围映入岛区。

Concept Analysis: abandoning the detached layout of the buildings to form a uniform system; the high storeys gradually incline to the northwest to form the enclosed shape with the five independent south buildings, which produce the focused landscape, while the overall design enriches the urban space.

Construction is not independent, and the full considerations to the scattered and continuous rolling architectural form create a diversed city space image as well as a relatively quiet interior space environment by enclosure. The unevenness of the entire strap-shaped construction creates a new landscape space. The central single independent building of office is surrounded by water and forests, creating the office environment of winding paths. The overall construction creates a diversed urban air landscape.

Dreamy Island: the central detached office is surrounded by water and forests, layers of balconies conneced through stairs form interactive communication space. The office area forms water island landscape, three wooden paths map into the island area from the periphery island.

宁波百丈公园

NINGBO BAIZHANG PARK

项目地点：中国 · 宁波　建筑面积：8500 m^2
建筑设计：宁波市建筑设计研究院有限公司
建筑师：翁建祥，唐莹，许建波

LOCATION: Ningbo, China　BUILDING AREA: 8500 m^2
DESIGN CORPORATION: Ningbo Architectural Design & Research Institute
ARCHITECTS: Weng Jianxiang, Tang Ying, Xu Jianbo

就像文化进程中的其他过渡定义和公式化的东西一样，公园（景观）也正变得僵化，中庸、平淡的情形需要被挑战和制止。百丈公园就是对此情形的回应。

公园：顾名是为人提供多样行为的场所，而不仅仅是观赏。它并不是一盆景，而是容纳人活动的空间。当前许多公园的设计缺乏激活人行为、吸纳人参与的考虑，更多停留在简单植物配置、平面图案化的设计层面上，从而导致许多公园缺乏人气。有的公园甚至由于低矮植物众多，形成许多阴暗角落，为犯罪提供场所。人们害怕进入这样的公园。它拒绝了人而成为了盆景。

百丈公园推广的是一种唤醒人各种行为冲动的公园环境。这使得公园更有活力、更有意义。在这一意义上，公园不仅仅是一种环境，更是一出舞台剧。人是演员，展示着各自不同的行为。所以百丈公园是一露天舞台，一个让演员可以表现他们天性的舞台。它是包含人们许多行为可能的场所——并非仅仅是看看花草，在树林中散散步。

随着汽车量的迅速增加，城市需要考虑留出更多的城市空间来方便停车。百丈公园就是宁波市为解决停车而建的公园式停车库。业主宁波市前期办希望在解决停车的同时建设一可供周边居民休闲、交流的城市公园。基地占地 8500 m^2，拟建一个 3500 m^2 的地下停车库（容纳 99 辆小汽车）和 8500 m^2 的地面公园。

设计中强调以整体的设计手法来解决地下建筑与地面景观的关系，即不是孤立地把地下车库埋入地下，再在地上孤立地考虑公园设计，而是把两者统一，整体地来考虑。设计采用了一木铺地系统，该系统不是简单地在地面铺木地板，而是一个把地下车库采光、汽车坡道、出地面疏散楼梯、公共厕所等地面建筑统一归纳到公园活动空间来设计的一个整体系统，在公园里看不到孤立的突出地面的汽车下地库坡道、楼梯间、公共厕所等建筑。突出地面的建筑成为木铺地系统的有机组成部分，它或抬或坡或斜，成为公园市民的活动空间、景观空间的一部分，形成高低错落的丰富的公园地面空间，木铺地局部抬、斜形成地下室的采光，同时也成为人可坐、可躺的空间。木铺地系统迂回穿插于公园的草丛和树林中，其中设计了一系列树池和木椅，树池种树可遮阳，树池周围木凳可供人休息。

设计是对传统公园景观平面式、图案式设计的反思和探讨，希望把竖向与横向的空间引入景观中，为人的行为提供多种可能性。另外是对整体式设计的一个研究——各元素不是孤立的，而是互相作用形成创造性和活力。

Like other transitional notions and formulations of the things in culture progress,the situation that parks (landscape) are also becoming rigid, moderate and insipid needs to be challenged as well as curbed. Baizhang Park is a response to this situation. Park, as it its name implies is a place for people to conduct various activities more than being visited. It's not just a bonsai but a container for human activities. Now the design of many parks lacks consideration for activating human behaviors and absorbing human participation, while stays more on the simple plant setting and the design of flat patterns, which results into the lack of human atmosphere in many parks, and some even cause dark corners providing a forum for the crime due to the overflow of boskage. People are afraid to enter these parks, and this rejection to man just makes them bonsais.Baizhang Park promotes the park environment capable of waking up peoople's impulse for various acts. This makes the park more dynamic and meaningful. In this sense, the park is more a stage play than an environment, where people are actors displaying their behaviors. So Baizhang Park is exactly an open-air stage,on which actors show their nature. It's a place containing possibilities of human actions, instead of just showing grass and flowers and walking of the visitors. With the rapid increasement of automobiles, the city needs more convenient parking space. Ningbo Baizhang Park is such a park-style garage to solve this problem. The owners, Ningbo Municipal Engineering Prophase Office,expected to solve the parking problem by building a leisure and communication park area for the residents. The base covers an area of 8500 m^2,which is proposed an underground garage of 3500 m^2 (holding 99 cars) and the ground park of 8500 m^2.

Design emphasizes the overall design approach to treat the relationship between the underground architecture and the ground landscape, that is not to isolate the garage underground to consider the park design independently, but to consider the two as a uniform. The design uses the wood floor system, which is not just a simple paving of wood, but an overall systematic design of underground garage lighting, vehicle ramp, ground evacuation stairs, bathrooms and other buildings. You don't see the vehicle ramp, staircases, or bathrooms, etc. , on the ground. All the overground buildings are integral parts of the wooden floor system which is actually a park space for public activities in various shapes. As a part of the landscape,it forms scattered ground park space,with local lift and ramp of the wooden floor to form the basement lighting as well as the rest area where people can sit or lie. The wooden floor system windingly interludes in the grass and trees of the park,where trees and wooden chairs provide shade and rest space for the public.

The design is the reflection and exploration to the landscape design and graphic design of the traditional park, hopefully the vertical and horizontal spaces can be introduced to the landscape to provide a variety of possibilities for human behavior. Moreover it is a research on the overall design—elements are not isolated, but interactive with each other to form creativity and vitality.

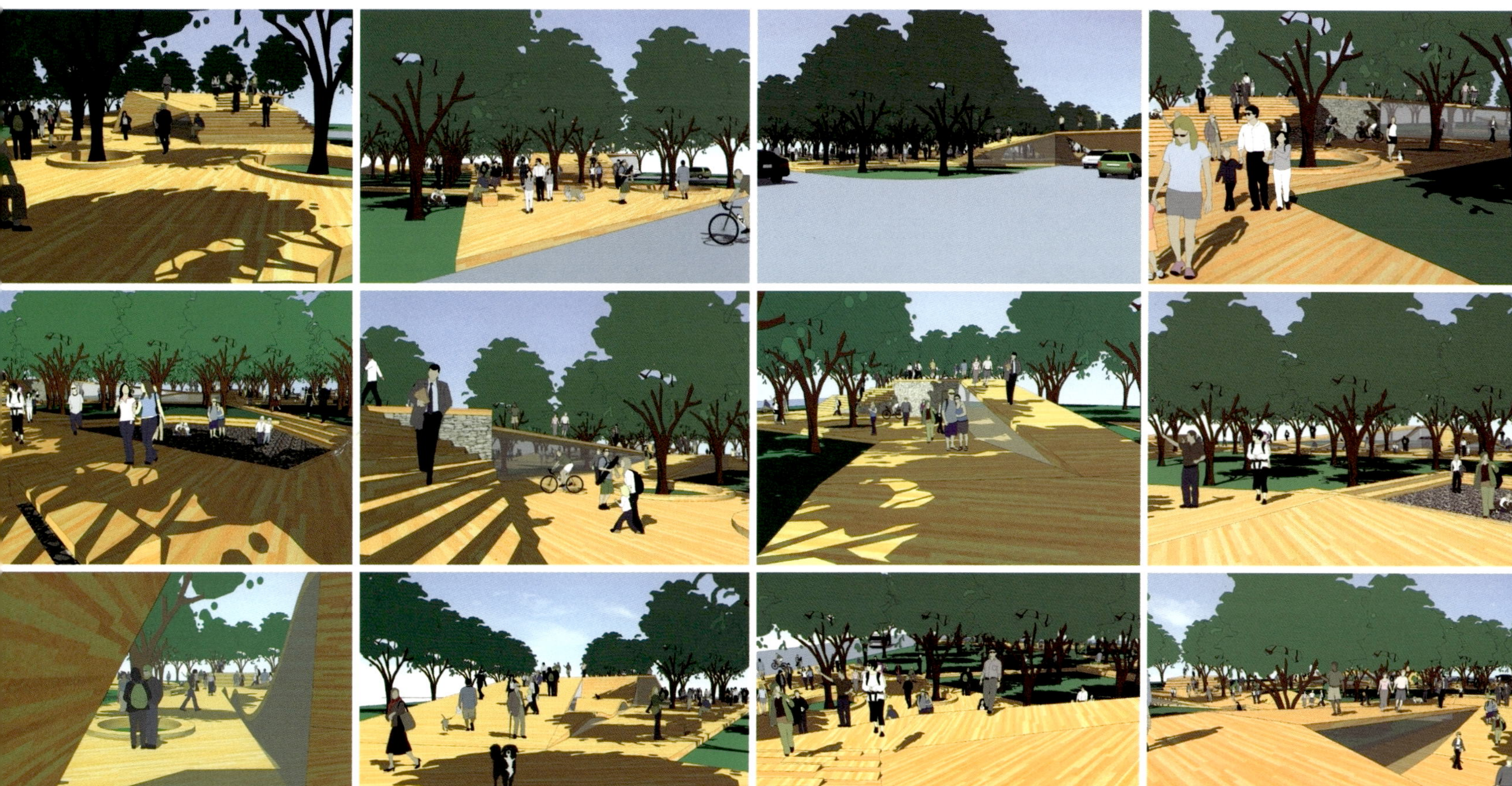

BAIOA ARCHITECTURE STUDIO

BAIOA般诺建艺工作室

刘强

冯保军

公司合伙人

-刘强　Banno　建筑师

单位：青岛北洋设计院　BAiOA 般诺建艺工作室（青岛）

青岛理工大学　建筑学学位

-冯保军　Kato　建筑师

单位：青岛北洋设计院　BAiOA 般诺建艺工作室（青岛）

TOUCH THE EARTH

项目地点：美国·得克萨斯州达拉斯
用地面积：72 000 m^2
建筑面积：14 664 m^2
建筑设计：BAiOA 般诺建艺工作室
建筑师：刘强，冯保军

LOCATION: Dallas Texas, USA
SITE AREA: 72,000 m^2
BUILDING AREA: 14,664 m^2
DESIGN CORPORATION: BAiOA Architecture Studio
ARCHITECTS: Banno, Kato

5 about concept foucsing on the INPUTS and OUTPUTS,the air,the water

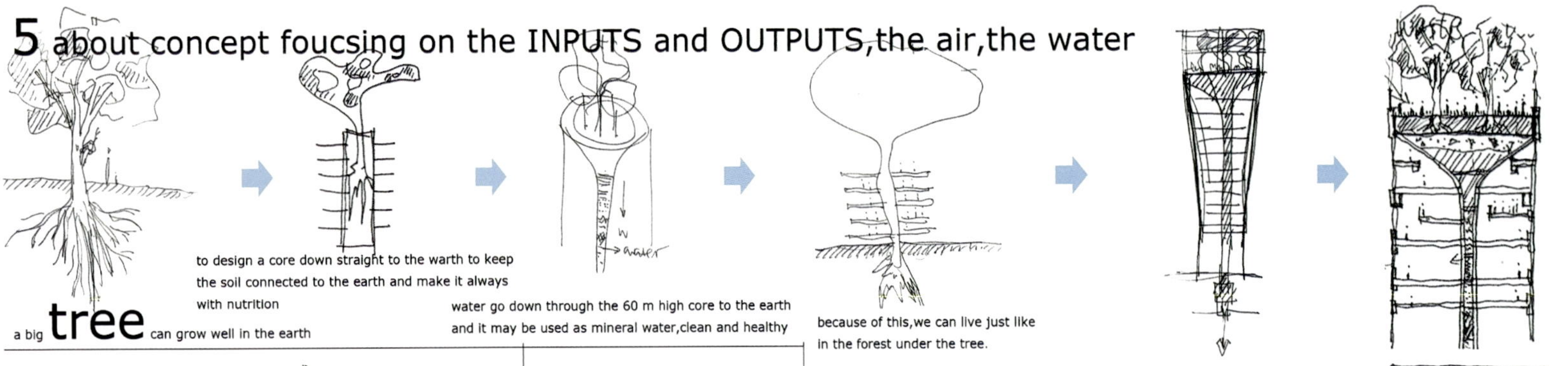

a big **tree** can grow well in the earth

to design a core down straight to the warth to keep the soil connected to the earth and make it always with nutrition

water go down through the 60 m high core to the earth and it may be used as mineral water,clean and healthy

because of this,we can live just like in the forest under the tree.

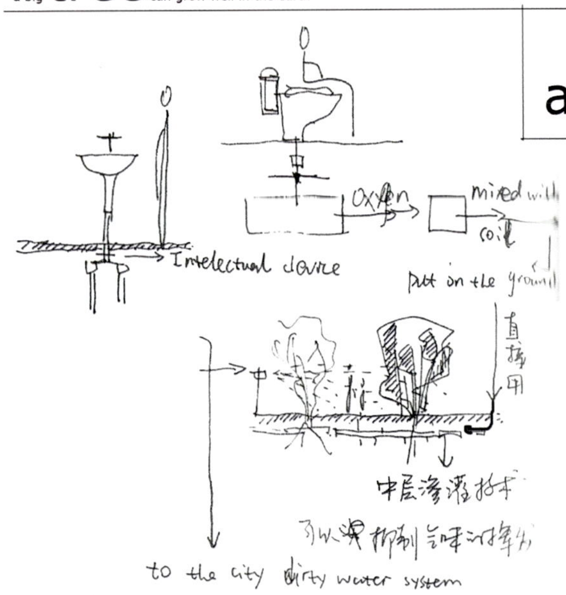

air conditioner

Intelectual device heated by the sun

then hot air in it will go up so it can move the air conditiona system

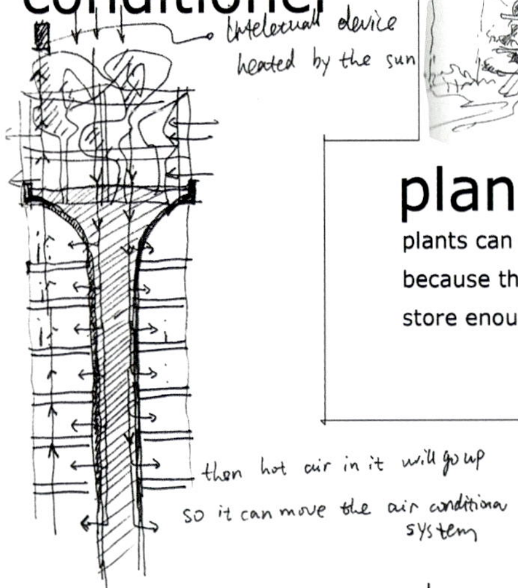

winter to heat the cold air in the soil
summer to cold the air in the soil.

plants

plants can grow well because the core can store enough water

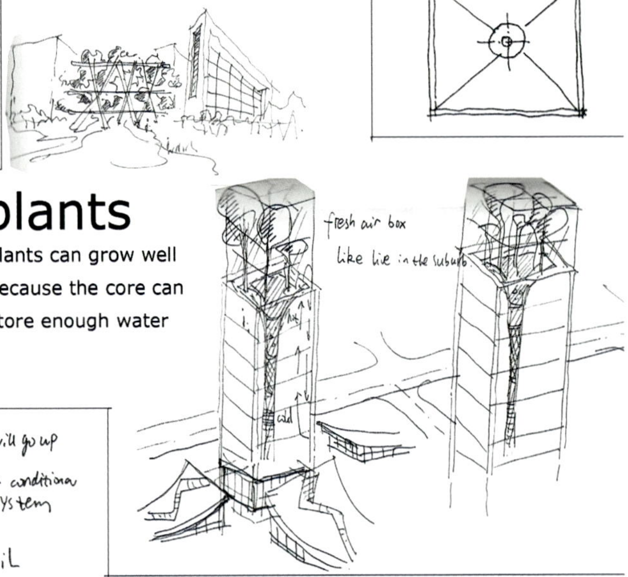

fresh air box like live in the suburb

water

no water will be throwed away as waste

Intelectual device

oxygen → mixed with soil

put on the ground

中层渗灌技术

to the city dirty water system

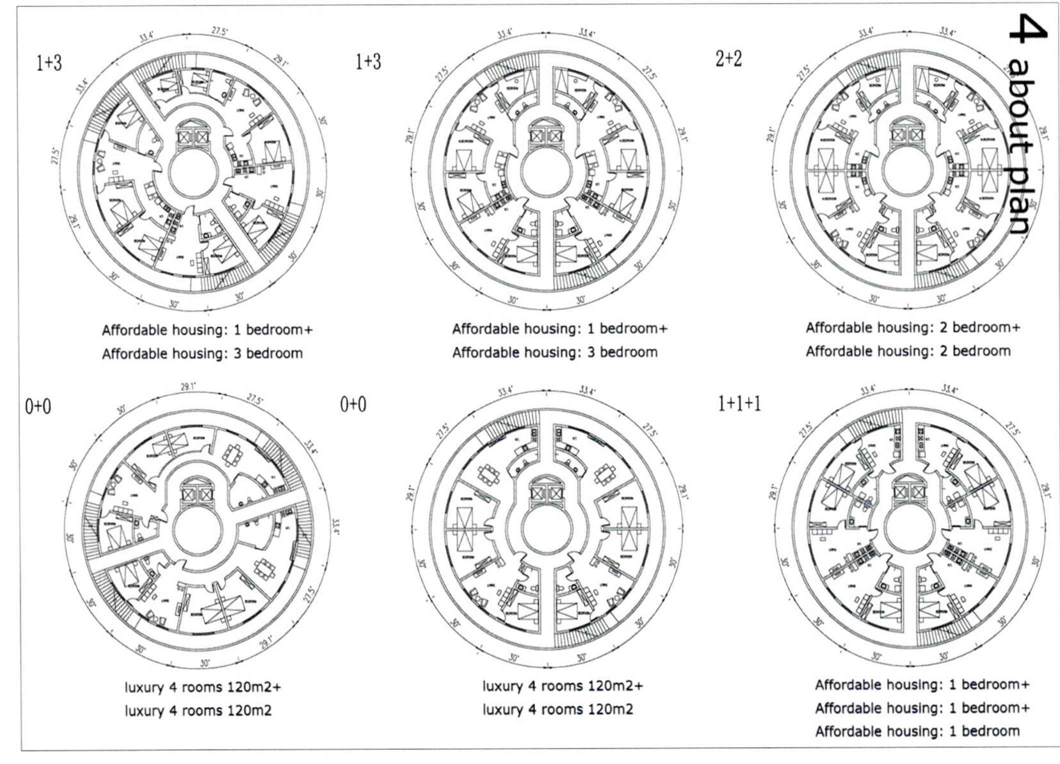

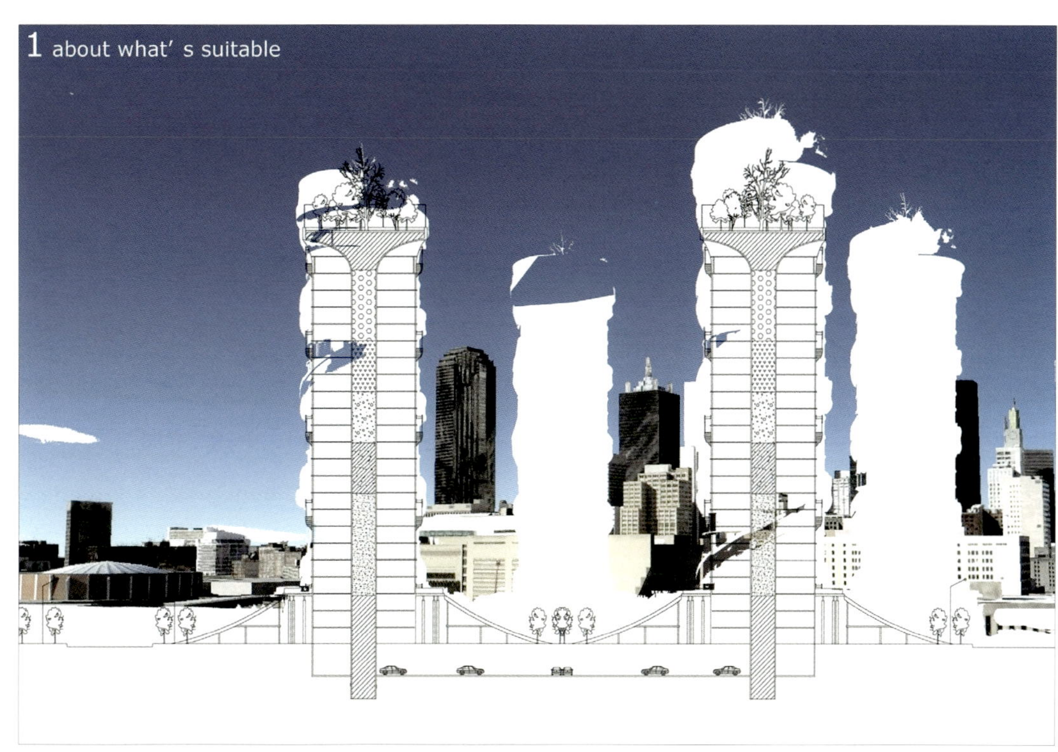
1 about what' s suitable

SHANGHAI INSTITUTE OF ARCHITECTURAL DESIGN & RESEARCH CO.,LTD.

上海建筑设计研究院有限公司

公司合伙人及主要设计师

- 姚陟　Yao Zhi

上海建筑设计研究院有限公司
主创建筑设计师
2000.7—2005.7 重庆建筑大学
2005.7至今 上海建筑设计研究院有限公司

无锡体育中心北侧商业地块概念方案

THE CONCEPT PROGRAMN OF THE NORTH COMMERCIAL LAND OF WUXI SPORTS CENTER

项目地点：中国 · 无锡 用地面积：44 500 m^2 建筑面积：120 000 m^2
建筑设计：上海建筑设计研究院有限公司
设计指导：吴文，程明生
建筑师：姚陟

LOCATION: Wuxi, China SITE AREA: 44,500 m^2 BUILDING AREA: 120,000 m^2
DESIGN CORPORATION: Shanghai Institute of Architectural Design & Research Co.,Ltd.
Design Directors: Wu Wen, Cheng Mingsheng
ARCHITECT: Yao Zhi

城市就是集聚。

紧扣“商业”主题，是城市生命活力和生活质量的体现。

商业文化的营造带来城市改造、更新的机遇。原来杂乱、无序、陈旧的片区，可以营造成整体完美、适应 21 世纪城市可持续发展的新街区，同时高品质的商业中心也需要适当地与原有城市环境拉开差距。在设计时，从以下三个方面切入。

• 变零乱无序为整体有序；
• 变破旧落后为新颖先进；
• 变暗淡无光为闪光景观。

以达到商业空间与公共空间的融合，建筑景观与生态景观的融合，商业氛围与文化氛围的融合。

购物中心可以采用以下四种空间形式。

①室内大开间。无街道，商铺间没有完全隔绝。

②露天步行街。上空无顶棚，空间开敞的步行街。

③回廊式步行街。步行街的两侧或一侧为回廊，步行街局部遮盖。

④室内步行街。步行街在室内，完全遮盖。

BOSCH

City means high density.

Closely related to "Business" theme is the reflection of the urban vitality and quality of life.

Business culture brings urban renewal and updated opportunities. The original messy old area can be refromed to a wholly perfect new neighborhood to adapt to sustainable urban development of the 21st century, while the high-quality business center also needs appropriate separation from the original urban environment. We started from the three aspects:

- Replacing the mess and disorder with order and integrity.
- Changing the dilapidated and shabby into the advanced and new.
- Changing the dim and dark view into gleaming view.

Thus, the following are achieved: the fusion of business space and public space, the fusion of architectural and ecological landscapes, and the fusion of commercial atmosphere and the cultural context.

The shopping center can take the four space forms:

Large indoor room. The shops are incompletely isolated and there are no streets.

Open-air pedestrian street. The open pedestrian space are without roof over the head.

Corridor-type walking Street. The corridor on both sides or one side of the pedestrian street is locally covered.

Indoor pedestrian street. The pedestrian street is completely covered in the space.

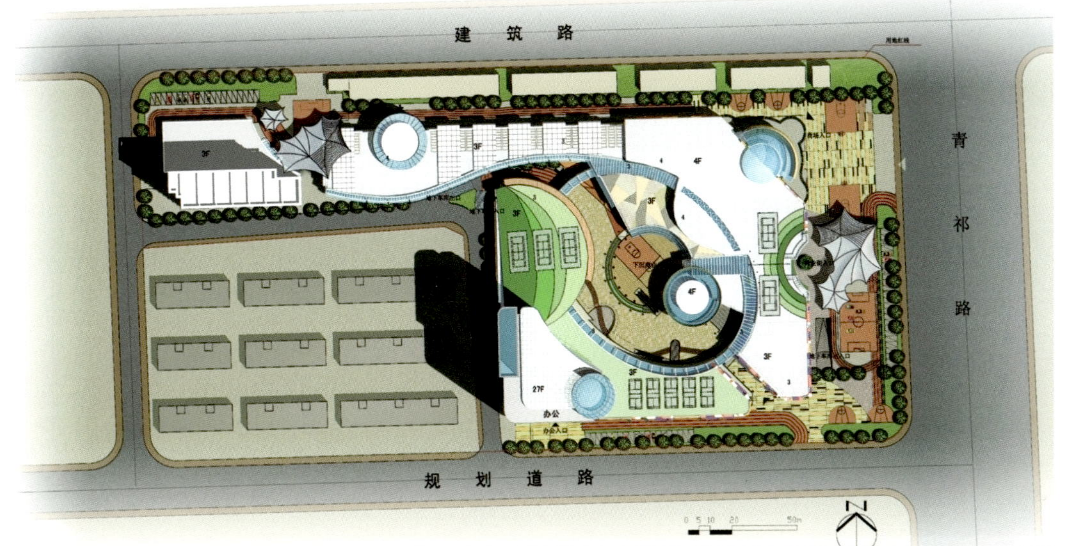

太原奥体中心酒店概念方案

TAIYUAN OLYMPIC HOTEL CONCEPT PROGRAM

项目地点：中国 · 太原　用地面积：23 809 m^2　建筑面积：92 774 m^2
建筑设计：上海建筑设计研究院有限公司　设计指导：吴文
建筑师：姚陟

LOCATION: Taiyuan, China　SITE AREA: 23,809 m^2　BUILDING AREA: 92,774 m^2
DESIGN CORPORATION: Shanghai Institute of Architectural Design & Research Co.,Ltd.　Design Director: Wu Wen
ARCHITECT: Yao Zhi

本方案以满足城市发展需求，丰富城市景观，传承当地悠久文脉三大板块内容为主要出发点，明确项目的预期定位。接下来将具体从功能规划布局，建筑主体设计，绿化布局，景观视线分析等多方面来进行设计说明。

根据整体建设背景，本方案力求在功能组织上做到合理，与体育中心的整体建设风格和使用需求以及未来的发展要求保持一致。

酒店、办公楼为相互独立的两栋建筑，分立于基地北部与西南角。东南角裙房则是高档餐饮休闲区，与酒店区和办公区共同构成围合布局。地块中央区则是主要交通枢纽区兼做休憩大堂，构成一个"中心"，一层设置酒店大堂接待处等相关辅助设施，便于集中式服务管理。

所有建筑的主楼与裙房部分均相互独立，便于建造，结构合理。建筑外观采用直线与曲线元素灵活结合的手法，主题鲜明，外观造型独具一格，与体育馆的建筑造型协调呼应，以进一步在所在地塑造旅游目的地的形象。主体立面与细节上结合地域文化进行设计，局部使用山西当地的文化元素、符号，以彰显本地深厚历史底蕴。

酒店入口大堂视野开阔，可以看到汾河美景，内部装修与当地历史文化紧密结合，烘托主题氛围。大堂内部设有三层通高的阳光长廊来活跃室内气氛，扩充室内的交流界面，提升空间的流动性。

南部下沉绿地结合景观与建筑使用功能要求进行设计，从道路边界往大堂方向逐层下降，既满足地下室采光通风的要求，又为一层大堂内的休憩空间提供了宁静、开阔的景观带，给人以舒适的感受。

酒店自然间不大于 600 间，85% 以上客房都保证了南向直接采光，保证了良好的日照条件和室内微环境，按白金五星的标准建造，采用智能化设计。

两栋高层视野开阔，彼此不存在视线干扰。由于采用了 V 形平面，使得入住宾客可往东南方向遥望汾河，或西南方向远眺宏伟的体训设施，既增添了视野趣味性，又提升了室内品质。

The program mainly aims to meet the needs of urban development, enrich urban landscape, passing a long context of local cultural content and to make clear the project's expected positioning. Then it will give introduction of design in aspects including the specific functional layout, the main building design, green layout and the landscape view analysis.

According to the whole building background,the program seeks to achieve reasonable and dignified functional layout, in line with the overall construction style, the using demands as well as future development requirements of the sports center.

The hotel and the office buildings are independent from each other,respectively set in the north and the southwest corner of the base. The annex in the southeast corner is the high-quality dining area,constituting the enclosed layout with the hotel and office buildings together. The central area of the block is the transport hub also offering open lobby to form a "center". The first floor is the hotel reception lobby and other related facilities for efficient centralized service management.

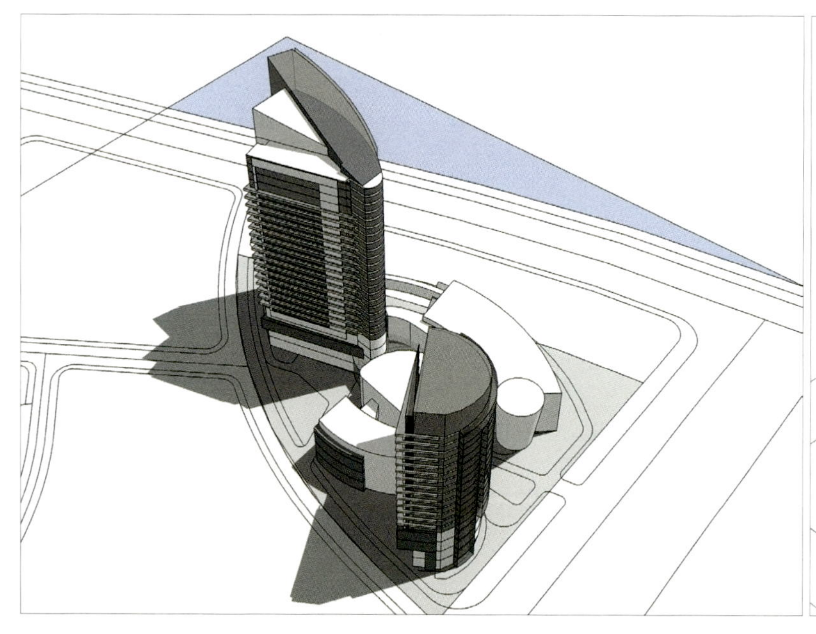

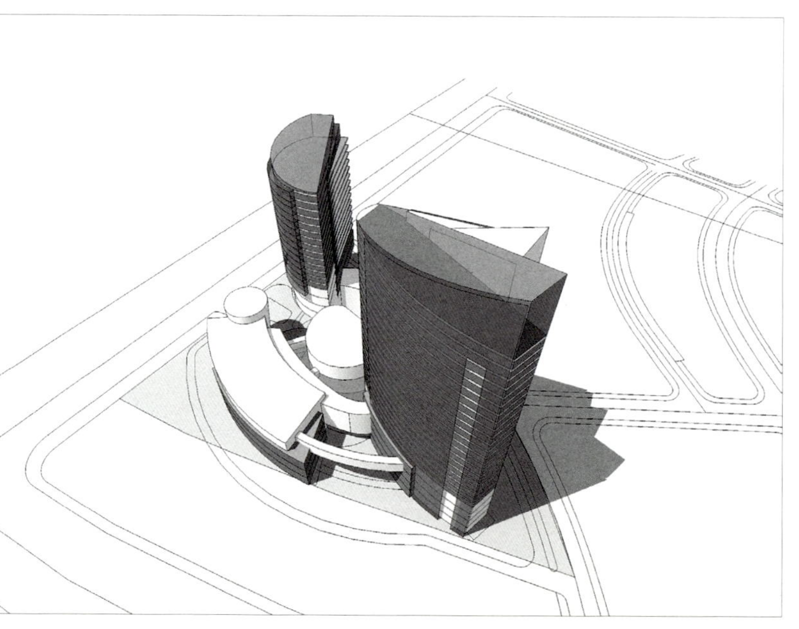

All the main buildings and annexes are independent, easy to build, with more reasonable structure. The building facades use the flexible combination of straight and cured lines with a clear theme and a unique shape, echoing with the building shapes of the stadium to further establish a tourist place in the site. The main facade and the details are designed according to the local culture, and partly use the local cultural elements and symbols in Shanxi to highlight the rich historical heritage of the local area.

With the wide view of the entrance hall,you can enjoy beautiful Fen River scenery. The interior decoration is closely integrated with local history and culture to largely interpret the theme . Inside the lobby there is the three-storey high sunshine corridor to activate the atmosphere, enlarge the communication interface and enhance the space mobility. The southern sunk green are designed to integrate into the landscape and meet the functional requirements. It gradually declines from the road boundary to the lobby layer by layer,both to meet the basement lighting and ventilation needs and to provide a quiet and open landscape fro the lobby as well as the comfortable feeling for the guests.

Among the less than 600 hotel rooms, 85% of rooms directly take natural light from the south to ensure good sunlight conditions and indoor micro-environment. They are all built according to five-star rate standard with intelligent design.

There are enough open space between the two high-rise buildings without sight block to each other. The use of the V-shaped plane enables the guests to look into the Fen River from the distance of the southeast, or magnificent spporting training facilities from the southwest, which both add fun to the view and enhace the indoor taste and quality.

腾冲大酒店概念方案

TENGCHONG HOTEL CONCEPT PROGRAM

项目地点：中国·腾冲 用地面积：62 727 m^2 建筑面积：45 999 m^2
建筑设计：上海建筑设计研究院有限公司 设计指导：吴文
建筑师：姚陟

LOCATION: Tengchong, China SITE AREA: 62,727 m^2 BUILDING AREA: 45,999 m^2
DESIGN CORPORATION: Shanghai Institute of Architectural Design & Research Co.,Ltd. Design Director: Wu Wen
ARCHITECT: Yao Zhi

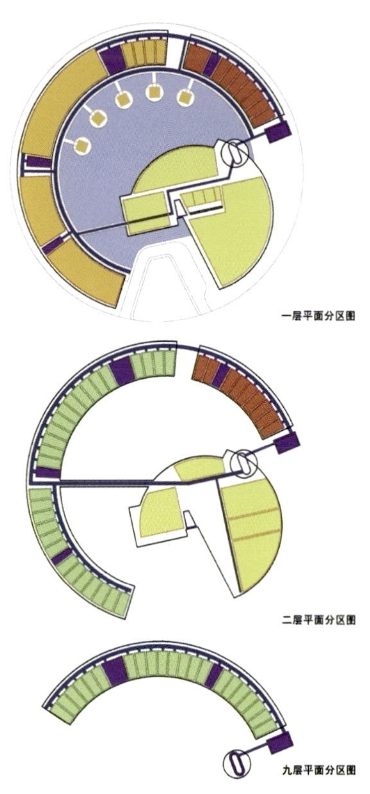

本方案定位为五星级酒店，并包含一个大型会议中心。以传承与创新的设计理念为指导，从简洁入手，理性与感性结合，用明快、大气的造型，创造出具有时代气息的酒店建筑。

该方案以超前的思维，务实的精神，创造新的城市品质，带给人们深刻印象，为城市带来新的生机，使其焕发出古老城市前所未有的活力与激情。

方案整体功能布局合理大气，主要分为会议中心、酒店。酒店主楼布置在地块的西南侧，酒店功能模块大致可分为公共部分及客房部分。公共部分包括门厅、餐饮、娱乐、会务等空间，其性质相对外向布置在一层；客房部分包括标准客房、套房，此模块性质较为私密，布置在二层以上。因为地块西北侧有公安小区的不利视线干扰，我们在客房的朝向设计中也作了充分的考虑，尽量让所有房间有好的景观朝向，豪华套房布置在地块北侧的一个小岛上，环境清幽，雅致。会议中心位于酒店用地东南侧，相对独立，但又通过连廊与酒店公共部分联系紧密，通过巧妙的设计安排，平时非会议时段又可用做宴会厅，高效率的空间使用为酒店自身提高了营业率，灵活经济。

整个建筑从当代建筑的美学角度出发，用简洁的设计手法，现代的建筑材料设计出极富张力与厚重文化内涵的建筑形态。圆形的总体构图契合和腾冲当地热海火山的大地之韵；酒店飘逸的主体仿佛一位姑娘跳着孔雀舞在百花丛中轻舞飞扬；酒店的面材以玻璃幕墙为主，以碧绿色为主色调，远远看过去就像一块无瑕的美玉。

我们一直在摸索和探讨一个问题：如何成为现代而又能回归自己的源泉，去追寻自己文脉的根；如何能够唤醒一个古老的、沉睡的文明，而又让它参与到现代的文明？

The program is positioned as a five-star hotel including a large conference center. It's guided by the innovative and traditional design concepts, starting from simplicity and the combination of sense and sensibility. It creates a hotel building with a sense of time with a crisp, and dignified modeling.

The program impresses people with the new urban quality it built with advanced thinking and pragmatic spirit, bringing new life to the urban development with unprecedented vitality and passion.

The overall function layout of the construction is reasonable and dignified, mainly including the conference center and the hotel. The main building was set in the southwest of the block. The functional modules are mainly the rooms and the public spaces. The former includes the standard rooms, suites,which are more private and arranged above the second floor. The latter includes the entrance hall, the dining, entertainment area, the conference hall and other spaces,which are relatively open and arranged in the first floor. Because the northwest view is sort of blocked by the public security section, we took full consideration of the construction orientation, in order to ensure all rooms have a good landscape view. The deluxe suites are set in an northern island of the block,with quiet and elegant environment. The conference hall is located in the southeast of the hotel site, relatively independent, but closely related to the public spaces of the hotel through the corridor. Due to the ingenious design, it can also be used as the banquet hall during non conference time. The efficient use of the spaces enhance business rates for the hotel economy.

The entire building was designed with full tension and cultural content on the basis of contemporary architectural aesthetic, simple design techniques and modern building materials.

The overall circular composition fits and rushes into the local volcanic earth rhyme;the graceful main body of the hotel is like a girl lightly dancing in the flowers;the facing materials of the hotel mainly use the glass curtain wall with aquamarine,just like a flawless piece of jade seen from the distance.

We have explored and discussed the question: how to become modern while returning to its source to pursue its own cultural root; how to restore an old and dormant civilization which is now involved in today?

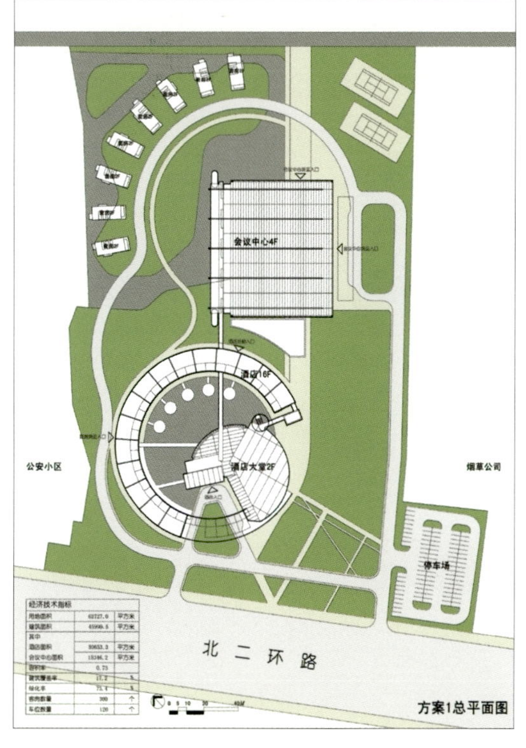

方案1总平面图

舟山 CBD 区超高层方案

ZHOUSHAN CBD SUPER HIGHRISE BUILDINGS SCHEME

项目地点：中国 · 舟山　用地面积：30 600 m^2　建筑面积：206 000 m^2
建筑设计：上海建筑设计研究院有限公司　设计指导：吴文
建筑师：姚陟，朱莉霞，张萌，张英斌

LOCATION: Zhoushan, China　SITE AREA: 30,600 m^2　BUILDING AREA: 206,000 m^2
DESIGN CORPORATION: Shanghai Institute of Architectural Design & Research Co.,Ltd.　Design Director: Wu Wen
ARCHITECTS: Yao Zhi, Zhu Lixia, Zhang Meng, Zhang Yingbin

随着社会环境和世界经济的发展变化，办公建筑面临着空间革新的挑战。发展商从单纯追求立面造型的新奇效果，改变为越来越注重空间和设备的高效性和经济性，使用方亦越来越注重环境和使用的舒适性和灵活性。本设计从建筑内部空间的有效利用到城市环境文脉的生成发展，结合业主的功能要求和设计长久以来积累的社会经验，力求创造一个全方位的智能化办公大楼。

1. 简洁、高雅的经典造型

在舟山 CBD 区纷繁错落的城市天际线中，设计一个简洁、高雅的经典超高层建筑，既能与现有 CBD 区的建筑风格相协调，又可以更加突出建筑的标志形象，体现出作为高层办公建筑稳定而端庄的个性。建筑造型以轮廓清晰的整体直线体块构成，并借鉴了三段式立面设计的经典手法。其端正的体量不仅在地理位置上，而且在视觉上也使建筑真正成为 CBD 区的一个重要标志。

2. 高效、舒适的智能空间

从办公楼层使用的便利性出发，采用近似于正方形的平面布置。方形大空间办公既保证了高效的使用率，又可以适应各种企业规模的灵活布置。3 m 的空间净高构成了一个开放宽敞的工作环境。

3. 丰富的空间组织和人性化的环境设计

大楼四层裙房的商业空间设有共享中庭，裙房与主楼之间设置有下沉式广场，西南部还设有景观步行广场，建筑通过中庭、下沉式广场和景观广场把建筑内外空间连成一个连续的环境线。裙房的功能布局沿环境线安排了大堂、商店、餐饮等各项综合公共设施。以风、光、水、影为主题的环境设计实现了立体的空间组织。

With the development of social environment and world economy, office buildings face the challenges of space innovation. More and more developers focus on the space efficiency and equipment economy instead of simply pursuing the novel effect of surface design, while more and more users emphasize the environment and the comfort and using flexibility. This design strives to create a full range of intelligent office building ranging from the effective use of the interior space to the formation and development of the urban environment and culture, combining the owners' functional requirements and accumulated experience in practice of the designer.

1. Simple and elegant classical style

To design a simple and elegant classic high-rise building in the numerous scattered city skylines in the Zhoushan CBD area is not only coordinated with the existing architectural style but also highlights the landmark image

of the building and reflects its stable and dignified nature as a high-rise office building. The building form is constitued by the clear outlined linear body masses with the classic approach Triade facade design. The decorous body volume not only lies in the location but also in that it makes the building the real landmark of CBD area in vision.

2. Efficient, comfortable and smart space

Office floors emphasize on the ease of use, with plane layout similar to square. Square large space both ensures efficient utilization and adapts to the flexible enterprise layout of various scales. The space of 3.0 m clear height provides an open and spacious work environment.

3. The diversed spatial organization and humane environment design

There is an atrium in the commercial space of the annex in the fourth floor, and between the annex and the main building there's a descented plaza. In the southwest there's also a landscape walking square. The three parts of the building constitue the continuous environment line, along which the lobby, shops, restaurants and other integrated public facilities are arranged in the annex functional layout. The environment design of the four themes of wind, light, water and shadow fulfilled the three-dimensional spatial organization.

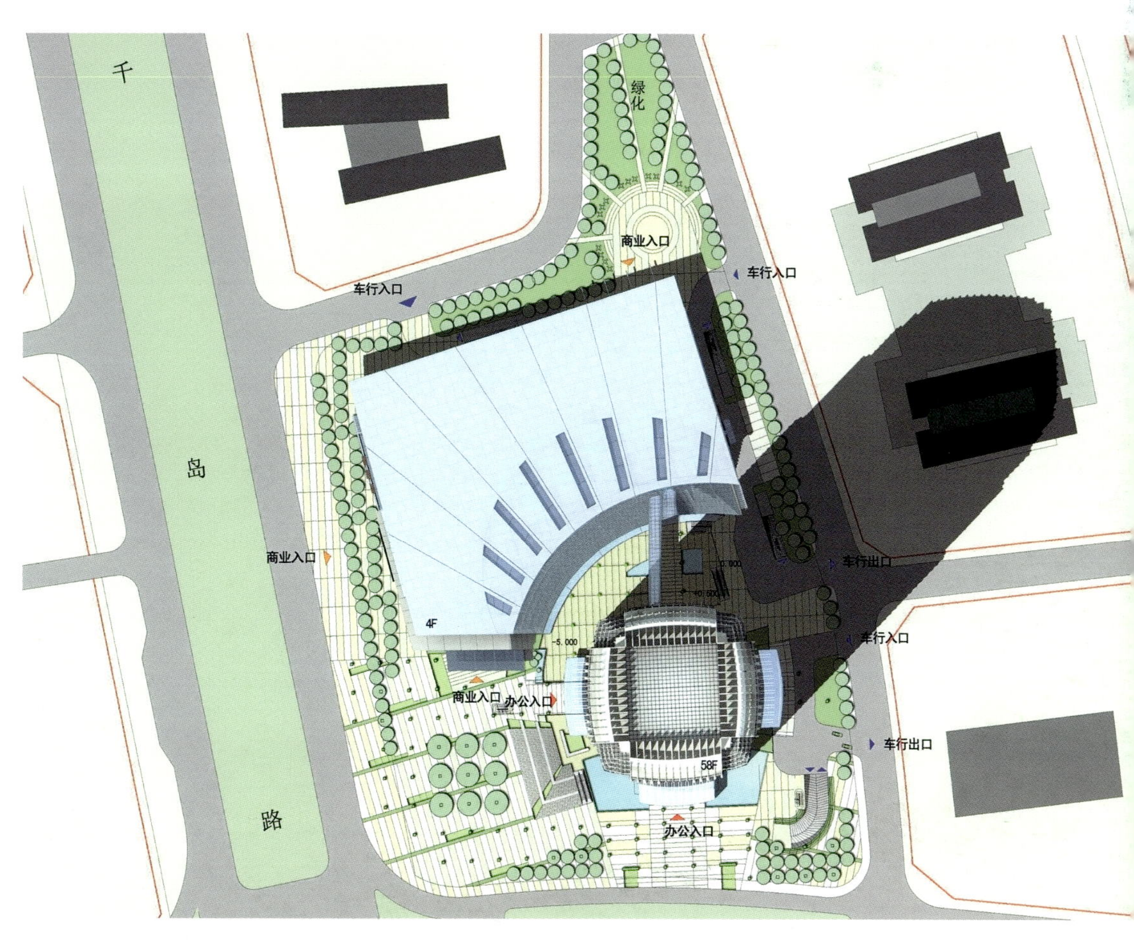

OTHERS STUDIO

沈阳别人建筑工作室

公司合伙人及主要设计师

-杨胤　Yang Yin

1972年出生
1995年毕业于沈阳建筑大学建筑系　留校执教至今
1998年创立别人建筑工作室
2005年作品入选《建筑中国》
2007年作品入选《中国建筑设计作品年鉴》
2007年作品入选《国魂——设计精英卷》
2008年作品入选《中国建筑竞标集成》
2009年作品入选《创意中国——设计卷》

- 高德战　Gao Dezhan

1995年毕业于武汉工业大学建筑系,学士
1999年加入别人工作室合伙人

- 张亦宁　Zhang Yining

1995年毕业于沈阳建筑大学建筑系,学士
2002年加入别人建筑工作室合伙人
2005年毕业于同济大学建筑与城市规划学院,硕士

我们总是在找不到自我的时候，想到别人，在别人的节日里寻找快乐。"在别人的思想中思想，在别人视线里生存"。——直到我们找到了自己的节日。那正是工作室成立之初的一种状态，所以我们命名工作室的名字为"别人"以纪念当时的心态。"别人"不是理论或口号，它是繁杂地交织在一起的建筑认识论和始终影响着我们创作心态的一个关键词，它时而是"地狱"，时而又成了"天堂"，它的解释的多样性正好说明了生活与建筑本身的多样性。
建筑设计时，不再考虑作品的内在含义，让体量与空间交织在一起，使形式变得鲜活，任由它们自在地表达。探索一种形式的语言，类似于音乐中的连续性与音调的和谐，也类似于写作中的文字排列，在那里，各个元素都受到内部相互关系的严格限制。

We always think about others and look for the happiness in the others' festival when we can't find ourselves."Thinking in the others' thought, living in the others' sight" - until we find our own festival. That was our state exactly when we just established our studio. So we named it "Others" to memorise the mentality."Others" is not a theory or a slogan. It is a complex architecture epistemology and a keyword to affect our creation. Sometimes it is "Hell", sometimes it changes to "Heaven". It can be explained by a lot of ways, and that just illustrates the life's and the architecture's diversity.
Duirng the architectural design,we no longer considered the inherent meaning of the work, just made the body intertwined with the space to enliven the forms and left them free expressions. To explore a form of language is similar to the pursuit of the continuity of the music in harmony with the tone or similar to the word arrangement of the text, where all elements are subject to restrictions on the internal relationship.

沈阳地铁营运控制中心

SHENYANG SUBWAY OPERATION AND CONTROL CENTER

项目地点：中国 · 沈阳　用地面积：6400 m^2　建筑面积：50 700 m^2
建筑设计：沈阳别人建筑工作室
建筑师：杨胤，张亦宁，高德战

LOCATION: Shenyang, China　SITE AREA: 6400 m^2　BUILDING AREA: 50,700 m^2
DESIGN CORPORATION: Others Studio
ARCHITECTS: Yang Yin, Zhang Yining, Gao Dezhan

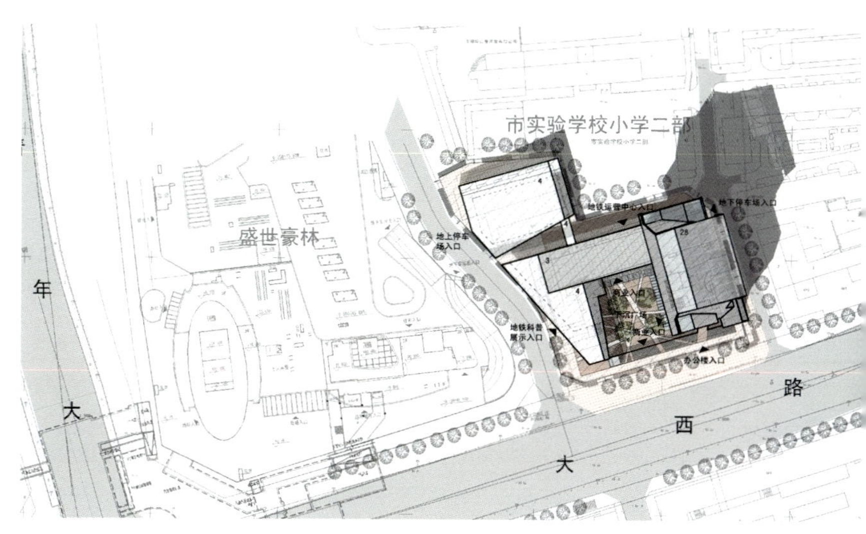

让城市体验运动

(1) 非常规体验——折线运动：常识告诉我们，躯干关节的弯折是人体运动的前提（如行走、跑步、跳跃），也是肢体形态变化的必要条件（如坐、卧、站、蹲）。同样，当我们已不满足于火柴盒式的呆板面孔，需要改变建筑静止姿态的时候，我们选择了折线。通过对折线和连续折线的运用和控制，让建筑开始行走和运动。具体策略是：基于基地的不规则平面形态，运用连续折线将高层和多层部分联结为一个整体，各线段在弯度、角度上均有变化，从而形成错动、片段化，而又浑然一体的整体运动效果，而这种线性运动又与地铁线性流动的特征相契合，从而能够激发人们的无限联想，并体现了地铁项目工程的标识性。

(2) 空中立方：方案并没有将折线只停留在二维层面上，在整个体量上基于对折线的控制，我们又采用整体切割、局部穿插的手法使整个建筑在各个方向均有丰富的变化，从而使城市各个角度都能获得不同的空间感受。侧斜的多个体量形成强烈的向外扩散的张力。空中舞动的体量以完全非对称的方式出现，带给人们的是完全陌生而又有强烈戏剧化的建筑体验，在震惊之余也许还会有一些“解放”的快感。

(3) 跳跃的节奏：让我们将镜头摇近，去感受那些具体的跳动的节奏。在高层主体建筑的顶部和西侧，一面由金属百叶构成的“折墙”由百米高空俯冲而下，直接插入地下，让人们感受到速度带来的快感。这种模糊的屋顶与墙面的做法，使建筑形态更加完整而有趣。顶部的百叶之下是屋顶花园。高层建筑的南侧，两个折动的体块以“俯仰”的姿态出现，它们相互交合又相互排斥，一虚一实，体现出“太极式”虚实相生的立体景观。

基地西侧临近规划路的低层部分，两个相互分离的体量，由于线条的延续和切割手法的一致性形成完整而又断裂的沿街形象，而其南端的架空和延伸，产生强烈的方向性和运动感，不仅增加了视线的通透性，而且产生强烈的视觉冲击力。

在整个建筑中看似不经意出现的洞口和缝隙，其实经过了反复的推敲和修改，它们是体量穿插、视线贯穿的结果。

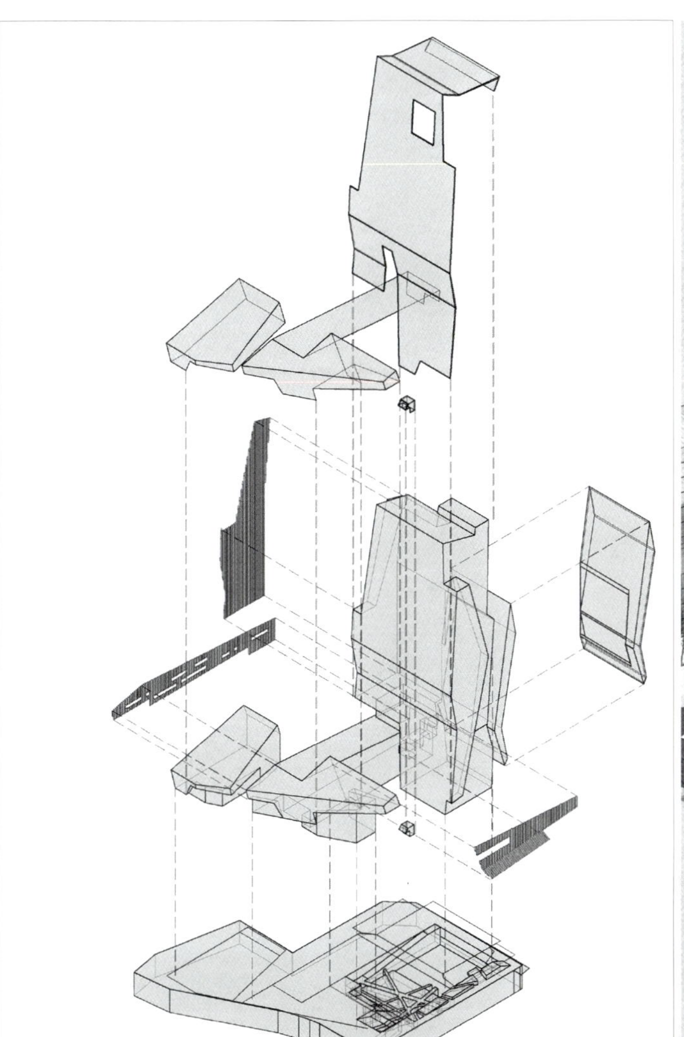

Let the city experience the movement

(1)An unconventional experience - foldline Movement: we all know that the bending of the trunk joint is the premise of human body movements (such as walking, running, jumping) and the prerequisite for changes in body shape (such as sitting ,lying, standing and squatting). Similarly, when we become unsatisfied with the rigid box-like faces and need to change the motionless posture we selected the foldlines. By the use and control of the foldlines or even the continuous foldlines we can make the construction seem to walk and move. The specific strategies are: as for the irregular surface shape of the base, the continuous foldlines connect the high-level and multi-level parts into a whole. There are various angel radians in every line to form dislocating and fragmented yet seamless overall moving effect. While this linear movement is in agree with the linear flowing characteristics of the subway so that it stimulates unlimited imagination of people and effectively reflects the identity of subway project.

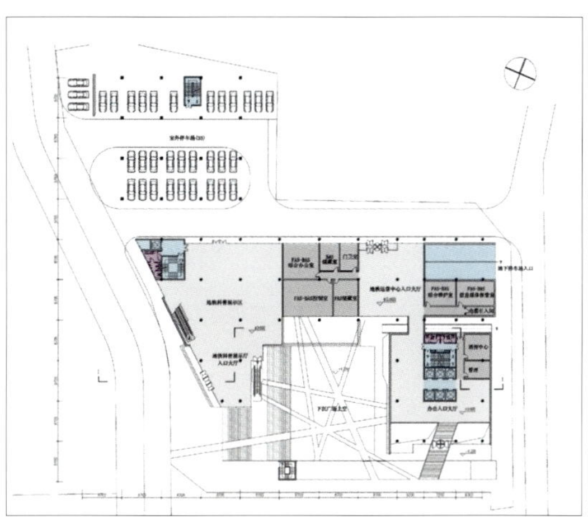

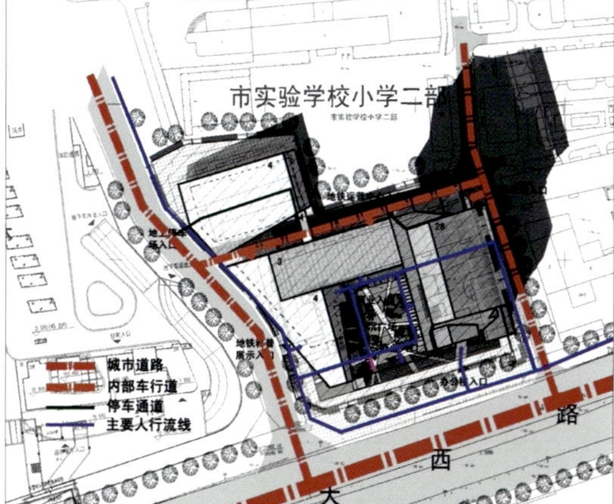

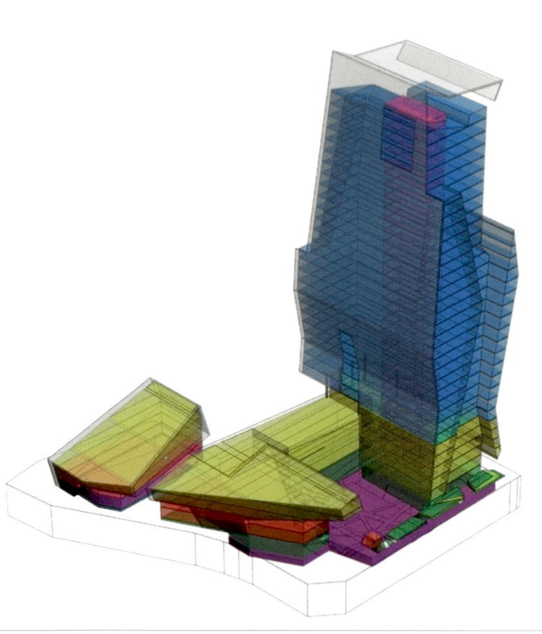

南立面图

(2)Air cube: the program has not only stayed in two-dimensional level. Based on the control of the foldlines in the entire block, we used the way of general cut but local interspersal to make the building have extensive changes in all directions, so that one can get different space feelings from different angles of the city , and a number of skewed blocks form a strong outward tension. The dancing blocks in the air appearing in a non-symmetrical way bring people a completely strange but a strongly dramatized architectural experience, perhaps in shock but without lose of pleasure of "liberation".

(3)Jumping rhythm: let us approach the lens closer to feel the concrete jumping rhythm. At the top and the west side of the main building, a "folded wall"constituted by the metal shutters rushes down from the 100 -meter altitude and dives directly into the ground, bringing people thrilling feeling of speed. This ambiguous approach of the roof and the wall complete the architectural form and make it more interesting. Under the metal shutters is the roof garden.

In the south side of high-rise buildings, two folded blocks popping up in "pitching" shapes. They are both interactive and mutually exclusive, with one virtual and the other solid, showing the "Tai Chi style" of three-dimensional landscape.

The west side of the base near the planning road,two separate blocks, due to the continuity of the line and the uniformity of the cutting method form an intact but fratured consistency of the street. While the southern frame work and extension produce a strong sense of direction and movement, not only increasing the transparency but also creating strong visual impact. Throughout the building the unwittingly appearing holes and cracks actually experienced repeated scrutiny and revision. They are the results of body interspersal and view penetration.

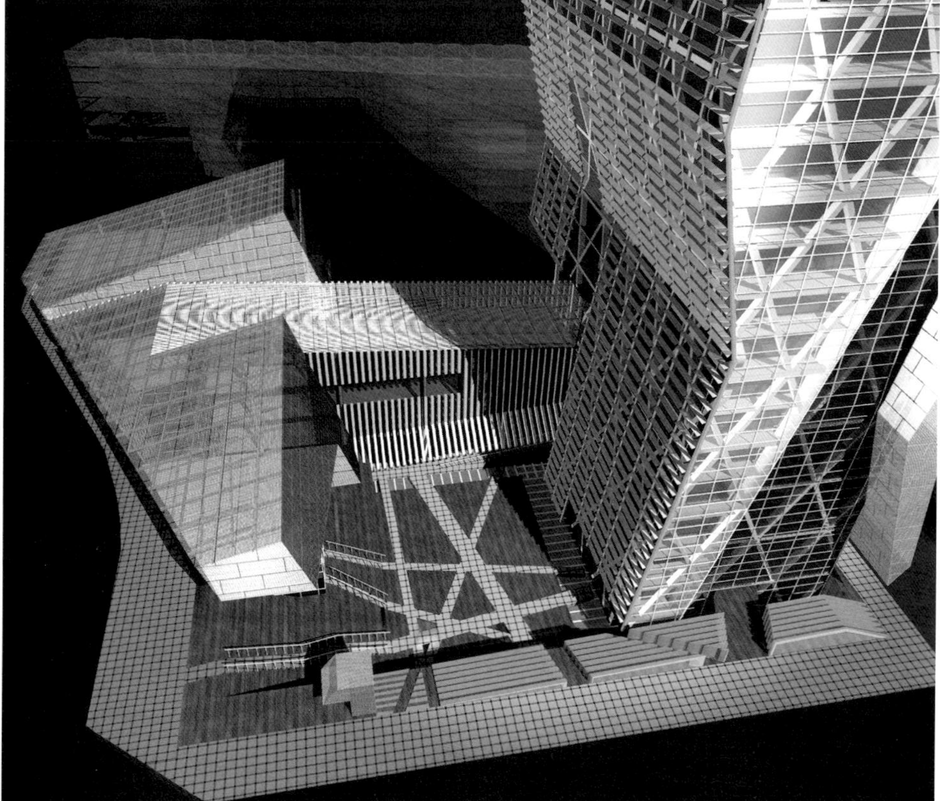

YG 艺术中心

YG ART CENTER

项目地点：中国 · 沈阳　**用地面积**：4700 m^2　**建筑面积**：64 423 m^2
建筑设计：沈阳别人建筑工作室
建筑师：杨胤，高德战，倪伟宁，洪亮

LOCATION: Shenyang, China　SITE AREA: 4700 m^2　BUILDING AREA: 64,423 m^2
DESIGN CORPORATION: Others Studio
ARCHITECTS: Yang Yin, Gao Dezhan, Ni Weining, Hong Liang

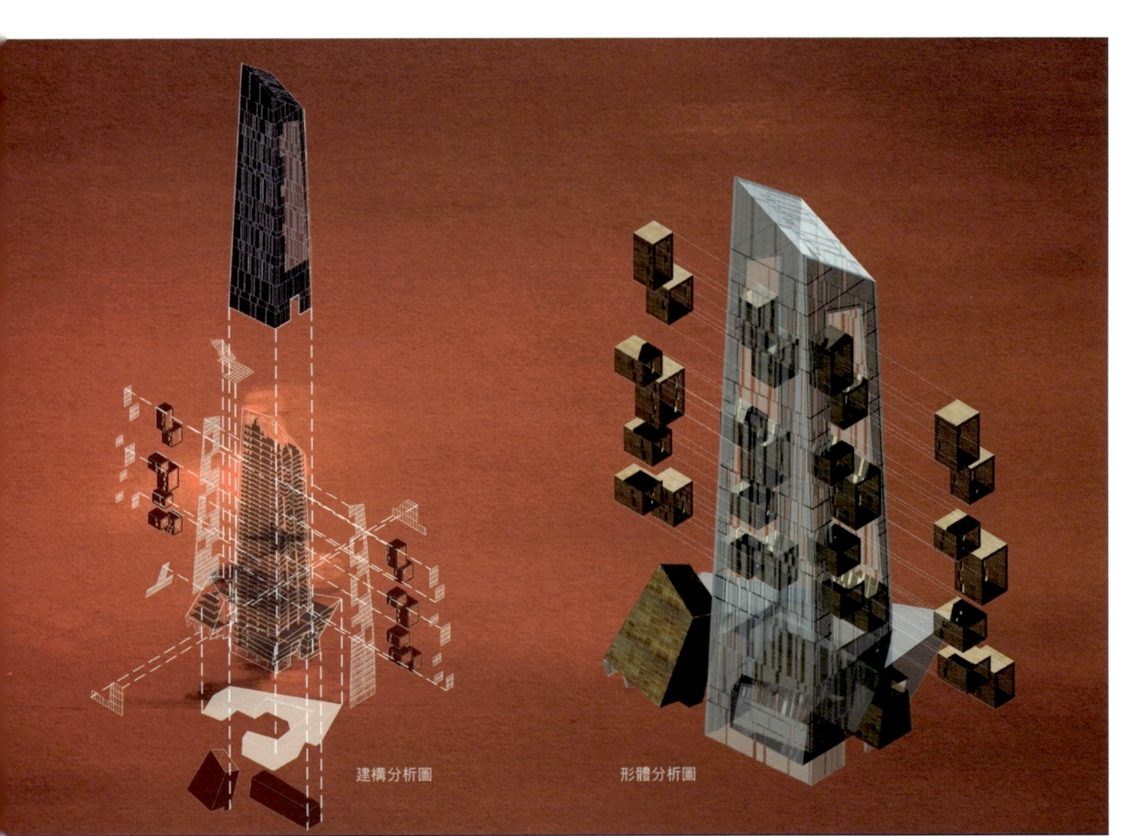
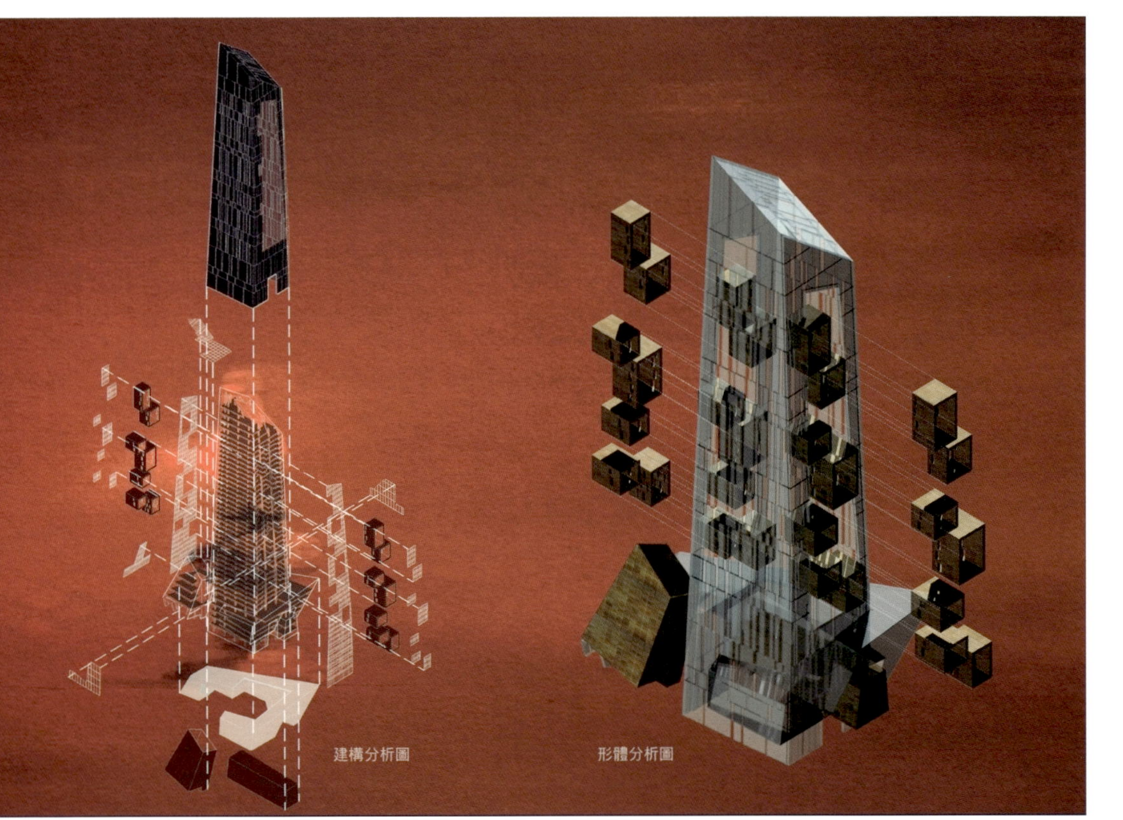

艺术展示类建筑可以有以下两种方式。

（1）建筑以背景方式呈现，用以容纳、烘托展品。

（2）建筑与展品共同表演，建筑融入到展品中 - 即具有强烈的个性、标志性、戏剧性，呈现出一种形体与空间的变奏，建筑是容器的同时也是展品。

此方案选择了后一种方式，它产生在这个喧闹的年代、躁动的都市，它不选择隐身，而是积极地参与。它要激活的不仅是一个区域、一个城市，更应该是一种创造性的心态和真实的人性的反思。

墨，在砚中游动，每次都沿着不同的轨迹，不经意的，不可预料的。

人，在生活中，每个人都有着不同的生存轨迹，所谓记录，其实都是片断的。

这里的建筑，整体上，涵盖了些许片断与不经意，这也反映了我们对这种世界的理解。

建筑以开敞的空间形态与原生态景观呼应，在原有植被全部保留的基础上，在树木间轻巧、谨慎地设置铺装，水景、草地灯、座椅、雕塑等，使其具有足够的可进入、可停留性。这片难得的宁静，不应该因为建筑的融入而被打破。

艺术中心入口空间采用雕塑手法，彰显空间张力的同时，又有效地把基地西侧的原生态景观收纳其中，这也是建筑在设计之初的主要概念之一。

静逸的内部展示空间，建筑的外部形态与内部的空间感受完全契合，磨砂玻璃外墙均匀稳定了随时变动的日光，金属百叶进一步调整了照度，随墙而上的楼梯既提供了垂直动线，又活跃了空间本身。

非匀质空间——建筑力求摆脱整齐划一的标准层模式，想法来自于人的生活本身就不应该是标准化的，在主体中，我们嵌入了尺寸、位置都不同的盒子空间，用这种方式打造不可预知的尺度变化，形成个性化的场所空间。

“建构”不仅显示结果，更是一个过程。我们的建筑突出体现了一个过程，强化了一种体块间的搭接、穿插的关系，还原建筑的本意。我们认为，过程永远重于结果。

关于景观：除了建筑给城市带来全方位的景观价值外，其内部空间同样运用景观设计手法，达成内与外、内与内的全景交融，建筑不仅是功能与雕塑形体调和的结果，同时也是各种具有空间张力的场景以空间序列的方式合理组织的结果。

关于流线：三种主要功能、五种流线相对独立，互不干扰，不仅是使用的需要，更是管理的需要。

艺术中心的展示空间分为上下两部分，顶部三层为永久展区，底部六层为临时展区，可根据面积、性质的不同需求灵活布展。

另外，在主体的 1 ～ 2 层设置了常年对外营业的画廊，在保障日常人气的同时，又可以分担一部分艺术中心的运营成本。

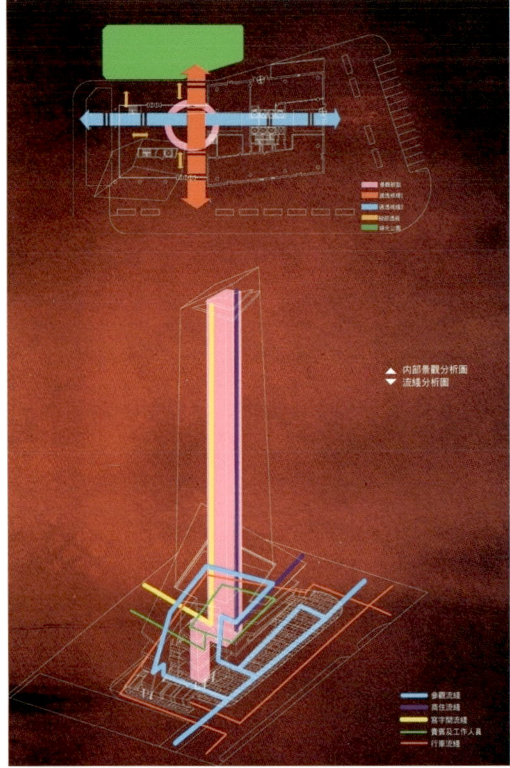

Art show constructions can have the following two ways:

(1)To present the building as the background to accomodate and set off the exhibits.

(2)The bulding and the exhibits show together with the building fused into the exhibits – which is strongly identified, symbolic and dramatic to show a variation of form and space.The construction acts both as the container and one of the exhibits.

This program have chosen the second way. It's born in this noisy age and restless city. It does not select to be invisible, but to actively participate. What it strives to activate is not only a region or a city, but a creative mind and a true reflection of human nature.

Ink, moving in the inkstone, along different paths, is casual and unpredictable.

People, in life, everyone has different life path, and the so-called records are actually fragments.

The buildings here,in the whole, contain some fragments and casualness, which reflect our understanding of this world.

The open space form of the building echoes the original ecological landscape. Remaining the original vegetation the building has sufficient accessiblity and ratainability with the light and meticulous decoration between the woods, the water features, lawn lights, seats and sculptures etc.

This moment of tranqility should not be broken by the participation of the building.

The Art Center entrance space takes sculptural techniques, not only dynamically showing the spatial tension but also effectivelly taking in the original ecological landscape in the west of the base,which is one of the main concepts at the beginning of construction design.

The tranquil internal exhibition space, and the external form well conform to the internal space feeling.The frosted glass wall well stabilize the changing daylight. The metal shutters further adjust the illumination. The staircase along the wall both provides a vertical moving line and enlives the space itself.

Non-homogeneous space – the construction seeks to break the standard uniform layer patterns. The idea derives from that human life itself should not be standardized. In the main body, we embed box spaces of different sizes and locations to create unpredictable scale changes and individualized space.

"Construction" is more a process than a result. Our building highlights a process to strengthen an overlap and interspersed relationship between blocks to restore the real meaning of a building. We believe that process is always more important than results.

About the landscape: in addition to a full range of landscape values brought by the building,the interior space also uses landscape design methods to reach panoramic effect indoor and outdoor or between the both. Architecture is not only the result of reconciliation between functions and appearances but also one of the rational organization of tension spaces in spatial sequence.

About the streamline: three main functions,five relatively independent streamlines without interference to each other are not only for use, but especially for management needs.

The Art Center exhibition space is divided into two parts by the top three floors of permanent galleries and temporary exhibition areas of six floors at the bottom,which can be flexibly used for different needs of sizes and features.

Moreover, in the first and second floors of the main body are galleries open to all people all year round, which shares part of the Art Center's operating costs in the protection of the daily populations.

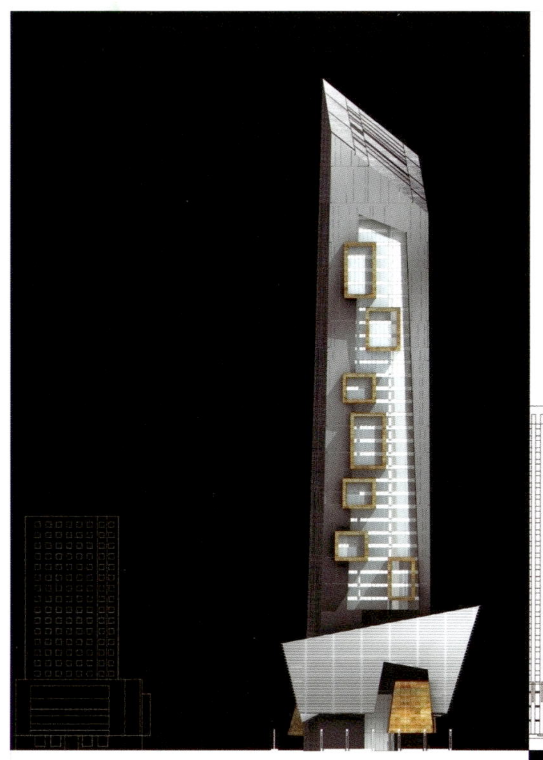

沈阳市音乐厅

SHENYANG MUSIC HALL

项目地点：中国 · 沈阳　用地面积：12 000 m^2　建筑面积：18 000 m^2
建筑设计：沈阳别人建筑工作室
建筑师：杨胤，张亦宁，高德战

LOCATION: Shenyang, China　SITE AREA: 12,000 m^2　BUILDING AREA: 18,000 m^2
DESIGN CORPORATION: Others Studio
ARCHITECTS: Yang Yin, Zhang Yining, Gao Dezhan

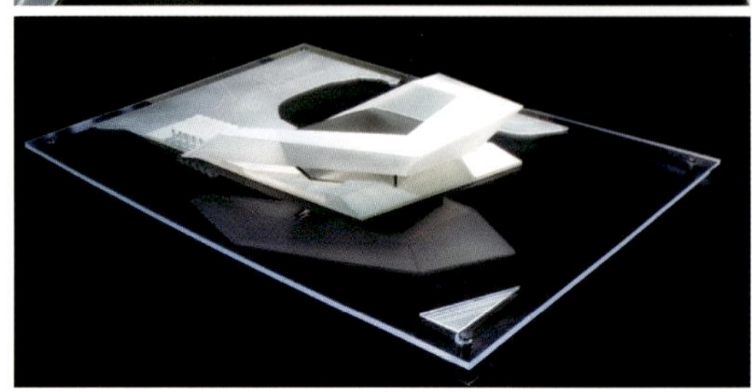

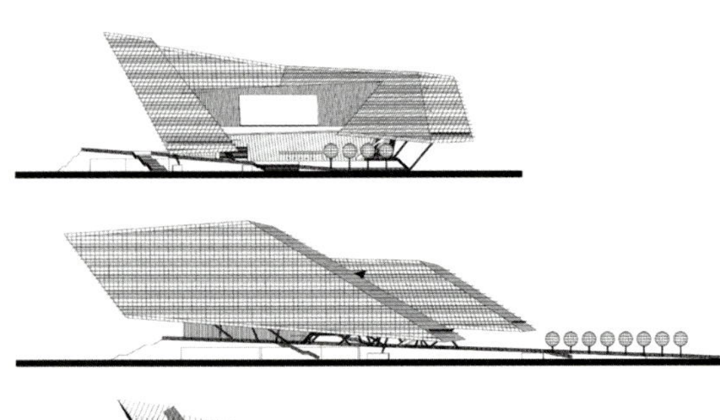

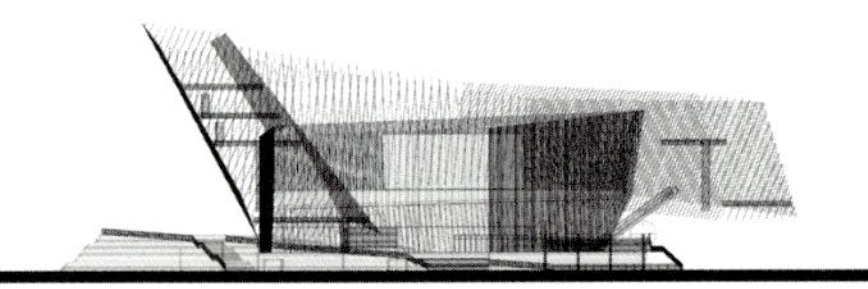

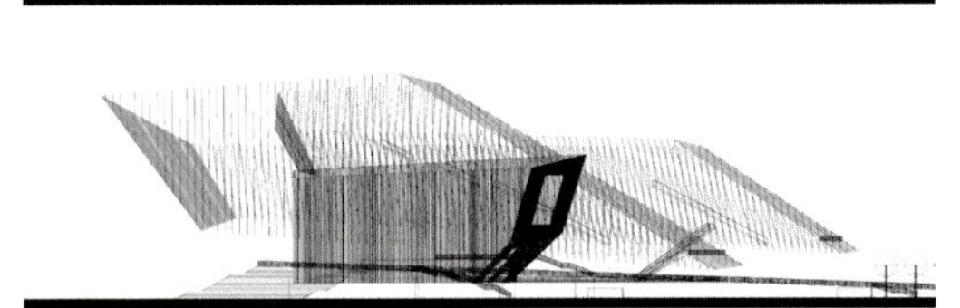

悬浮、轻盈——上虚下实。

通透——内实外虚。

动态系统：(1) 视觉动态循环系统，通过多维度、多方向的形态产生不规则的建筑外表，而其形式却经过了高度解析，从城市各个角度都能获得动态的建筑感受。(2) 动态流线系统：通过立体交叉的流线设计，产生连续、清晰的动线系统。(3) 动态空间系统：倾斜的墙面、屋面、地面打破了传统意义上对建筑边界的认识，内部空间与外部形态相互呼应，产生具有冲击力的空间效果。

交叉意向：通过对功能和体积的复合交叉，使建筑从两个层面上进行交叉。(1) 功能交叉。(2) 体量交叉：体量交叉与功能交叉相对应，底部平台为实体，平台上部公共空间部分为虚体，虚体包围的中间音乐空间部分为实体，形成上虚下实，内实外虚的交叉体量。

开放性和参与性：(1) 通过大平台的设置，可提供露天影院、大型现场演唱、演奏会场所，大屏幕的设置更是为演出提供了良好的舞台背景。(2) 展览、餐饮、音乐教育等开放式公共空间的设置为广大市民提供了非正式的音乐体验空间。(3) 大平台的设置使底部停车成为可能，解决了青年公园内部停车难的问题。

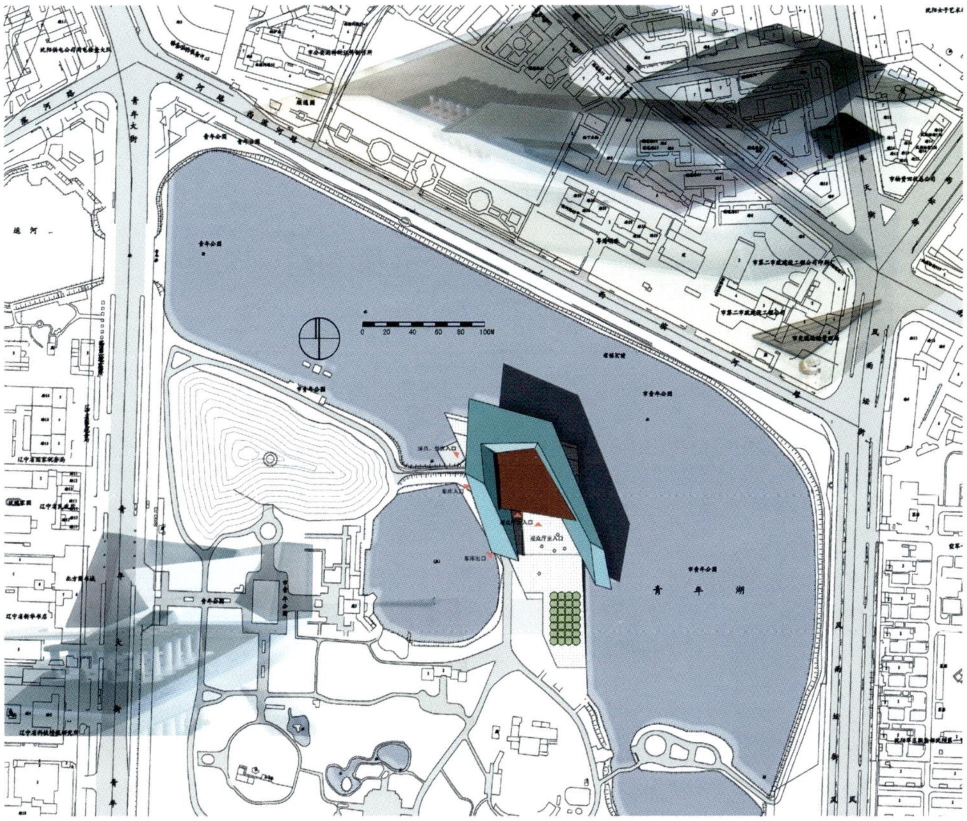

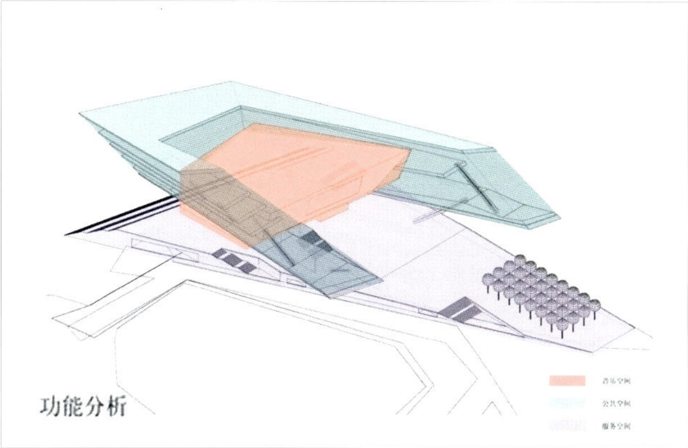

功能分析

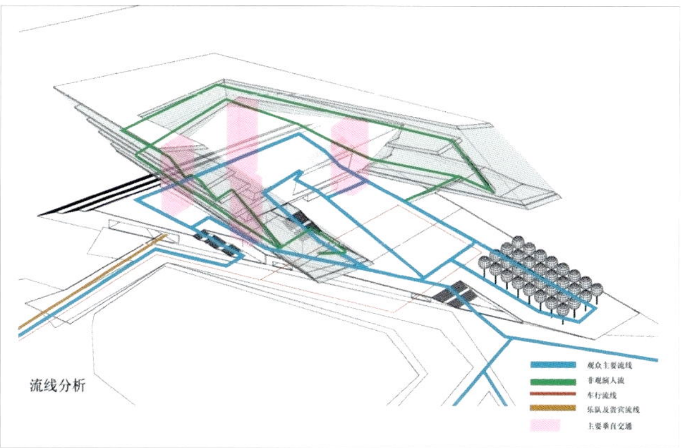

流线分析

Suspension, lightweight - void in the upper and solid in the lower.

Transparent -- void inside and solid outside.

Dynamic Systems: (1) Visual dynamic circulatory system: the multi-dimensional, multi-directional shapes produce irregular building exterior appearance,while its form has been highly analytical. It shows dynamic construction views from various angles of the city. (2) The dynamic flow line system: the intersectional flow line design produces continuous and clear dynamic line system. (3) Dynamic spatial systems: obliquitous walls, roofs and the grounds broke the traditional knowledge of construction boundaries. The interior and exterior spaces echo with each other, generating spatial experience with strong impact.

Interweaving Intention: The construction uses compound interweaving on two levels,the functions and blocks. (1) Functional interweaving. (2) Bulk intercross: bulk intercross corresponds to the functional interweaving. The bottom platform is solid while the public area above is void. The central music space surrounded by the void area is solid. So it totally forms the bulk intercross, void in the upper and solid in the lower, hollow outside and solid inside.

Open and Participatory: (1) The setting of the large platform provides an open-air theater and a large live concert venue. The large screen provides an even better stage background. (2) The open public spaces, such as exhibition, food, music education areas, provide an informal music experience space for the public. (3) The setting of the large platform makes the bottom parking possible to solve the parking problem of the Youth Park.

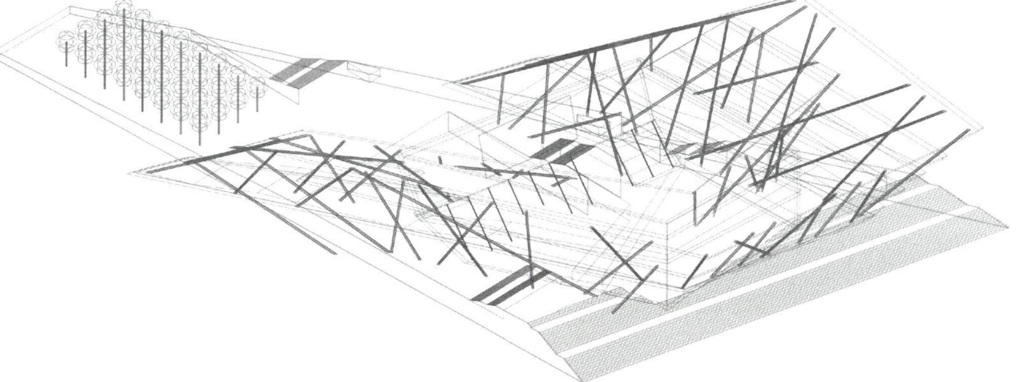

沈阳文化艺术中心概念性设计

SHENYANG CULTURE & ARTS CENTER CONCEPTUAL BESIGN

项目地点：中国 · 沈阳　用地面积：45 000 m^2　建筑面积：42 000 m^2
建筑设计：沈阳别人建筑工作室
建筑师：杨胤，高德战，张亦宁

LOCATION: Shenyang, China　SITE AREA: 45,000 m^2　BUILDING AREA: 42,000 m^2
DESIGN CORPORATION: Others Studio
ARCHITECTS: Yang Yin, Gao Dezhan, Zhang Yining

关键词：振颤、律动、轻盈、通透、开放

动态系统：（1）视觉动态循环系统：通过多维度、多方向的形态产生不规则的形态外表，而其形式却经过了高度解析。从城市各个角度，都能获得动态的建筑感受。（2）动态流线系统：通过立体交叉的流线设计，产生连续、清晰的动线系统。（3）动态空间系统：倾斜的墙面、屋面打破了传统意义对建筑边界的认识，内部空间与外部形态相互呼应，产生具有冲击力的空间效果。

交叉意向：（1）功能交叉：文化艺术中心包含有音乐厅和美术馆两个大的功能板块，但又不是音乐厅和美术馆的简单集合，而是包含有多种公共文化设施的综合性文化场所。音乐空间包括：舞台、观众厅、室内乐演奏厅等。美术空间包括：各类展厅、美术研究室等。

公共空间包括：内部有音乐展览厅、餐饮部、音乐教育室、音乐体验中心、音乐家工作室、视觉体验中心、美术家工作室、艺吧等。外部有下沉广场、入口平台、室外舞台等。（2）体量交叉：体量交叉与功能交叉相对应，音乐厅和美术馆两部分体积相互分离又相互交叉，共同形成起伏跌宕、错落有致、浑然一体的整体意向。

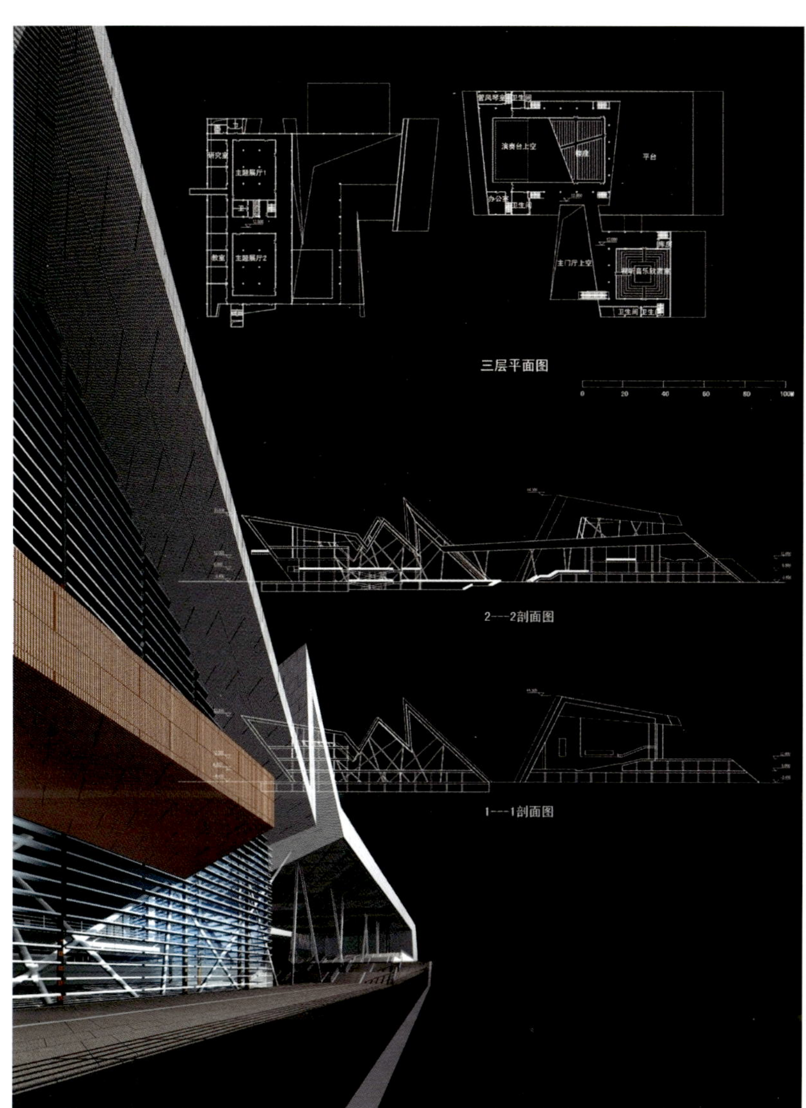

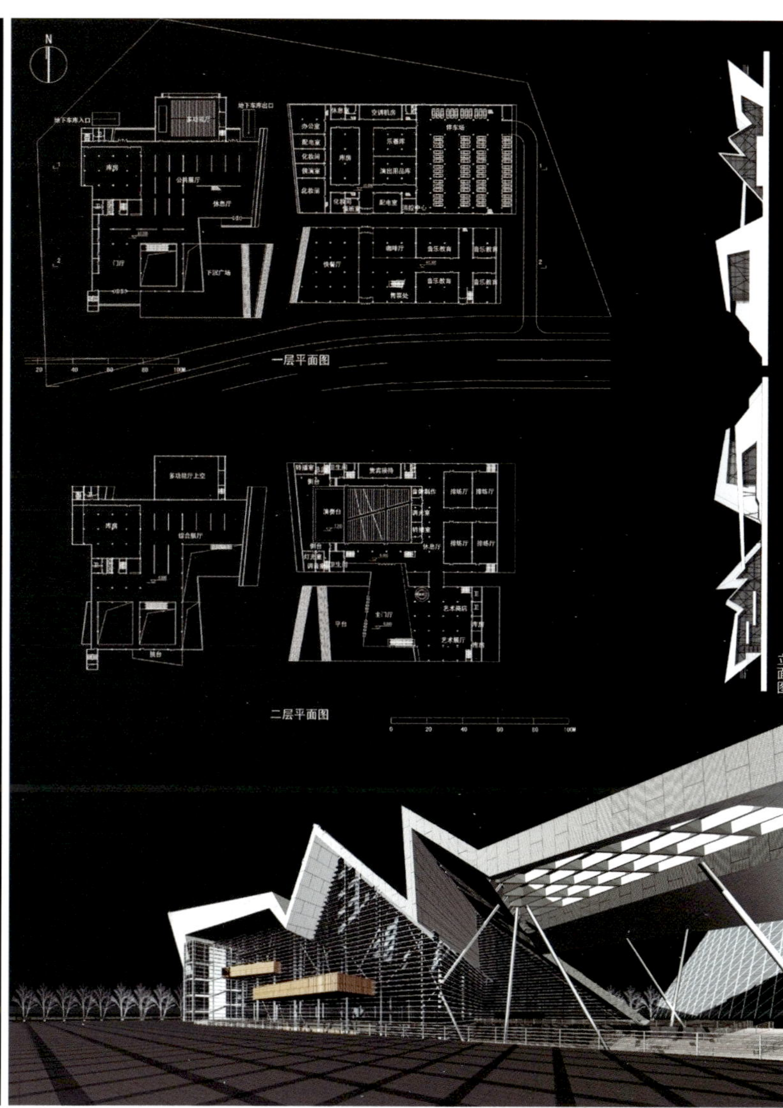

Key words: vibration, rhythm, lightness transparency and openness

Dynamic Systems: (1) Visual dynamic circulatory system: the multi-dimensional, multi-directional shapes produce irregular building exterior appearance,while its form has been highly analytical. It shows dynamic construction views from various angles of the city. (2) The dynamic flow line system: the intersectional flow line design produces continuous and clear dynamic line system. (3) Dynamic spatial systems: obliquitous walls, roofs and the grounds broke the traditional knowledge of construction boundaries. The interior and exterior spaces echo with each other, generating spatial experience with strong impact.

Interweaving Intention: (1) Functional interweaving: Cultural and Arts Center includes large functional blocks of Concert Hall and Art Gallery, but not the simple combination of the two. It's comprehensive cultural venue of a variety of public cultural facilities.

Music space includes stage, auditorium and chamber music hall etc. Art space includes: all kinds of exhibition halls and art research.

Public space, includes: internal - music exhibition, catering, music education, music experience center, musicians' studio, visual experience center, art studio and art bar etc..

External - sunken plaza, the entrance platform and outdoor arenas etc.

(2) Bulk intercross: bulk intercross corresponds to functional interweaving. The bulks of the concert halls and galleries are both separate and intercross with each other to form the whole image of intact up- and-down patchwork.

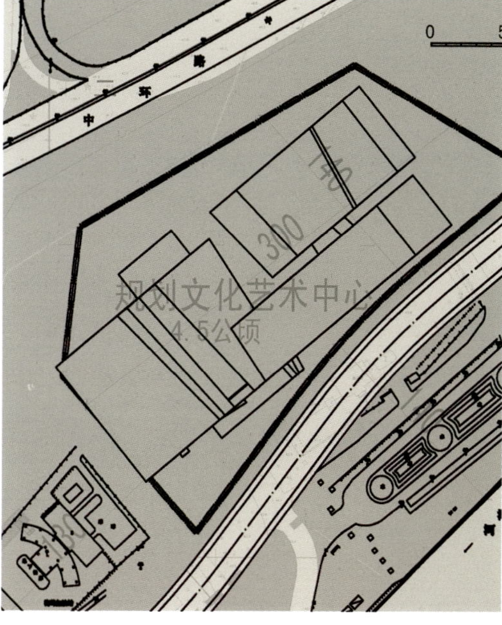

演通寺

YANTONG TEMPLE

项目地点：中国 · 黑龙江　用地面积：48 000 m^2　建筑面积：48 000 m^2

建筑设计：沈阳别人建筑工作室

建筑师：杨胤，张亦宁，高德战

LOCATION: Heilongjiang, China　SITE AREA: 48,000 m^2　BUILDING AREA: 48,000 m^2

DESIGN CORPORATION: Others Studio

ARCHITECTS: Yang Yin, Zhang Yining, Gao Dezhan

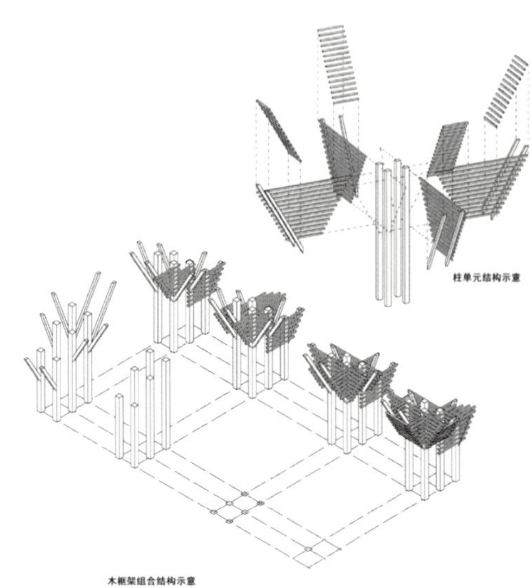

木榧架组合结构示意

禅宗认为：空间存在于物质消失的界域。也就是说，只有在物质的具体属性如内外、上下、体积、边界趋于消失的地方，空间才可能出现。间意味着不在这里，也不在那里。它是自心停留的场所。

我们的方案从某种角度上说其实就是对“间”的理解、阐释和试验。通过简易的折叠来整合、净化空间。在传统的寺庙建筑中，折叠是单个建筑屋顶（坡屋顶）形成的主要手段。我们所做的是采用连续折叠的方式，将单个庙宇连接起来，使整个寺院成为连绵山峦的一部分。在这个大的自然域中，众生平等，一切归于平静。

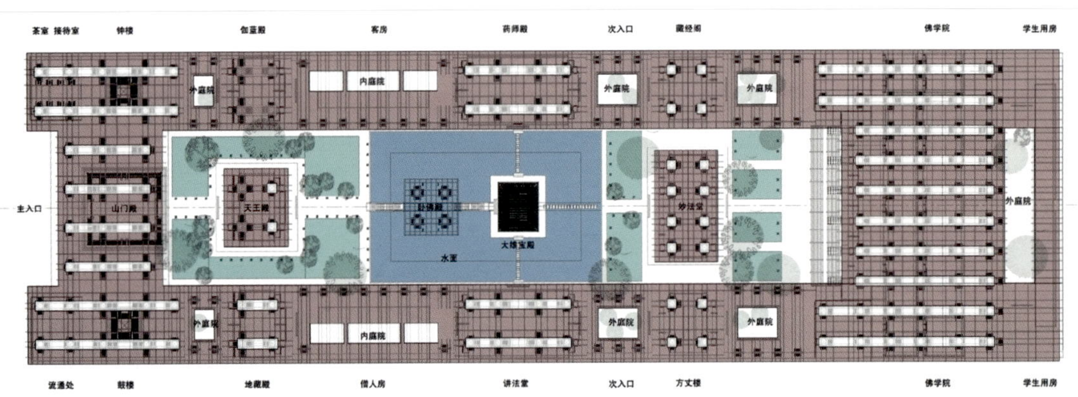

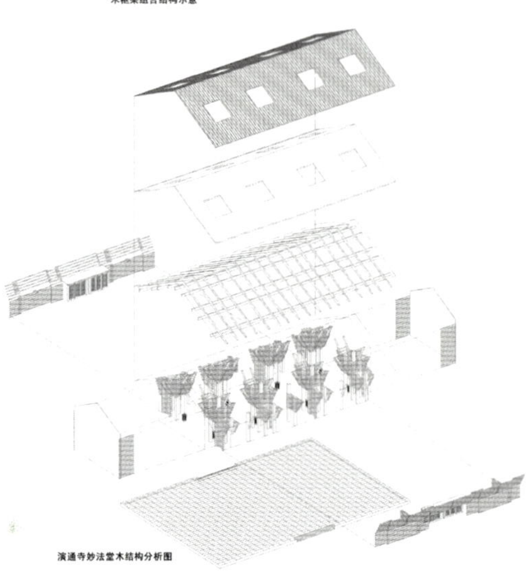

演通寺妙法堂木结构分析图

Zen said: space exists in the domain of physical disappearance,that is, can space emerge only when the specific properties of material tend to disappear, including inside and outside, up and down, volume and boundary."Jian"means nowhere but just a place for your heart to rest. In fact our program is the understanding, interpretation and experimentation to this meaning, and to integrate and purify the space through the simple folding. In traditional temples, folding is the primary form of the single building roof (sloping roof). What we did is to fold in a continuous way, in which the buildings are intactly connected to make the whole temple integrated into rolling hills. In this big natural domain, everyone is equal, everything was tranquil.

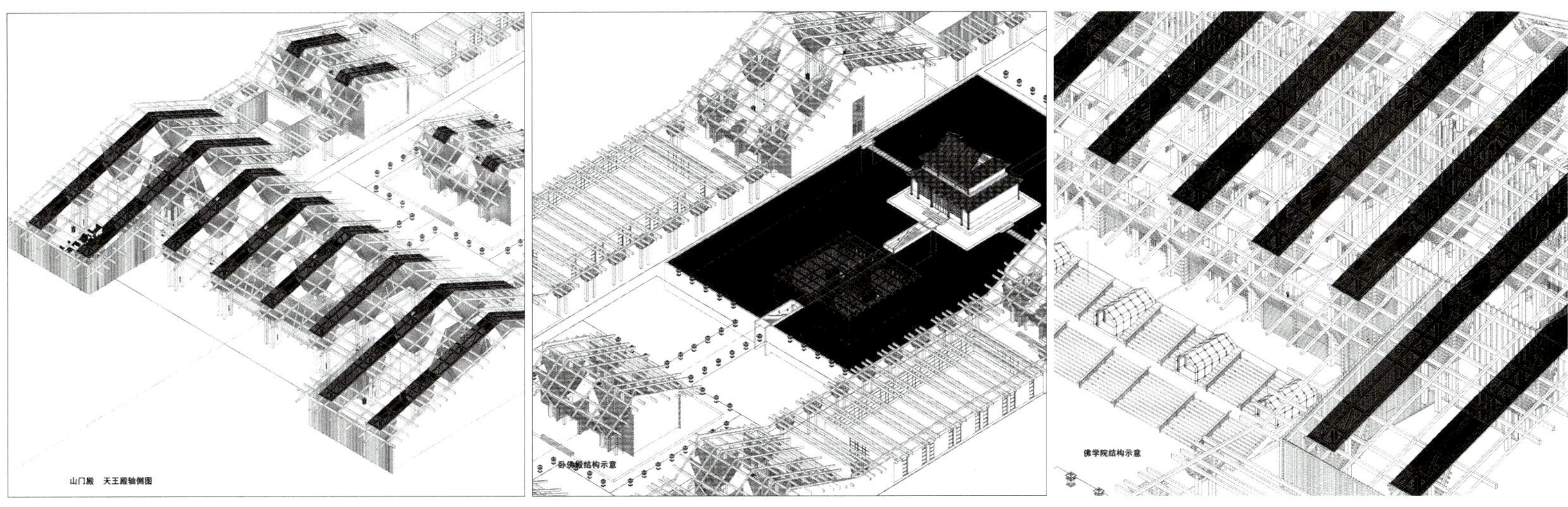

SIC ARCHITEKTEN GMBH

德国SIC建筑设计责任有限公司

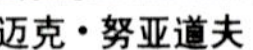
迈克·努亚道夫

斯蒂芬．亚斯伯

杨·古特木特

[sic!]成立于1983年，1988年总部迁往德国科隆至今，是德国科隆最大的集工程咨询、项目管理、规划、建筑、景观、室内设计为一体的综合性工程咨询公司。

[sic!]的经营哲学具有开放性以及全面性。[sic!]认为建筑服务应该是全方位的，它不仅仅包括设计和规划，工程的计划编制和施工服务也应涵盖其中，而如今只有为数不多的建筑公司持有这种理念。因此，从构思草图到工程完工，[sic!]重视项目整个过程中的每一个细节。此外，在方案设计的阶段[sic!]就会充分考虑与后期施工相关的细节，从而确保施工阶段的顺利实施，最终工程质量能达到相当高的标准。

对于复杂的、高标准的工程设计，[sic!]具有相当丰富的经验。在德国，[sic!]已为为众多大型企业及机构进行了建筑设计：德国保险公司（Gerling DEVK）、Deutsche银行、德国科隆会展中心（Koelnmesse）、德国电视广播公司（RTL and ZDF）、福特汽车公司（Ford）、科隆大学（University of Cologne）、土耳其驻德国使馆（Turkish Embassy）等。除了在欧洲国家开展业务，[sic!]还开拓了亚洲市场。

[sic!]在中国参与的第一个项目可以追溯到20世纪80年代末期，位于深圳的深圳国际会展中心，[sic!] 现任的两位总裁，均参与了当时的项目建设。此后，[sic!] 开始尝试对中国市场的研究，与其主要的合作伙伴，著名的深圳市建筑设计研究总院，以及其他的一些中国国内的建筑师事务所陆续展开了合作。近年来，[sic!]已为南京市政府、宁波市政府、徐州市政府、绿地集团、河南新闻出版集团、湖南广电集团、贵阳科协等重要客户设计了众多的建筑方案，赢得了广泛的赞誉。

目前，[sic!]深圳代表处主要负责其在中国的项目，所提供的服务主要集中在建筑设计和建筑节能方案上。其中包括：

建筑设计：市政公共建筑；商业建筑；酒店建筑；医疗建筑；办公楼宇；住宅区域规划及建筑设计（多层、高层、联排别墅 及高档别墅等）；旧建筑或旧区域的翻新改造设计；大型项目的概念性规划；旅游地产策划及规划。

建筑节能：

提供成熟的德国建筑节能方案（由德国节能方面的工程师提供支持与咨询）；为中国客户选择和推荐最切合实际情况的节能解决方案。

此外，[sic!]在做中国项目的过程中，能够得到中方合作伙伴方的各个专业工程师的充分配合与支持，进而能保证在项目的后期，与工程各专业及中国建筑规范相关的方面能顺利进行，从而确保为客户提供高质量的工程精品。[sic!]的建筑师同时也会履行与在德国参与工程的同样职责，参与到工程相关的每一个环节中，本着高度负责的态度与德国式严谨，对工程的质量做出全程的监控。

在未来，[sic!]希望把更多更好的已成功在德国采用的节能技术带到中国，为降低建筑维护成本、节约能源做出贡献。

[Sic!] was established in 1983, and since 1988 its headquarter has been in Cologne, Germany. It is the largest comprehensive engineering consulting firm set in Cologne, Germany integrating engineering consultancy, project management and planning, landscape and architecture as well as interior design.

[Sic!]'s management philosophy is open and comprehensive. It believes that the construction services should reach every aspect including not only the design and planning but also the project layout and building services, which is only a belief held by a handful of construction companies today. Therefore, throughout the process from the sketch to completion of the project [sic!] pays attention to every detail of the project. In addition, at the program design stage [sic!] takes full consideration to the details relevant to construction to ensure the smooth implementation of the construction so that the quality of the final project can reach very high standards.

As for complex and high-standard project design, [sic!] is a very experienced company. In Germany, [sic!] has completed architectural design for many large enterprises and organizations, such as German insurer (Gerling DEVK), Deutsche Bank, Cologne Exhibition Center (Koelnmesse), the German broadcasting company (RTL and ZDF), Ford Motor Company, University of Cologne, the Turkish Embassy in Germany.In addition to doing business in European countries, [sic!] also explores Asian market.

The first item [sic!] involved in China dates back to the late 1980s,i.e.Shenzhen International Convention & Exhibition Center. The two incumbent presidents both participated in the project at that time. Since then, [sic!] started trying to research the Chinese market and gradually expanded the cooperation with its main partners, the famous architectural design unit, Shenzhen Research Institute, and several other architectural firms in China. In recent years [sic!] has designed many construction programs for many important customers including Nanjing Government, Ningbo Government, Xuzhou Government, Greenland Group, Henan News Publishing Group, Hunan Broadcasting Group, Guiyang Scientific Association, etc, and has won widespread reputation.

So far, [sic!] Shenzhen Office is mainly responsible for its projects in China, providing services mainly in architectural design and energy efficiency building programs,including.

Architectural design:

Municipal public buildings; Commercial buildings; Hotel buildings; Medical Buildings; Office buildings; Residental area planning and architectural design (multi-storeyed, high-rise, townhouses and luxury villas, etc.); Retrofit design of the old buildings or areas; Conceptual planning of large projects; Tourism and Estate Planning;

Energy efficiency:

Providing mature German energy efficiency building programs (support and advice provided by energy saving engineers from Germany); Recommending and selecting the most realistic energy-saving solutions for the Chinese customers .

Moreover in the process of doing projects in China [sic!] will get the full support and cooperation from all professional engineers of the Chinese partners, which then can guarantee the smooth conduction of related aspects of the professional engineering and the Chinese building regulations in the later stage of the projects, to ensure refined high-quality projects for our customers. Architects of [sic!] will also implement the same duties as they do in Germany to every relavant detail of the projects, adopting a highly responsible attitude and German rigorousness to thoroughly monitor the quality of the project.

In the future, [sic!] hopes to bring more and better energy-saving technologies successfully used in Germany to China to contribute to the reduction of building maintenance costs and energy conservation.

深圳市观澜版画基地美术馆及交易中心

SHENZHEN MISSION HILLS PRINT BASE MUSEUM AND TRADING CENTER

项目地点：中国 · 深圳　占地面积：31 802.22 m²　建筑面积：13 784.43 m²
建筑设计：德国 sic 建筑设计责任有限公司
建筑师：杨 · 古特木特

LOCATION: Shenzhen, China　SITE AREA: 31,802.2 m²　BUILDING AREA: 13,784.43 m²
DESIGN CORPORATION: Sic Architekten Gmbh
ARCHITECT: Jan Gutermuth

项目总体设计构思：观澜版画基地是中国印刷艺术起源的一个历史中心区，同时也承载着历史人物陈烟桥的艺术生命，因此美术馆与交易中心将作为一个“划时代”的地标性建筑，重新诠释观澜版画基地。概念设计的主导思想来自于一个印刷艺术所带给我们的象征意义。

美术馆设计构思：第二个功能建筑即版画美术馆的设计。作为一个象征，代表着中国最先进的印刷术，其悠久的历史，正是从她的受限的方形正交网格的形状中跳跃出来，自由地站立在地上，并且不断地盘旋向上延伸，象征着新的希望以及活力，也寓意着艺术的自由发展。

交易中心设计构思：交易中心是一个以技术和交流为主要导向的功能场所。而印刷艺术的特点则把这些特质完全地反映在了图书印刷中。因此交易中心的设计采用了严格的正交系统，方形网格的布置体系象征着传统的中国文字。

整个项目用地将通过交易中心的方形网格结构重新规划排布，分布在美术馆螺旋上升的体块周围。北部地块中的立方体块规划为公园般的景观空间。该交易中心不仅在平面上而且在竖向上也呼应着历史的沿袭。原有的历史性建筑碉楼，矗立在交易中心的入口正门处，注视着美术馆，似乎是在和她相互交流，令人兴奋，同时更展现了美术馆作为观澜版画基地“里程碑”的地标性建筑的风采。

在项目用地的东南面保留了现有的河塘景观以及客家传统建筑排屋。整体上，南北景观互相呼应，结合中央广场地带，形成一个完整的围合区域。

变异性和灵活性，不止是对于美术馆，同时也适用于交易中心，是这个方案设计出发的核心。

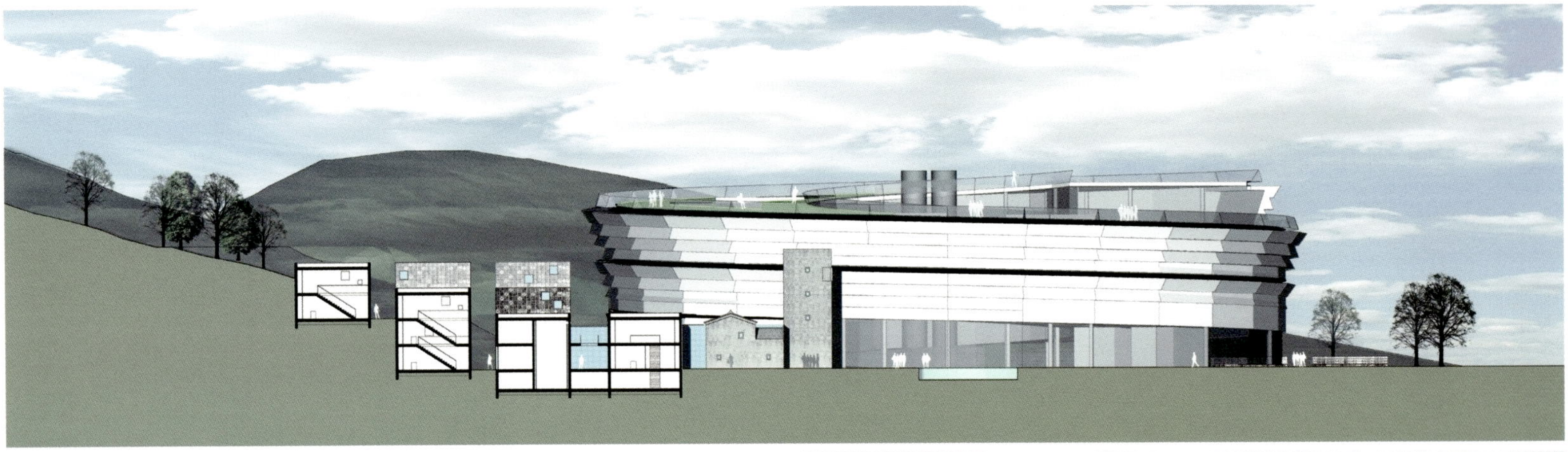

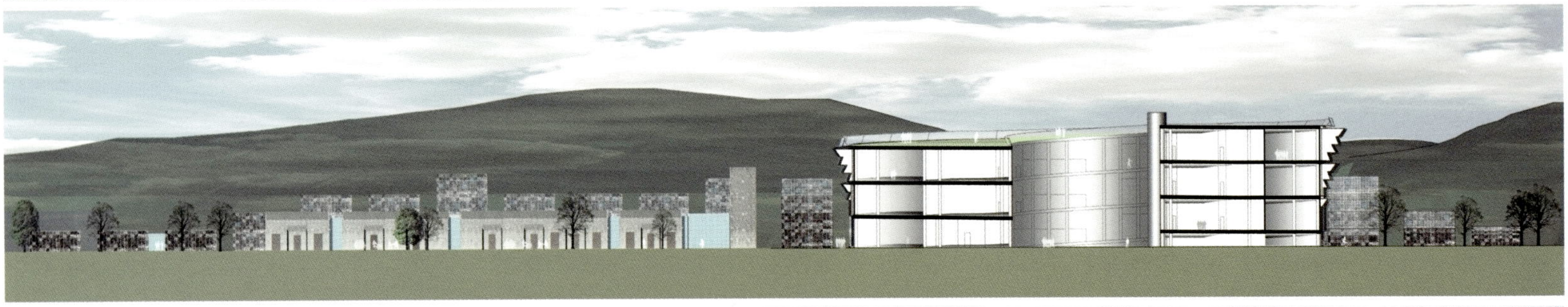

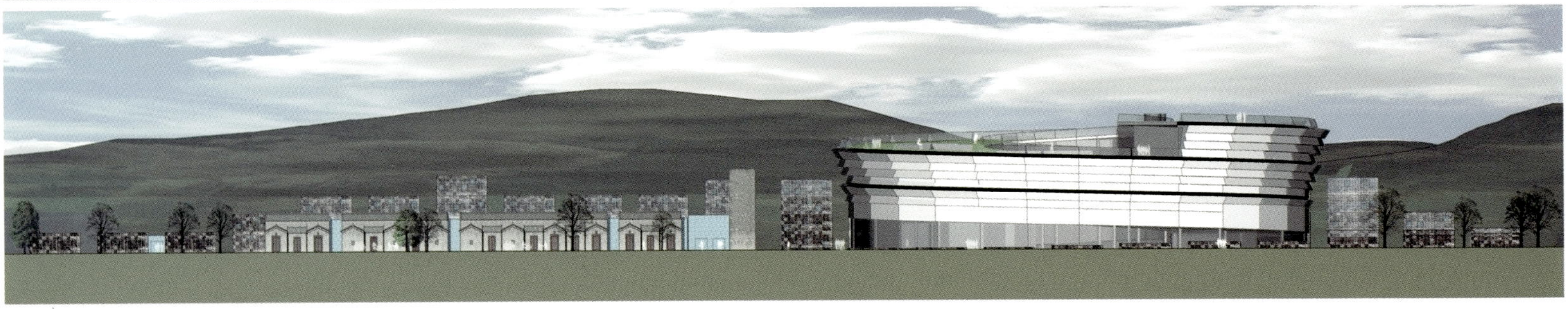

The overall design concept: Mission Print Base is a historical center of the printing art origin. It also carries historical figure Chen Yanqiao's artistic life,therefore the Art Museum and Trading Center will serve as a landmark architecture to reinterpret the Mission Print Base. The dominant design concept comes from a symbolic meaning brought by the printing art .

Museum Design Concept: the second functional building design is the Print Art Museum Design. As a symbol, it represents the most advanced printing.The long history just jumps out from her limited orthogonal grid square to stand on the ground without any constraints. It constantly spirals upward , which implies the free development of art, also symbolizing the new hope and creative energy.

The trading center design concept: the trading center is mainly a technical and communicational functional area. While the feature of the printing art is to reflect it in book printing.Therefore the design of the trading center uses strict orthogonal system, and the shape of the square grid symbols the traditional chinese characters. The area of the entire project will be rearranged through the square grid structure and distributed aroud the spiral chunk of the art museum. The cube chunk in the north is a garden like landscape.The trading center echoes with the historical development not only in the horizontal but in the vertical dimension. The origianal historical watchtower stands at the front door facing directly to the art museum, making you excited and meanwhile displaying the museum as a landmark of the printning base.

In the southeast of the project remains the existing pond landscape and the traditional Hakka architecture-row house. As a whole the landscapes of North and South echo with each other, and combined with the Central Square area to form a complete enclosure area.

Variability and flexibility is applicable not only to the museum but also to the trading center ,which is the design core of the project.

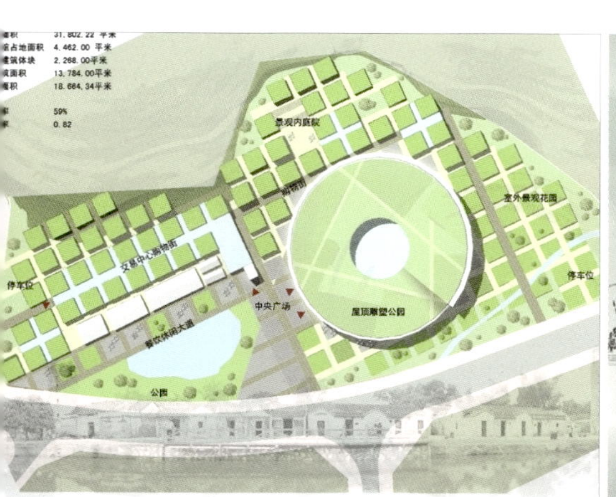

柏林自由团结纪念碑广场

THE LIBERTY AND UNITY MONUMENT SQUARE IN BERLIN

项目地点：德国 · 柏林　建筑面积：3000 m^2
建筑设计：德国 sic 建筑设计责任有限公司
建筑师：杨 • 古特木特

LOCATION: Berlin, Germany　BUILDING AREA: 3000 m^2
DESIGN CORPORATION: Sic Architekten Gmbh
ARCHITECT: Jan Gutermuth

象征统一与自由的纪念碑。

新的纪念碑把主体定位放在“人”上面，尤其以“追求自由和统一的人”作为中心。它强调了众所周知的基座，我们是民众，我们是一个民族。

从 1848 年的汉巴赫节革命到 1989 年的和平革命，始终是由个人的微小火花汇聚在一起从而形成的为争取自由和统一且充满鼓舞精神的炽热火焰！

前景的图案是为了纪念发生在 1989 年的“莱比锡星期一示威”活动。这个事件在德国的历史上拥有着独一无二的地位。在刚开始的时候，在一片似平静海面一样的火焰中仅有一些“羞怯而犹豫不决”的烛光，最后在这样的海中，原德意志民主共和国的狭隘政权沉没了，而一个分裂的德国却因此统一并站立起来。

前民主德国人民议院前总统辛德曼（Sindermann）曾说过：“对除了死亡以外所有可能发生的事都有所准备。”

The monument symbolizes unity and libety.

The new monument poses "human" as the main theme,esp. it focuses on "the human in pursuit of freedom and unity". It emphasizes the well-known base-we are the people, we are one people.

From the Hambach Festival of 1848 to the peaceful revolution in 1989, it was always a history of encouraging spirit blazing flame that was centralized by tiny individual spark of fight for freedom and unity. The visionary design is to mark the 1989 "Leipzig Monday demonstrations". This event occupies unique status in the history of Germany.

At the beginning,like in a calm sea, there just shines some "timid and hesitant " candle light in the flame just like a peaceful sea where the narrow regime of the former German Democratic Republic sank, while a divided Germany was unified and stood up.The former President of People's chamber Sindermann once said:" we are prepared for everything that might happen apart from the death."

贝纳通伊朗旗舰店及办公楼

BENETTON IRAN'S FLAGSHIP STORE AND OFFICE BUILDING

项目地点：伊朗 · 德黑兰　占地面积：1500 m^2　建筑面积：10 625 m^2
建筑设计：德国 sic 建筑设计责任有限公司
建筑师：杨 · 古特木特

LOCATION: Iran, Tehran　SITE AREA: 1500 m^2　BUILDING AREA: 10,625 m^2
DESIGN CORPORATION: Sic Architekten Gmbh
ARCHITECT: Jan Gutermuth

设计该方案的目标是：通过建筑本身来传递出该建筑功能的独特性以及其地理位置的显著性。

贝纳通的 CI“全色彩贝纳通”（Benetton 是世界驰名的休闲类服饰品牌）通过一个处于不唐突环境中的媒介性色彩，来体现该产品的主题“各种色彩”，并展现出它们的效果。建筑的外立面在外形上富有表现力，但是在色彩上却相对适度，这衬托并加强了玻璃幕墙后的彩光投影效果。

该建筑的外型及功能布局是建立在传统的伊朗百货类建筑的基础上，外立面受到伊朗传统建筑元素即“拱顶”的影响，采用传统百货建筑中的双拱廊，创新性地阐释并体现了该建筑的功能——“当代的百货店”。访客可从双拱廊的后面进入到一个巨大而空旷的空间，从该空间垂直延伸的中庭与所有的楼层相接，同时此处也是通往各楼层的主入口，而这样的空间也同样来自于伊朗传统百货建筑的元素。

创新：这个独一无二的设计为该建筑的使用者提供了一个与众不同的地标。它独特的外型与创新性的传统主题，体现了“旗舰店”建筑应有的品质。

弹性设计：功能上的组织要求有相当的弹性。一层与二层包括三个商业性的单位，可分开设置，也可合并或隔断为较大的单位，而在上层的各办公楼层中也遵循着同样的系统：三个办公室空间可合并为一个较大的空间，或者通过简单的施工隔断为所需要的且合适的办公空间。

The design goal: to deliver the the unique function of the building and significance of its location through the building itself.

Benetton's CI "full color Benetton" (Benetton is the world famous leisure clothing brand.) reflects the theme of the product-various colors,through the medium color in the unobtrusive environment and shows their effect. Building's facade is expressive in shape, while the color is relatively modest, which sets off and strengthens the projection and luminalart effect behind the glass curtain wall.

The appearance and function layout of the building is based on the traditional department store category architecture of Iran: the facade is subject to the traditional architectural element of Iran – the dome. The use of the double arcade in traditional department store building innovatively displays and interprets the building's function - "a modern department store." Visitors can enter a huge and empty space from the back of the double arches, where the vertically extending atrium are connected to all the floors, and besides it's the main entrance leading to all floors. This space also comes from Iran's traditional department store building elements.

Innovation: the unique design of the building provides users with a unique landmark. Its unique appearance and innovative traditional theme reflect the quality of a "flagship" building.

Flexible design: functional organization requires considerable flexibility. The first and the second floors include three commercial areas which can be set separately or merged into a larger partitioned area. In the upper floors of the office follows the same system: three office spaces can be combined into a larger space or an appropriate office space through simple construction of partition as needed.

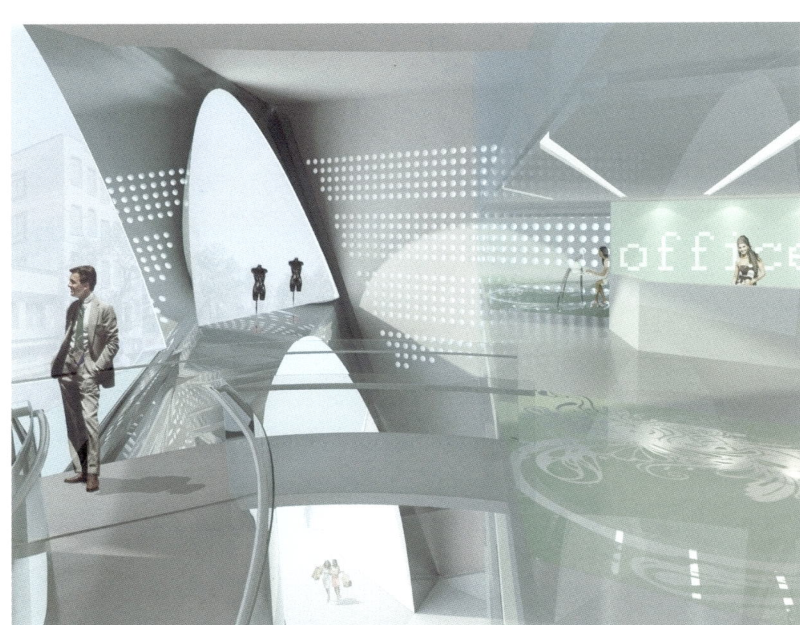

德国广播电台总部

THE GERMAN RADIO STATION HEADQUARTERS

项目地点：德国 · 科隆　建筑面积：20 000 m^2
建筑设计：德国 sic 建筑设计责任有限公司
建筑师：麦克尔 · 努因道夫

LOCATION: Cologne, Germany　BUILDING AREA: 20,000 m^2
DESIGN CORPORATION: Sic Architekten Gmbh
ARCHITECT: Michael Neuendorff

本建筑包含一栋 24 层高的楼房以及一栋 4 层高的行政大楼（地下有三层车库）。因为这是一个翻新改建工程，所以在进行整体设计和施工的过程的同时，还必须保证电台正常运营以及传送节目。除了对原有建筑进行整体翻新和布局，还新建了一个员工餐厅、一个内部庭院以及一个主传送中心。

The building contains a 24-storey building and a four-storey administrative building (beneath which is 3-storey garage). Because this is a renovation project, during the overall design and construction we must ensure the normal operation of radio and program transmission. In addition to an overall renovation of the original construction and layout, we also created a new cafeteria, an internal courtyard and a major transmission center.

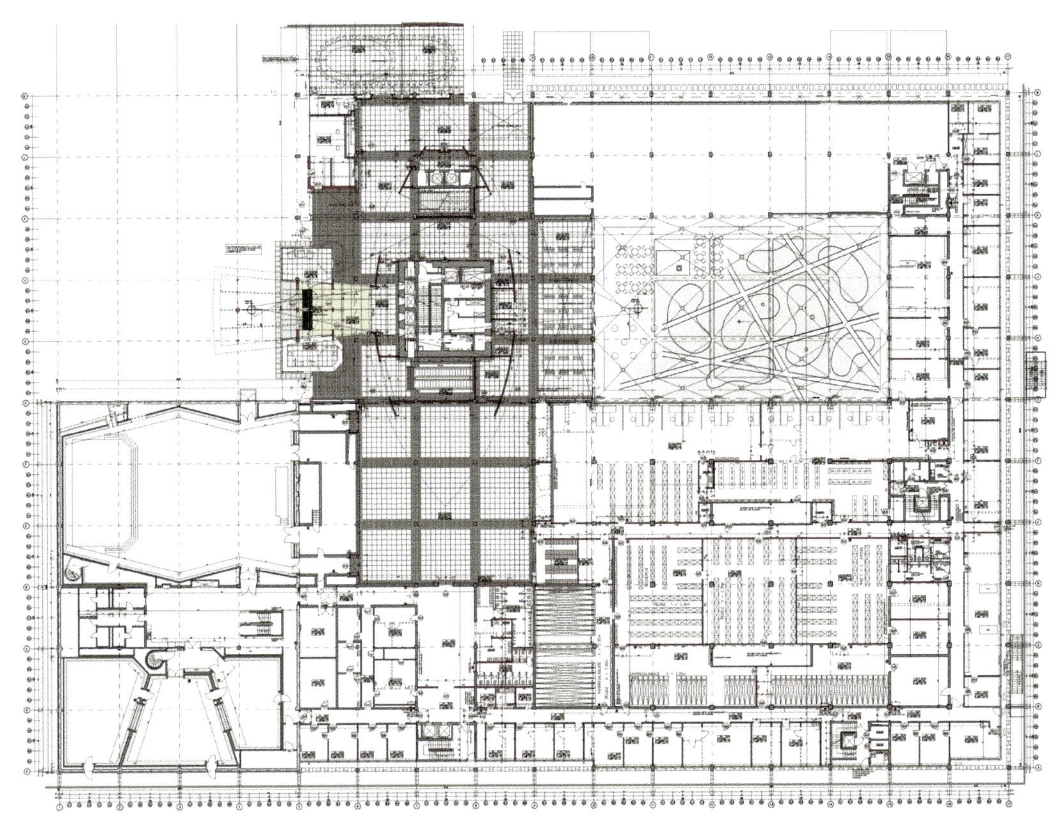

移动、游弋的建筑

A MOBILE, CRUISING CONSTRUCTION

项目地点：德国 · 勃兰登堡市　建筑面积：50 m^2（每栋）
建筑设计：德国 sic 建筑设计责任有限公司
建筑师：史蒂芬 · 亚斯伯

LOCATION: Brandenburg, Germany
AREA BUILDING: 50 m^2 (per building)
DESIGN CORPORATION: Sic Architekten Gmbh
ARCHITECT: Uwe Stephan-Jasper

勃兰登堡市的劳济茨湖泊地区将由一个工业景观带发展成为一个自然休闲的景观带。

设计的主要构思是将设计本身转换到对事物的塑造上去。

这个酷似水滴形状以及象征鸡蛋外壳的理想设计，将建筑本身转变成了一个“文化景观”。

建筑的形态适应周围的环境，并具备一系列的优点。它提出了一个新的设计尺度，为劳济茨湖泊景观增添一个恰当的独特功能；它非常适合于当地的气候条件，同时也显示了可持续性生活的思想和亲近了解大自然的概念；高科技的设备能够为建筑达到尽可能广泛的自己自足的补充，这也是创作思想的决策依据以及方案确定的强有力支持。作为“劳济茨湖泊地区”整体改造的一部分，对于漂浮建筑的更好理解将会有助于确定湖区总体规划的基准，而这个基准也将会继续补充直至完善。

The oberlausitz District Brandenburg City will be developed into a natural leisure landscape from an industrial landscape.

The main design concept is to convert design itself to the shaping up of things.

This water droplet shape and the the ideal design symbolizing egg shell transformed the building itself into a result of cultural landscape.

The shape of the construction which adapts to the surrounding environmentand has a range of advantages: it proposes a new design standards, and adds a unique and appropriate function to the landscape of oberlausitz; it's quite suitable for local climate and in the meantime shows the concepts of sustainability of life and close understanding of the nature; high-tech equipment can achieve self-contained complement as much as possible for the building, which is also the base of the creation and strong support for the program establishment. As an integral part of the renovation of oberlausitz district a better understanding of floating architecture will help to ascertain the base of the overall plan of the lake area, which will continue to be complemented to reach the perfection.

德绍大师之家

THE MASTER'S HOME DESSAU

项目地点：德国 · 德绍　占地面积：6000 m²　建筑面积：1000 m²
建筑设计：德国 sic 建筑设计责任有限公司
建筑师：杨 · 古特木特，凯南 - 多马斯

LOCATION: Dessau, Germany
SITE AREA:6000 m²　BUILDING AREA: 1000 m²
DESIGN CORPORATION: Sic Architekten Gmbh
ARCHITECTS: Jan Gutermuth, Kennan – Domas

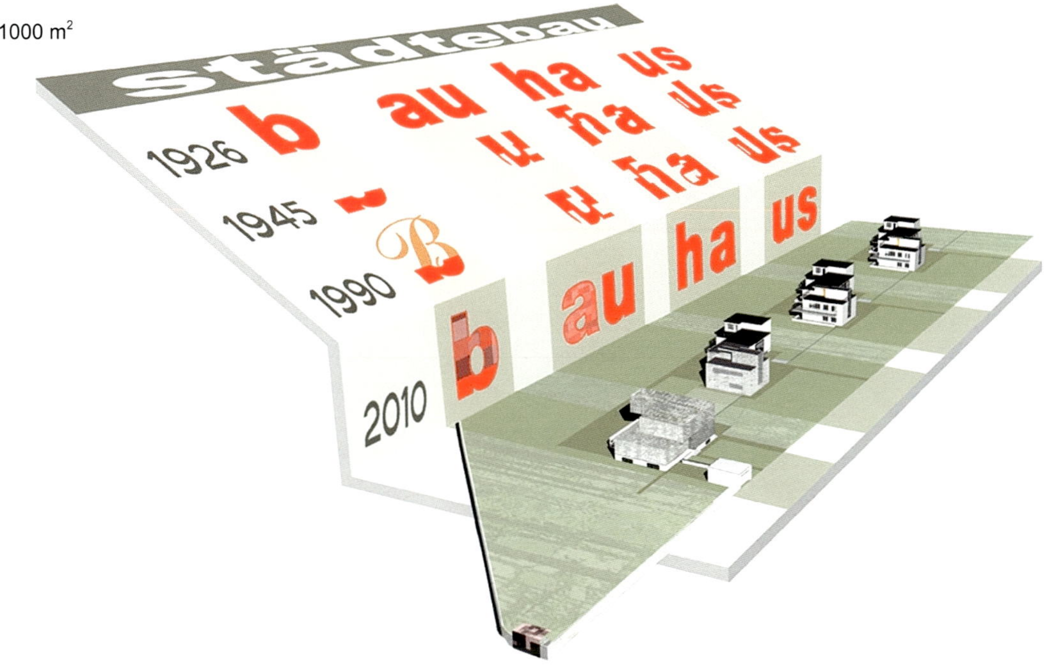

在魏玛和德绍的包豪斯建筑是现代建筑史上重要的见证，并于 1996 年列入世界文化遗产名录。

到目前为止德绍的包豪斯以及大师之家的三栋房屋需要进行维修和改造。

通过组织一次两个阶段的建筑设计竞赛将对在第二次世界大战中被炸毁的整个大师住宅区进行城市规划层面上的评估和修复，另外拓展和加强其在博物馆性质、科学研究性质和旅游性质的功能。

Bauhaus architectures in Weimar and Dessau's are the witnesses of the history of modern architecture, and were registered as world cultural heritage sites in 1996. So far, the Bauhaus Dessau and the three houses of Master's home need conservation and reconstruction.By organizing a two-stage architectural competition we'll give assessment and repair to the entire master residential area that have been destroyed in World War II on the urban planning level, and moreover strengthen its function in museum, tourism and scientific research.

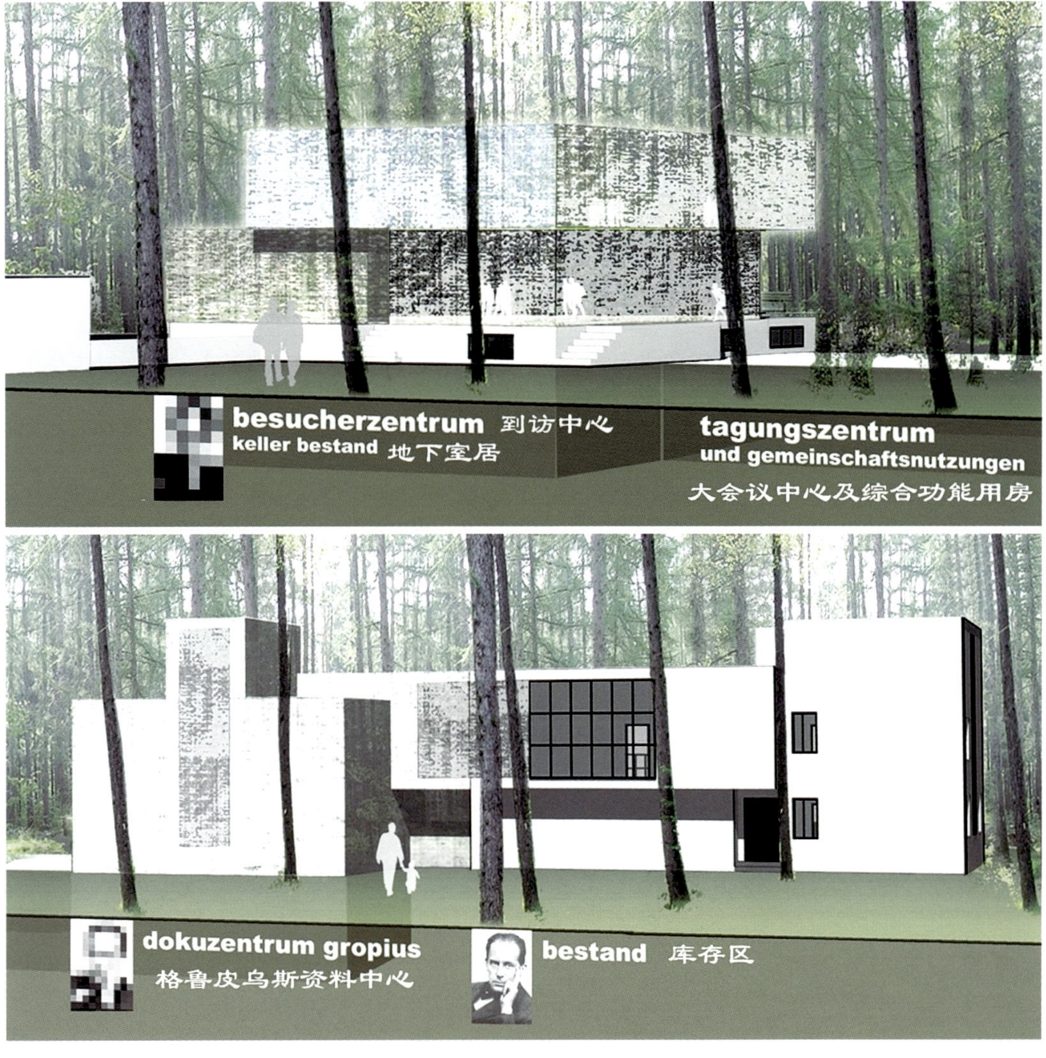

上海外滩鼎鼎金融大厦外立面设计

SHANGHAI BUND FRIEDRICH FINANCE BUILDING FACADE DESIGN

项目地点：中国 · 上海
建筑设计：德国 sic 建筑设计责任有限公司
建筑师：杨·古特木特

LOCATION: Shanghai, China
DESIGN CORPORATION: Sic Architekten Gmbh
ARCHITECT: Jan Gutermuth

拥有悠久历史的上海外滩将以一栋新的地标性建筑即鼎鼎金融大厦而得到新的诠释。

鼎鼎金融大厦的地理位置优越，位于外滩的南面收口上，紧邻古城及豫园，北面是上海西外滩中心与摩天楼，东面是浦东金三角。在现代建筑形式中形成的三角地块，以及来自豫园及外滩历史建筑的影响，决定了该建筑是现代与历史的一个交合点。

鉴于以上的特征，设计的主题定位在“使上海城市发展规划与历史区域内的新建筑元素在这栋建筑上得到统一”。在基于采用建筑石材开窗洞的外墙立面的基础上，依靠周边的历史对其重新塑造，创立一个新的历史范畴。在中心部分设计一个现代的钢筋结构框架，凸显上海“新”的一面，同时在建筑的顶部，塑造一个玻璃制的蓓蕾，成为外滩的标志。

Shanghai Bund,with a long history,will accept a new interpretation by a new landmark - Friedrich Finance Building.

Friedrich Finance Building is strategically located in the southern convergence, close to Old Town and Yuyuan Garden. The north to it is the west of Shanghai Bund Centre and the skyscraper, and the east to it is the Pudong Golden Triangle. The triangular block formed in the modern architecture and the impact of the Yuyuan Garden and the historic buildings of the Bund ensure its position as a merging point of modern and history.

In view of the above features the design theme was fixed as "to harmoniously unite the Shanghai urban development planning and the new construction elements within the historic region in this building." Based on the use of stone window in the external facade and reshaping the historic surrounding a new historical context is created. In the central part is a modern steel structure framework, highlighting the newness of Shanghai, while the glass bud on the top of the building becomes a symbol of the Bund.

深圳市民中心广场改进规划及水晶岛地标概念设计

IMPROVEMENT PLAN OF SHENZHEN CIVIC CENTER PLAZA AND THE CONCEPTUAL DESIGN OF THE CRYSTAL ISLAND LANDMARK

项目地点：中国 · 深圳 占地面积：45 ha
建筑设计：德国 sic 建筑设计责任有限公司
建筑师：史迪文

LOCATION: Shenzhen, China SITE AREA: 45 ha
DESIGN CORPORATION: Sic Architekten Gmbh
ARCHITECT: Stephanus Joseph Klabbers

让该区域成为南北整条轴线的核心区域和市民停留区。
通观整条轴线，从文化板块到会展板块，缺乏让人停留的活动空间，单纯的人行轴线并不能满足人的活动需求，所以，有必要在该区域营造一个真正以人为本的空间，在这里，轴线不再仅仅是线性的，而是放射性的，将不同的功能空间向四周放射，让人进行自由沟通和活动。

Make the region be the core region and a public stay area of the north-south axis.
Giving a whole view of the entire axis, from the culture section to the exhibition section, there's a no space for people to stay. Just a pedestrian axis can not meet people's needs, therefore it's necessary to create a truly people-oriented space, where the axis is no longer just a linear, but radioactive to provide different functions in all sides so that people can communicate and move freely.

通往 JAHNPLATZ FORUM 入口的屋顶

THE ROOF LEADING TO THE ENTRANCE OF JAHNPLATZ FORUM

项目地点：德国 · 比勒费尔德

建筑设计：德国 sic 建筑设计责任有限公司

建筑师：杨 · 古特木特

LOCATION: Bielefeld , Germany

DESIGN CORPORATION: Sic Architekten Gmbh

ARCHITECT: Jan Gutermuth

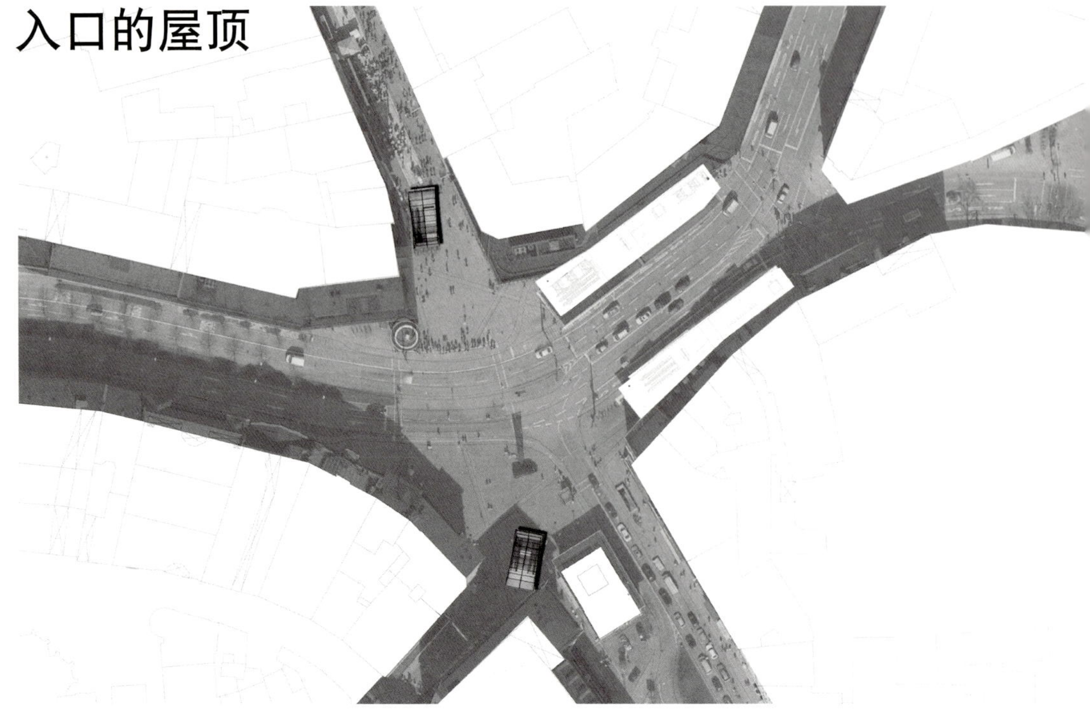

城市空间：该方案为 Jahnplatz Forum 的入口设计了两个完全透明一致的钢与玻璃的结合体，这就是在 Jahnplatz Forum 前所缺乏的一个醒目的城市空间。

Jahnplatz Forum 通过自身严谨而不张扬的建筑形态表达与现有的城市环境和谐共生，与 Jahnplatz Forum 周边地区的那些建筑互不干扰。

游客导向：通过一个类似于“洞门”的入口，访客可由此进入商场。三角形的建筑形态，为访客提供了非常醒目的指引标示，指引他们通往地下区域。

广告：在入口屋顶结构的底部，有一个非常有创意的多媒体屏幕。它能在一方面保证建筑的透明感，另一方面则能够在这样的一个通道处，提供个性化、多样性的广告空间，由建筑的内部向外部投射影像。在一些特别的场合，通过不同的影像或者灯光幻彩，这个屏幕还能够起到渲染气氛的作用，灵活多变的应用方式，无须考虑其结构或者周围环境的约束。

从现在的效果图可以看出，通道处的多媒体屏幕上投影的信息正展示出“游客导向”的建筑核心概念。

Urban Space: The program was designed two same completely transparent combinations of steel and glass for the Jahnplatz Forum entrance, which is a striking urban space that Jahnplatz Forum lacked. Through harmonious co-existence of its rigorous and low-key architectural expression and the existing urban environment Jahnplatz Forum brings no interference to those surrounding buildings.

Tourist guide: through a "Portal" entrance, visitors enter the mall. Triangular building form provides a very eye-catching label, guiding them to the underground area.

Advertizing: at the bottom of the roof structure of the entrance is a very creative multimedia screen. It provides a sense of transparency, and also a personalized and diversified advertising space out of such a channel, projecting images from the interior to the outside of the construction.In some special occasions the colors change by different images or lights, so the screen can also play a role in rendering the atmosphere with the flexible application methods without regard paid to its structure or environment constraints.

From the current design sketch we can see that multimedia screen in the channel is projecting "tourists oriented" , the core architectural concept.

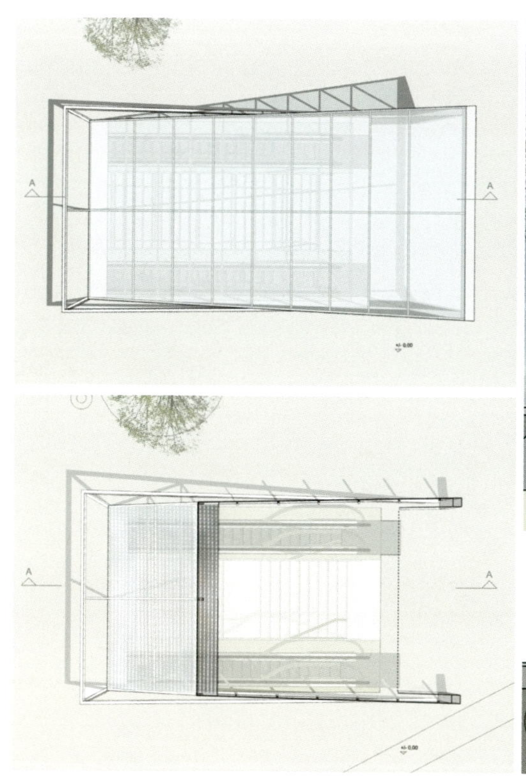

中国海油深圳新大厦

NEW BUILDINGS OF CNOOC IN SHENZHEN

项目地点：中国 · 深圳　占地面积：12 712.51 m^2　建筑面积：200 000 m^2
建筑设计：德国 sic 建筑设计责任有限公司
建筑师：史迪文

LOCATION: Shenzhen, China　SITE AREA: 12,712.51 m^2　BUILDING AREA: 200,000 m^2
DESIGN CORPORATION: Sic Architekten Gmbh
ARCHITECT: Stephanus Joseph Klabbers

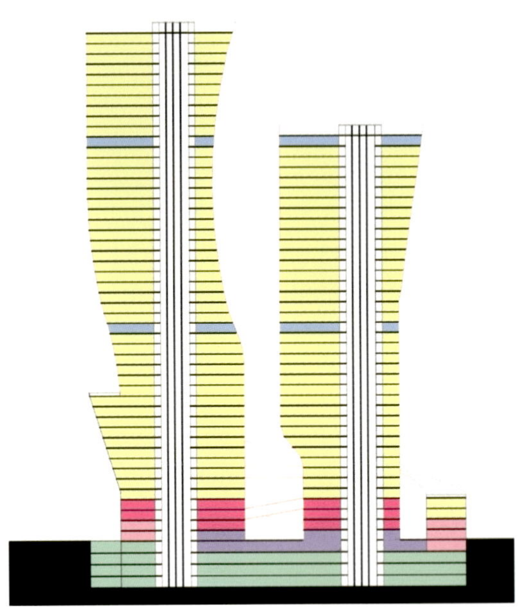

整体概念来源于海上采油的钻井平台的形象。两个火炬由大地拥抱，象征着可持续性与自然能源之间的特殊纽带。通过这样的联系，周围的景观与建筑之间的联系变得更亲近，在视觉上没有被分隔开来的效果。裙房部分带有“向上爬升”的趋势，而在建筑内部的空间则有反方向的旋转，交通经过商铺街往地下空间延续，终止于地铁站点。

螺旋飘带接地的部分（“屋顶”与地面相接处）可设置为供人休憩处，视为公共空间。在“屋顶”的较高处，可作为室内空间的延续，成为私有的室外场地。

The whole design concept comes from the images of offshore oil drilling platform. The two torches embraced by the land symbolize the special bond between the sustainability and natural energy. Through this connection, the intimacy between surrounding landscapes and the architecture becomes strengthened without the visual effect of being separate. Annex has an upward trend, while inside the space of the building shows a counterclockwise rotation. The traffic extends to the undergroud space through the shoping street and terminates at the MTR site. The point where the spiral ribbon is connected to the ground(the "roof" reaches the ground) can be set as a public rest area. The higher part of the "roof" can be used as a continuation of interior space and becomes a private outdoor space.

SINO-SUN ARCHITECTS & ENGINEERS

北京东方华太建筑设计工程有限责任公司

公司合伙人及主要设计师

- 徐建伟　Xu Jianwei

1981-1985年上海同济大学建筑系，学士
1993-1995年比利时LEUVEN大学建筑学，硕士
1985-1993年中国有色工程设计研究院
1993-1995年比利时DRIC设计事务所
1996年至今北京东方华太建筑设计工程有限责任公司

东方华太是由一批海外留学人员主持的国家甲级建筑设计单位，经过10多年的辛勤耕耘，已发展成具有良好品牌的优秀设计团队，也成为在国内建筑设计界有一定影响的设计公司。公司以建筑设计为主，同时涉及规划、景观及室内设计。

设计理念

东方华太坚持艺术与技术相结合的创作设计理念，尊重环境、技术和经济等客观条件，追求建立在理性基础上的个性和创新。

东方华太遵循"质量第一、服务至上"的原则，强调质量是设计工作的灵魂，服务是设计工作的保证。通过有效的管理体系、创新的设计手法来实现项目品质和价值的最大化。

东方华太追求：创新+质量+服务。

学术交流

公司在进行技术创新的同时，也注重学术研究与交流，积极鼓励员工在建筑学术杂志上发表专业论文，对设计工作实践及时进行总结和提高，用新的设计理念来指导自身的实际工作。公司积极参与各种建筑论坛和展览，加强业界的学术交流，推动国内建筑设计水平的共同提高。公司员工在各专业杂志上发表的学术文章有《建筑形体与设计创意》、《北师大珠海校区设计》、《"过渡"建筑》、《船与海的联想》、《空间·色彩和质感的艺术》、《与城市共生》、《再现童话故事》、《简约建筑的人性化》、《走中国特色的建筑设计之路》等。

国际合作

东方华太开展广泛的国际设计合作，在合作过程中，学习国际先进的设计理念和手法，不断提高自身的设计水平和质量。在与日本山本理显设计工场合作设计建外SOHO的过程中，感受到简约建筑的人性化；通过与法国AP3设计公司合作设计的中国银行金融研修院项目，了解了建筑"过渡"性理念；与加拿大ABCP设计公司合作设计的南昌浙大科技园区，共同创造了科技园建筑的亲自然品质。到目前为止，其合作的境外设计公司还有澳大利亚DEM设计公司等。

Sino-Sun Architects & Engineers is a first-grade state architectural designer under the leadership by a group of Chinese students ever studied abroad. With the hard work of over ten years, the company now becomes an excellent design team enjoying high reputation, and plays an important role in the domestic architectural design field. The company is mainly engaged in architectural design, and concurrently deals in planning, landscape and interior design.

Design Concept

Sino-Sun architects & Engineers sticks to the design concept of combining art with technology. It focuses on such objective conditions as environment, technology and economy, and is in pursuit of personality and innovation on the basis of rationality.

Sino-Sun Architects & Engineers follows up the principle of "quality and service first", which means that quality is the soul of design and service is the guarantee of design. The company optimizes the quality and maximizes the value of designed works through effective management system and innovative design methods.

The aim of Sino-Sun Architects & Engineers: innovation + quality + service.

Academic Communication

The company pays attention to academic research and communication while exploring technical innovation, and encourages its staff to publish professional thesis on architectural academic magazines actively. The company also makes a timely summary and tries to improve the design practice, guiding its own practice with new design concept. The company takes an active part in various architectural forums and exhibitions, with the purposes of strengthening academic communication and promoting the improvement of domestic architectural design field. The academic articles that are written by the staff of the company and have been published on various professional magazines include: Architectural Form and Design Creation, Design for Beijing Normal University Zhuhai Branch, "Transitional Buildings", Legend of Boat and Sea, Arts of Space, Color and Texture, Growing with the City, Fairy Tales Reappear, Personalization of Simple Architecture and Architectural Design with Chinese Characteristics etc.

International Cooperation

Sino-Sun Architects & Engineers has been involved in wide international design cooperation, during which it has learned international advanced design concepts and methods so as to improve its design level and quality constantly. During the process of designing Jianwai SOHO jointly with designer riken yamamoto, the company has felt the personalization of simple architecture. Besides, the company has understood the concept of transitivity of architecture when it co-designed the project of Finance Research Institute of Bank of China. When it cooperated with Canada ABCP Design Company in the design of Science and Technology Park for Zhejiang University in Nanchang, they added close-nature property to its architecture. In addition, the company also cooperated with Australia DEM Design Office and Taiwan Li Zuyuan Design Office etc.

清华坊

TSINGHUA FANG

项目地点：中国 · 中山　用地面积：130 327.3 m^2　建筑面积：51 136.5 m^2

建筑设计：北京东方华太建筑设计工程有限责任公司

合作单位：广州集美组室内设计工程有限公司

LOCATION: Zhongshan, China　SITE AREA: 130,327.3 m^2　BUILDING AREA: 51,136.5 m^2

DESIGN CORPORATION: Sino-Sun Architects & Engineers

PARTNER:Guangzhou Newsdays Interior Design & Construction Co.,Ltd.

清华坊以现代中式为设计风格，是现代化的中国文化的居住体，功能布局以内向性的庭院为居住核心。

本地块依山傍水，绿树成荫，鸟语花香，拥有得天独厚的自然环境。

该设计是一低层低密度的高尚住宅区，均为联排式，在每户的底层均设有独立的车库和私家花园，户主可以根据个人的喜好对私家花园进行个性化设计，与整个小区的绿化景观和中心湖泊相呼应，使人感到犹如住在公园中，充分体现了“以人为本”的设计理念。

处处有庭院、绿地、树，为人们提供一个安静、私密、清洁、美观、文明的居住与交流的环境。

蘭畦

Tsinghua Fang designed in modern Chinese style is a modernized Chinese cultural living body.The courtyard functions as the residence core in the whole layout.

It stands among the clear mountains and rivers where birds sing among lush trees and flowers in a unique natural environment.

This is a low-rise highly densified residential area. They are all townhouses. At the bottom of each house is equipped with a private garage and a private garden which can be individualized on the basis of your favor. These gardens echo with the central lake and the green landscape of the district making you feel as if living in the park, which fully reflects the "people-oriented" design concept.

The gardens, green space and trees everywhere form a beautiful, tranquil and civilized living and communication environment.

SOM自1936年成立以来已经在世界上50多个国家完成了各种类型项目的设计。SOM因在很紧的进度和预算内成功地完成具有挑战性和复杂的项目而闻名。因为在各种规模和类型项目上的成功记录，客户可以相信公司是在聘请经受过系统培训、具有丰富经验的专业人士在完成公司的项目。

在一个史无前例的决定中，由于SOM杰出的建筑设计，美国建筑师协会再度授予SOM该协会最高荣誉1996年“最佳建筑设计事务所奖”。1962年，SOM成为第一个“最佳建筑设计事务所奖”得主，是唯一两次获此殊荣的建筑设计事务所。SOM对优秀设计的不懈追求使得SOM完成了众多具有想象力的、实用的建筑。自成立以来，SOM因为其高质量和新颖的设计获得了1400多个奖项，其中的125项是在过去的10年中获得的。

SOM完成的各种类型及各种规模的项目包括办公楼、银行、商业建筑、工业建筑、教育建筑、宾馆、会展中心、博物馆、影剧院、医院、实验室、机场、火车站、地铁、娱乐建筑、宗教建筑和住宅等。室内设计和城市规划也是SOM专业领域中不可缺少的一部分，大范围的城市规划项目包括改造规划、影响评估和交通规划分析。

SOM分部

SOM在世界各地的分部为当地的项目提供服务，也为全国性和国际性的客户提供专业设计服务和资源。SOM的事务所分别于芝加哥、纽约、旧金山、华盛顿、伦敦和上海设立分部。

综合性事务所

SOM开创了将各设计专业融为一体的先河，为客户提供完全的、综合性的服务。这样的综合性使公司发展了跨越建筑设计和工程设计界限的创新的技术方法。SOM专业服务的实践使公司可以通过一个共同的、整合的管理体系提供单源的设计责任。

SOM的综合性服务领域包括建筑设计、结构工程、机电工程、土木工程、可持续性设计、总体规划、城市设计、室内设计、开发策划、造型设计和标志设计。其可持续性设计是深远的并跨越所有上述专业。

该事务所除了内部设置的这些专业以外，也擅长和其他事务所进行合作。SOM在与其他建筑师、业主的顾问以及自己聘请的顾问专家合作方面具有悠久的历史。从这些合作中产生的交流和对话加强和丰富了每个项目的构思和想法。SOM对所有团队成员的要求是对优秀设计和客户服务的共同承诺。

Ever since SOM was set up in 1936, it has completed designs of various types of projects in more than 50 countries in the world. SOM is famous for completing challengeable and complicated projects successfully within tight schedule and budget. Due to its successful records in all sizes and types of projects, customer may believe that the company is employing a systematically trained and experienced professional to perform the project of the company.

In a unprecedented decision, American Institute of Architects (AIA) again gave "Premium Architectural Design Office Award", the highest honor of the institute to SOM for its outstanding design. In 1962, SOM became the first winner of "Premium Architectural Design Office Award" and is the only architectural design office that has won the laurel twice. The indefatigable pursuance of excellent designs makes SOM achieved numerous buildings with imagination and practicability. SOM has won more than 1400 awards for its high quality and novel designs since its founding, of which 125 have been won in the past ten years.

The various types and sized of projects performed by SOM include: office buildings, banks, commercial buildings, industrial buildings, educational buildings, hotels, conference and exhibition centers, museums, theatres, hospitals, laboratories, airports, railway stations, underground railways, recreational buildings, religious buildings and residential buildings. Interior design and urban planning are also a part and parcel of SOM's professional field. Large urban planning projects include modification planning, impact evaluation and traffic planning analysis.

SOM Branch Office

SOM branch offices all over the world provide service for local projects and also provide professional design service and resource for nationwide and international customers. SOM has its offices in Chicago, New York, San Francisco, Washington, London and Shanghai.

Synthetic Office

SOM has initiated the very first example in integrating all design specialties into one body so as to provide the customers with total and synthetic service. Such a synthetics makes the company developed the innovative techniques that have surpassed the demarcation lines between architectural design and engineering design. The practice of SOM's professional service makes the company able to provide monophyletic design responsibility through a common and integral management system.

The synthetic service field of SOM includes architectural design, structural engineering, electromechanical engineering, civil engineering, sustainable design, general planning, urban design, interior design, development planning, style design and logo design. Its sustainable design is far-reaching and crossing over all the above specialties.

The office is good at cooperating with other offices besides our internal specialties. SOM has a long history in cooperating with other architects, customer consultants and the consultant specialists invited by ourselves. The exchanges and dialogues produced in such a cooperation have enhanced and enriched the conception and notion of each project. SOM requires that all the members of the team should provide a common commitment for excellent design and customer service.

新保利大厦

NEW POLY PLAZA

项目地点：中国 · 北京　基地面积：65 000 m^2　项目面积：100 000 m^2
建筑设计：SOM　楼层数目：24 层　建筑高度：110 m

LOCATION: Beijing, China　SITE AREA: 65,000 m^2　PROJECT AREA: 100,000 m^2
DESIGN CORPORATION: SOM　NUMBER OF STORIES: 24　BUILDING HEIGHT: 110 m

新保利大厦的设计异常独特，大厦平面呈三角形，24 层的办公空间从两侧将 90 m 高的中庭围合其中，余下一侧则是世界最大的索网支撑玻璃幕墙，8 层高的悬挂式“灯笼”是博物馆空间，其依靠四条平行钢绞斜拉索悬于中庭之内。

The unique design incorporates 24-storey office space built around a 90-meter-tall atrium whose one side is enclosed by the world's largest cable-net-supported glass wall. The museum is an eight-storey hanging"lantern" suspended in the building atrium by means of four parallel strand bridge cables.

保利国际广场

POLY INTERNATIONAL PLAZA

项目地点：中国 · 广州 基地面积：57 565 m^2 项目面积：180 000 m^2
建筑设计：SOM 楼层数目：36 层 建筑高度：150 m

LOCATION: Guangzhou, China SITE AREA: 57,565 m^2 PROJECT AREA: 180,000 m^2
DESIGN CORPORATION: SOM NUMBER OF STORIES: 36 BUILDING HEIGHT: 150 m

集办公和零售于一体的综合性保利国际广场，其设计充分考虑了当地气候及毗邻珠江的关键地理位置。表现型结构和偏移式核心筒诠释了保利国际广场的高能效设计；结构龙骨支撑着大面积开放式楼板及玻璃核心筒，从建筑楼层内放眼望去，江岸美景一览无余。

This mixed-use office and retail complex makes full use of the local climate and the location significance of the Pearl River. The expressive structure and offset core define this energy-efficient design; the structural spine supports the large open floor plates and glass core, which capture dramatic views of the waterfront.

南京绿地广场

NANJING GREENLAND PLAZA

项目地点：中国 · 南京　基地面积：28 294 m^2　项目面积：308 000 m^2
建筑设计：SOM　楼层数目：66 层　建筑高度：450 m

LOCATION: Nanjing, China　SITE AREA: 28,294 m^2　PROJECT AREA: 308,000 m^2
DESIGN CORPORATION: SOM　NUMBER OF STORIES: 66　BUILDING HEIGHT: 450 m

绿地广场的三角形平面与其所处地块息息相关，大厦拥有极佳的景致视野，附近山色、湖景和历史建筑尽收眼底。大厦的递进攀升具有功能性，办公、酒店与零售功能位于地上，餐馆及一个公共观景台位于大厦顶部，其上冠以螺旋天线。

The greenland plaza's triangular form relates to its site, its tower providing stunning views of the nearby mountains, lake, and historic buildings. The tower's stepping is functional: office, hotel, and retail are located above the ground while restaurants and a public observatory are housed at the top of the tower, which is crowned with a spire.

郑州绿地广场

ZHENGZHOU GREENLAND PLAZA

项目地点：中国 · 郑州　基地面积：30 478 m^2　项目面积：240 169 m^2
建筑设计：SOM　楼层数目：56 层　建筑高度：279.30 m

LOCATION: Zhengzhou, China　SITE AREA: 30,478 m^2　PROJECT AREA: 240,169 m^2
DESIGN CORPORATION: SOM　NUMBER OF STORIES: 56　BUILDING HEIGHT: 279.30 m

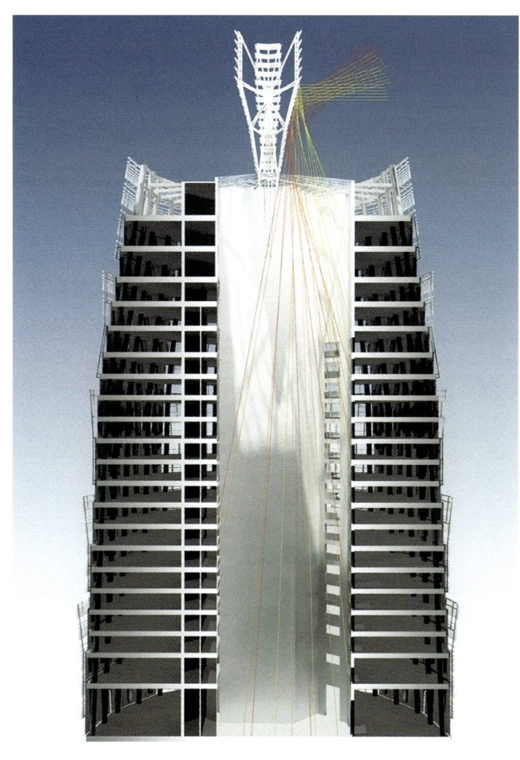

郑州绿地广场核心部分高层大厦的竞标获胜设计方案，结合先进的建筑设计、工程设计以及可持续设计，创造了一座56层的高能效建筑。SOM的设计通过前所未有的技术应用，满足不断变化的季节及环境供暖、制冷和照明需求。大厦功能布局包括酒店、办公、零售及娱乐设施，将造就活力四射的全天候“垂直都市”。

The competition-winning plan for the high-rise at the heart of Zhengzhou Greenland Plaza combines advanced architecture, engineering, and sustainable design to create an energy-efficient, 56-storey structure. The SOM proposal is unprecedented in its ability to use technology to effectively meet changing seasonal and environmental requirements for heating, cooling, and lighting. The program, which includes hotel, office, retail and entertainment, will constitute a vertical city that is active throughout the day.

温州鹿城广场——地标大厦

WENZHOU LUCHENG PLAZA— LANDMARK TOWER

项目地点：中国 · 温州 **基地面积**：132 325 m² **项目面积**：492 000 m²
建筑设计：SOM **楼层数目**：75 层 **建筑高度**：350 m

LOCATION: Wenzhou, China SITE AREA: 132,325 m² PROJECT AREA: 492,000 m²
DESIGN CORPORATION: SOM NUMBER OF STORIES: 75 BUILDING HEIGHT: 350 m

SOM 为地处瓯江入海口的温州市设计了一个广场及一座地标大厦。该开发将重塑温州的天际线，并为滨水带吸引其他新兴开发。SOM 以一贯的"功能性形式"手法设计该座综合性地标大厦，通过优雅的比例和不锈钢遮阳系统，将经典的柱式摩天楼提升到新的境界。不锈钢遮阳系统将控制热增益，并反射斑驳光影穿过玻璃幕墙。建筑中酒店、办公、以及住宅单元的功能布局直接造就了其形式——一座 75 层晶莹剔透的大厦——其顶部酒店空中大堂则扮演着"城市灯笼"的角色。

For the City of Wenzhou, at the mouth of China's Oujiang River, SOM designed a plaza and landmark tower that will reshape the city skyline and attract new development to the riverfront. The mixed-use tower draws on SOM's history of functional form, promoting the classic columnar skyscraper with elegant proportions and stainless steel sunshades that regulate heat gain, reflect dappled light across the glass curtain wall. The building's program of hotel, offices, and residential units led to its shape, a sleek 75-story tower capped by a hotel lobby that acts as a lantern for the city.

SPACEWORK ARCHITECTS

在场建筑

刘宏伟

钟文凯

公司合伙人及主要设计师

- 刘宏伟　Liu Hongwei

美国纽约州注册建筑师

2001年美国哈佛大学设计学院，建筑学硕士

1995年中国天津大学建筑学院，建筑学学士

- 钟文凯　Zhong Wenkai

美国纽约州注册建筑师

1998年美国加州大学伯克利分校环境设计学院，建筑学硕士

1997年美国莱斯大学建筑学院,建筑学学士

1995年艺术及艺术史学士

在场建筑2004年成立于美国纽约，是一家专业性建筑设计事务所。两位合伙人均先后在中国和美国接受建筑学教育，并在国际知名的建筑师事务所工作多年，积累了丰富的实践经验，熟知从项目策划、建筑方案到施工图及施工配合管理等各工程阶段。两位合伙人均为美国纽约州注册建筑师。

中国在现阶段正经历着政治、社会、经济、文化的深刻变革。前所未有的城市化进程为业主和建筑师们提供了宝贵的机会，但同时也提出了严峻的挑战。面对全球化与地域性、当代思想与传统价值、扩张与保护、质与量等众多无可回避的重要问题，我们有责任不断地进行开创性的思考，并努力参与实践。

建筑是营造场所的艺术。每一个建筑物的地点和内容都是独一无二的。建筑的意义最终必须通过建筑物的物质存在来实现，包括使用功能的合理、对场地和气候的回应、空间和光线、材料和质感，对施工质量和细部的追求。

技术进步不仅给我们的日常生活带来了巨大的效益，同时也造成了许多令人忧虑的后果。在新的世纪里，环境的可持续性问题正在日益引起人们的重新关注。针对每个设计任务的特殊性，我们将努力探寻在技术上创新、在经济上可行、同时又具有环境责任感的解决方案。

建筑是一项汇集众多领域的专业知识和群体智慧的工作，需要业主和使用者、建筑师、工程师、规划师、室内设计师、艺术家、建筑科技研究人员、施工单位、产品制造商等共同参与。成功的建筑只有通过所有这些专业人士之间的密切交流、默契配合、高效管理才得以实现。团队精神也同样体现在事务所内部分工合作、亲密无间的工作模式上。

Initially founded in New York in 2004, SPACEWORK Architects has now established its office in Beijing to fully engage the opportunities and challenges in China's dynamic process of urbanization. The two partners both received architectural education in China and the US. After graduation, they worked in internationally recognized US firms for extended periods of time, gaining valuable experiences in a variety of planning and building projects for institutional, academic, corporate, and individual clients. Both of them are licensed architects in New York State.

Today's China is undergoing fundamental political, social, economical, and cultural transformations. The unprecedented process of urbanization is offering tremendous opportunities as well as posing serious challenges to both clients and architects. Inevitably faced with such important issues as globalization versus locality, contemporary versus traditional values, expansion versus preservation, quantity versus quality, it is our responsibility to think creatively and practice rigorously.

Architecture is the art of place making. Each building is unique in its location and program. Architecture is ultimately fulfilled through the physical presence of building – appropriateness of usage, correspondence to site and climate, space and light, materials and textures, attentions to craftsmanship and details.

Technological advancements have brought great benefits as well as unwelcome consequences to the world we live in. At the beginning of the new century, we are witnessing a renewed consciousness to the long-term sustainability of our environment. Based on the specifics of each project, we strive for design solutions that are technologically innovative, economically feasible, and environmentally responsible.

Building is a collective activity that requires the knowledge and contribution of practitioners from diverse backgrounds: clients and users, architects, engineers, urban planners, interior designers, artists, building science researchers, contractors, and manufacturers. Successful buildings can only be realized through the intensive communication, vigorous coordination, and efficient management among all these professionals. The teamwork spirit is equally cultivated in the collaborative and collegiate working process within the office.

凤凰山峰顶建筑

FENGHUANG MOUNTAIN SUMMIT BUILDINGS

项目地点：中国 · 安吉　建筑面积：1375 m^2
建筑设计：在场建筑
建筑师：刘宏伟，张义海，孙婧祎

LOCATION: Anji, China　BUILDING AREA: 1375 m^2
DESIGN CORPORATION: SPACEWORK Architects
ARCHITECTS: Liu Hongwei, Zhang Yihai, Sun Jingyi

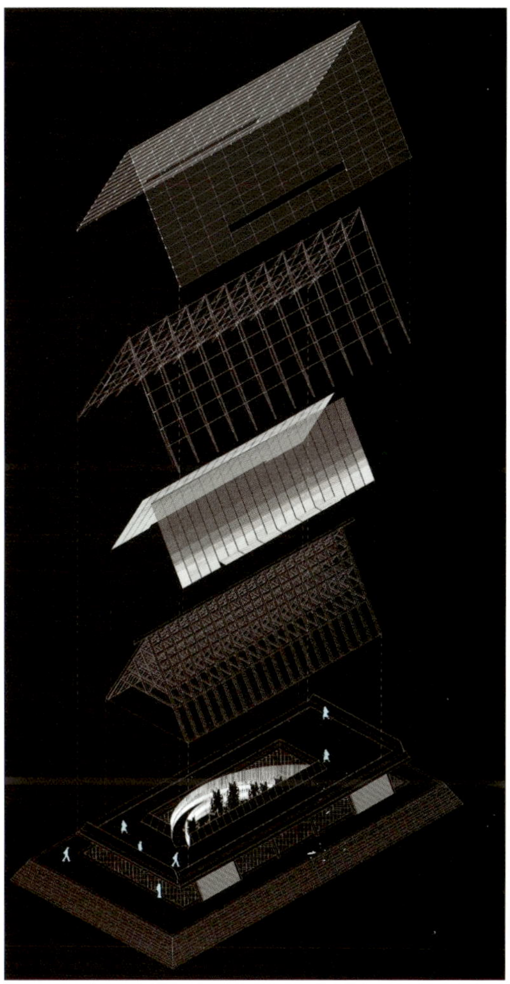

凤凰山峰顶建筑位于浙江省的竹乡安吉。作为城市的标志性景观建筑，设计的主旨在于重新塑造中国江南的地方精神，于主峰与次峰之间形成主次完整丰富，节奏轻重有致的景观体验：阁、栈道、桥、廊、台、亭等空间序列像一幅长卷渐次展开。建筑形式在时代精神与中国传统建筑文化之间寻求平衡，在外部形体、内部空间、结构形式以及构造细部方面保持着简练、清晰的现代设计逻辑，同时在这些层面上寻求诠释中国传统建筑与自然和谐共生的关系以及细腻优雅的构造品质。

The Fenghuang Mountain Summit Buildings are located in Anji, Zhejiang, a town known for its bamboo forests. As the town's symbolic scenic buildings, the design is intended to redefine the local spirit in an area south of the Yangtze River. Among the mountain's main and lower peaks are a series of scenic experiences that are rich and complementary to each other, with delicate rhythms and variations — the spatial sequence of pavilions, paths, bridges, galleries, terraces, and kiosks, unfolded like a long-scroll painting. The architectural language seeks to establish a balance between contemporary spirits and traditional Chinese architecture. The conciseness and clarity of modern design logic is rigorously pursued in the exterior forms, interior spaces, structure, and construction details, while at the meantime, the harmonious relationship between building and nature as well as the refinement and elegance in construction quality are reinterpreted from traditional Chinese architecture.

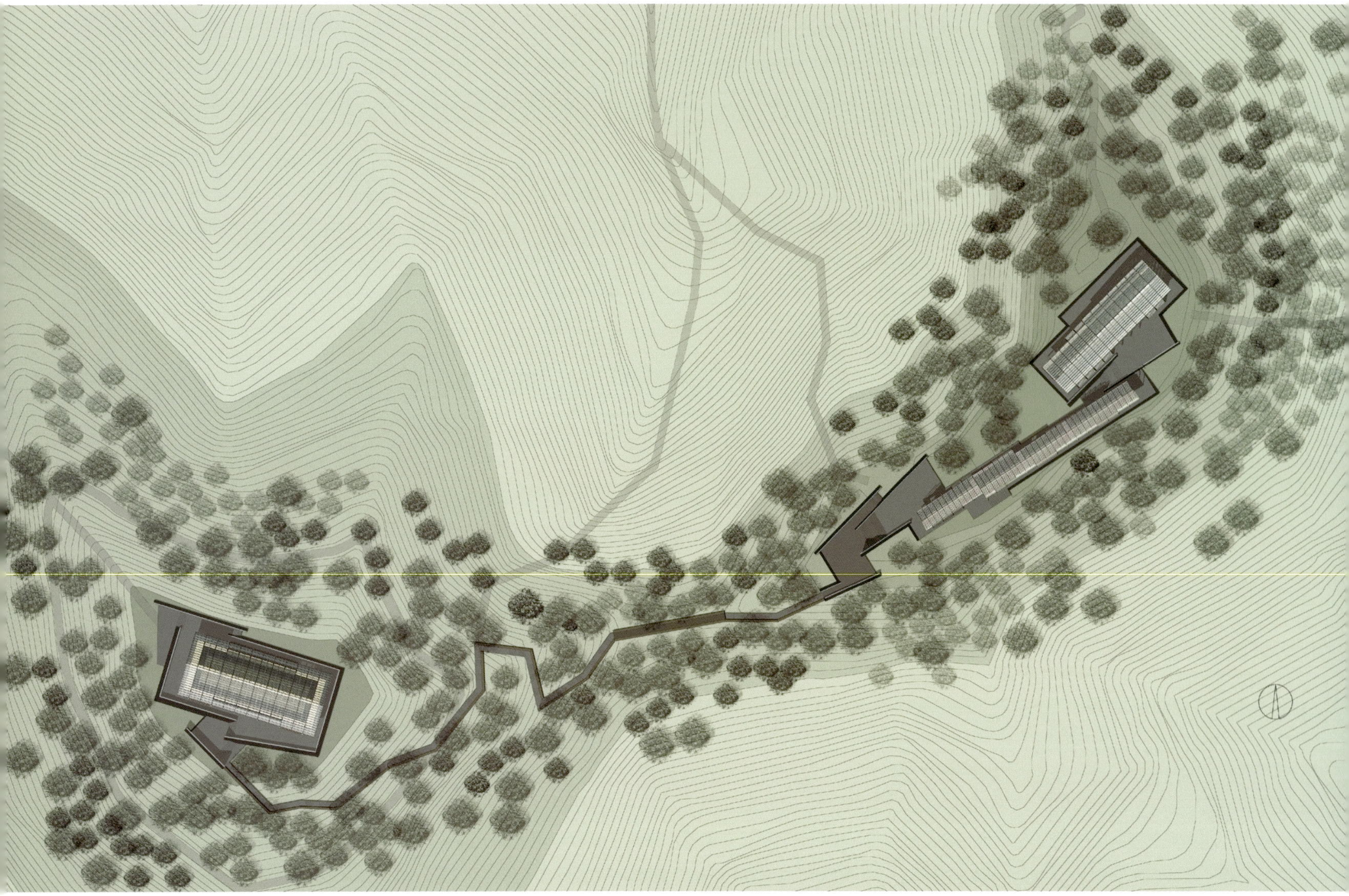

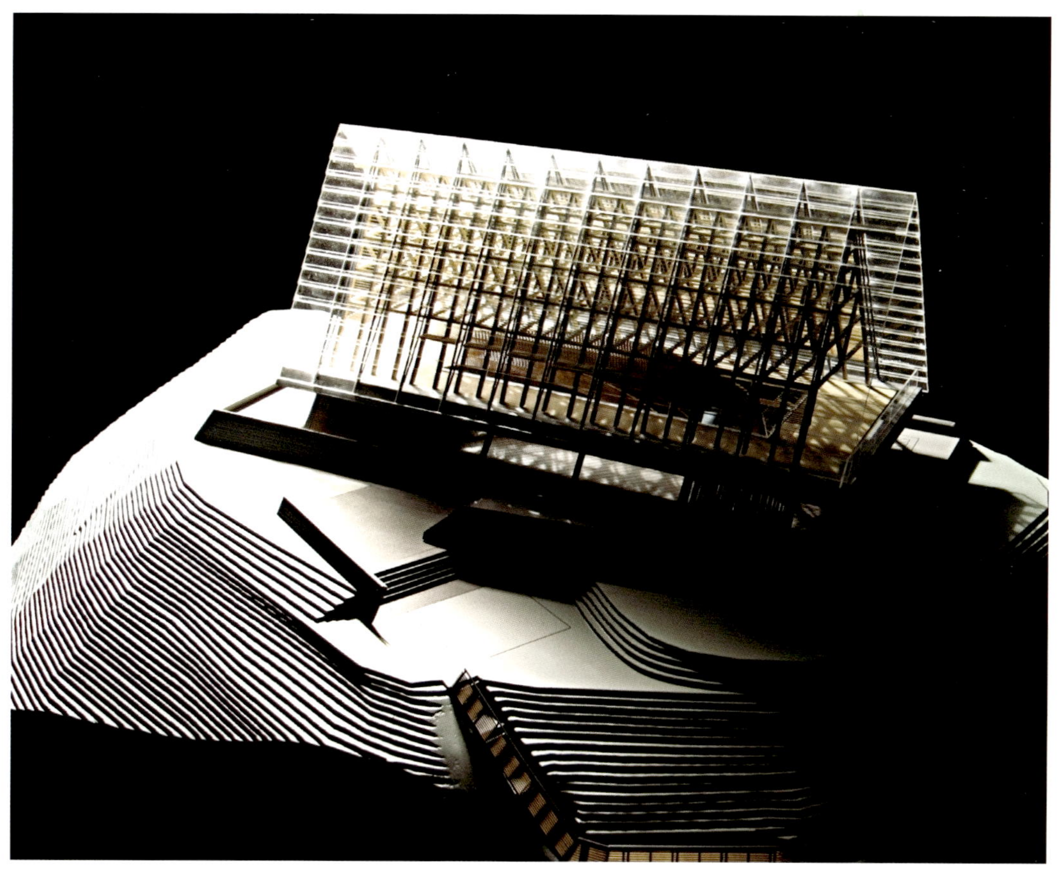

世博会万科馆方案

VANKE PAVILION FOR SHANGHAI EXPO

项目地点：中国·上海 **基地面积**：5000 m² **建筑面积**：3294 m²
建筑设计：在场建筑
建筑师：徐千禾，钟文凯，刘宏伟，高玉婷，孙婧祎，李然，刘治邑

LOCATION: Shanghai, China SITE AREA: 5000 m² BUILDING AREA: 3294 m²
DESIGN CORPORATION: SPACEWORK Architects
ARCHITECTS: Hsu Chienho, Zhong Wenkai, Liu Hongwei, Gao Yuting, Sun Jingyi, Li Ran, Liu Zhiyi

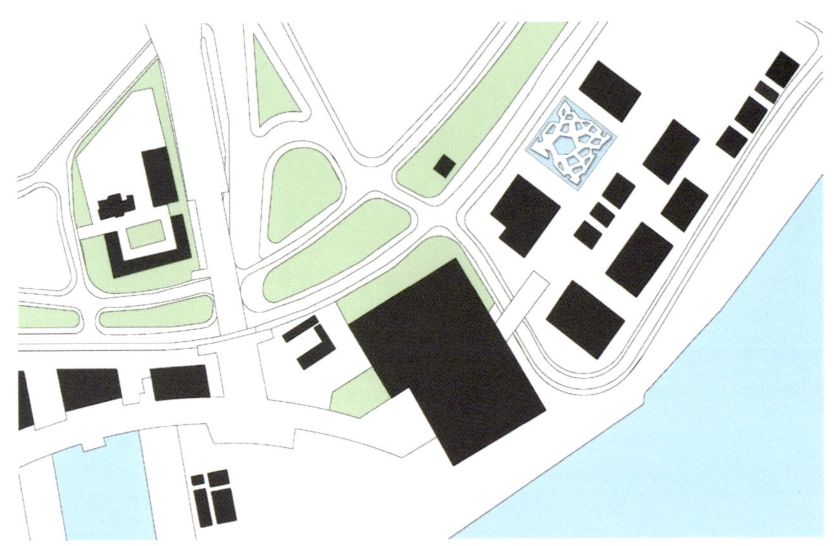

世博会是一场新思想相互交融、争相绽放的盛宴。它不仅仅是一个展示新技术新产品的场所，更是一个新思想得以传播并进入公众意识的场所。路易斯·康说：学校起源于一个人在树下……

世博会关注自然环境，倡导人类社会的“绿色生存”，希望将人类自身的生生不息融入自然界的造化之中。在都市生活中，人们渴望与大自然的重新邂逅，不管那是一片天空、一缕阳光，还是一阵风、一场雨……

万科是一个以为千万中国百姓建造家园为使命的企业。世博会中能否有那么一方所在，可以休息停留，既不在展厅中也不在去展厅的路上，能够遮蔽夏季炎热的阳光或躲过一场突如其来的雷雨？

一个新鲜而温暖的意象逐渐地呈现——树。建筑体量被托举到空中向四周伸展，地面的广场和景观则留给公众；人们从四面八方来到“树”下，没有大门或墙面的阻隔，可以自由穿越、停留；“树冠”所带来的荫翳、空气的对流、水景的降温效应，以及从二层的孔洞穿透下来的阳光创造了宜人的“微气候”；“树干”内是两层通高的主题展厅，并将参观者带离地面，进入“枝叶”；二层展厅如同四通八达的网，形成叶脉般的参观路径；墙面采用轻质薄膜结构，外面悬挂着一层由金属叶片制作的表皮，如轻盈的树叶一般随风飘动……

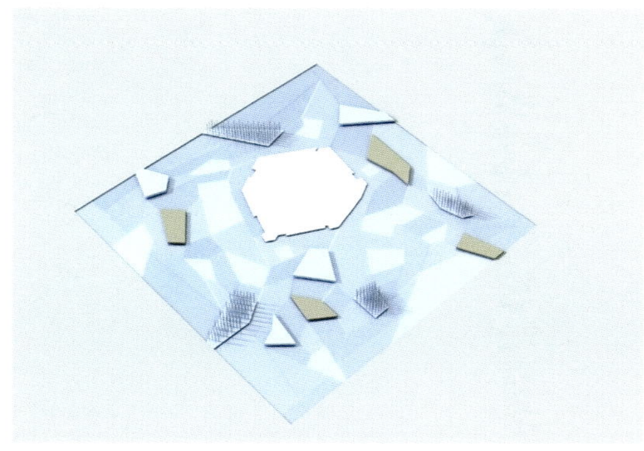

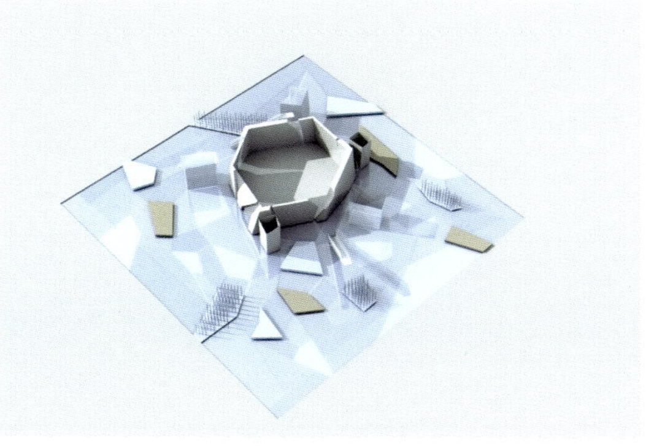

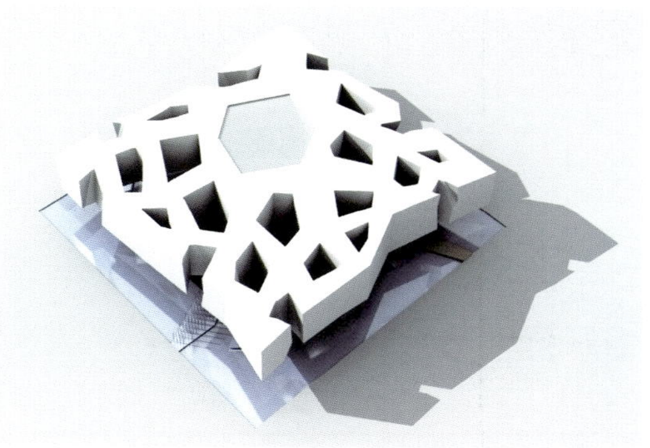

The World Expo is a festive arena where new ideas blossom and interact with each other. It is not simply a place where latest technologies and products are displayed, but also a place where new ideas spread out and enter into the public's consciousness. As Louis Kahn once said: "Schools began with a man under a tree…"

The World Expo focuses on the natural environment and advocates a "green" way of living in human society, with the anticipation that propagation of humankind can coexist with creations of nature. In urban life, people have been yearning for the chances to reconnect with nature, such as a piece of sky, a ray of sunlight, a gust of wind, or a moment of rain…

Vanke is an enterprise whose mission is to build homes for millions of ordinary Chinese. Within the vast Word Expo, could there be such a place where one could take a break or stay around – it's neither an exhibition hall nor on the way to an exhibition hall, where one could take shelter against the excruciating summer sun or a suddenlly coming thunderstorm?

A refreshing and heart-warming image gradually takes shape: the "tree". The building volume is lifted into the air and spread out, while the plaza and landscaping on the ground plane is reserved for the public; people arrive under the "tree" from all directions, and without the cumbersome walls and doors, they are free to pass through or stay; the shades provided by the "canopy", the cross ventilation of air, the cooling effects of water features, and sunlight penetrating through the porous second floor volume, combine to create a pleasant "micro-climate"; the "tree trunk" houses a double-height main exhibition hall, while carrying visitors off the ground and into the "branches"; second-floor exhibition spaces are interconnected like a network, generating a circulation pattern resembling those of leaf veins; walls are made of light-weight membrane structure, with a layer of hanging metal pieces forming an outer skin – they follow the movement of wind, like tree leaves…

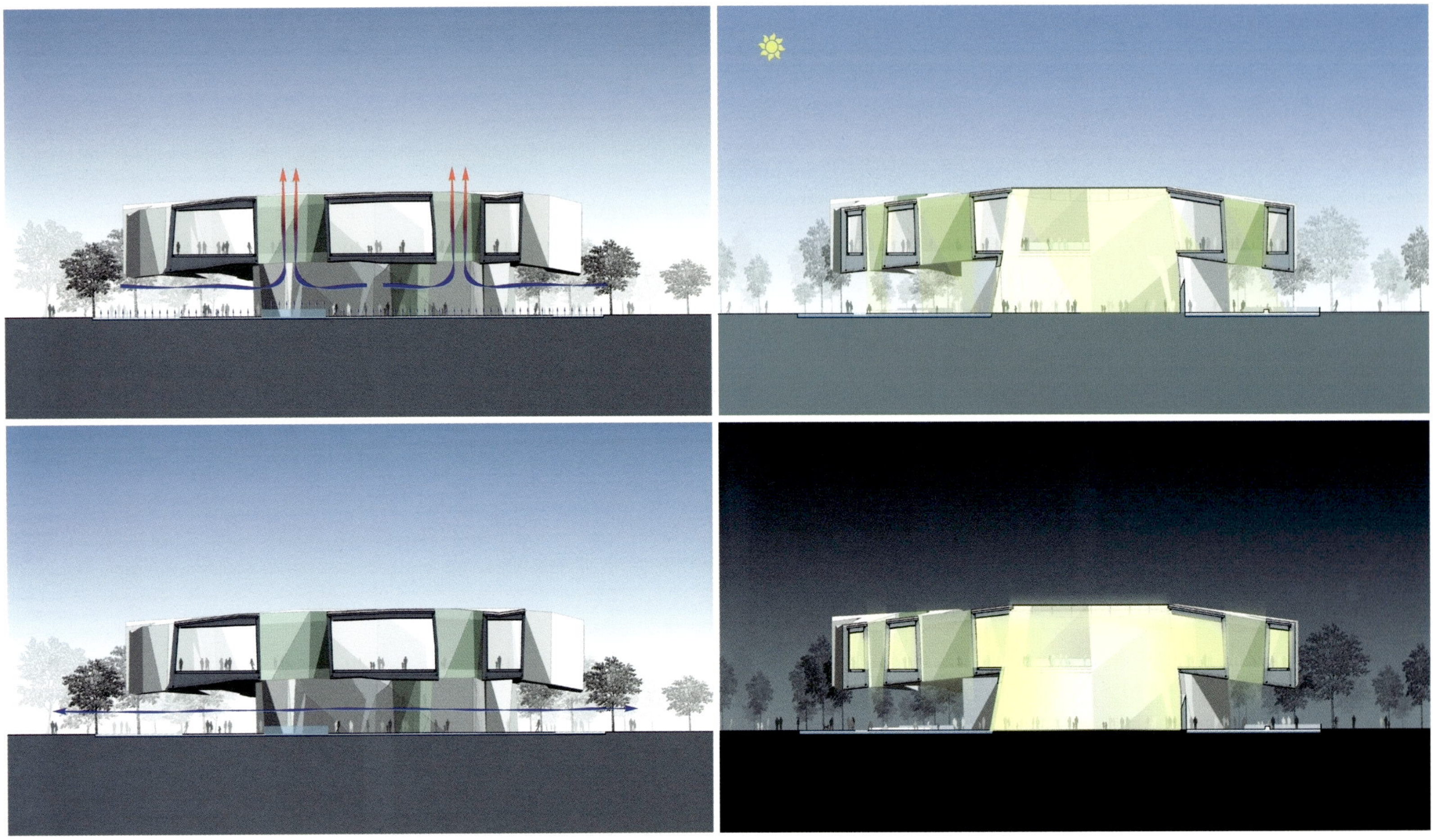

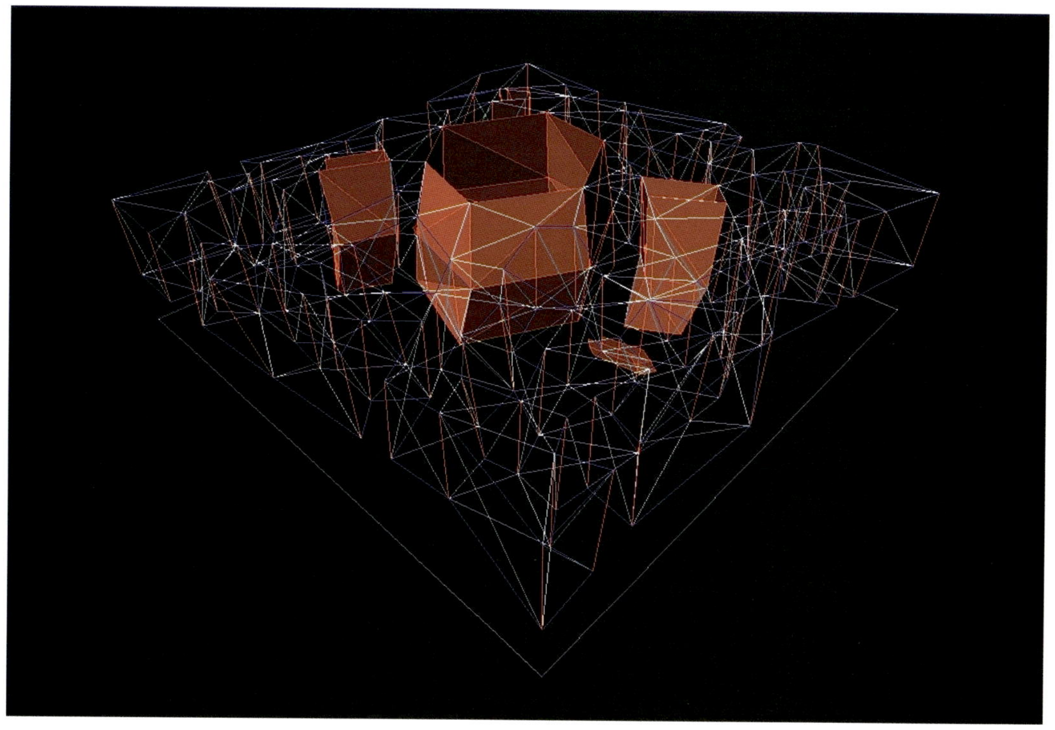

STEVEN HOLL ARCHITECTS

斯蒂文・霍尔建筑师事务所

斯蒂文・霍尔

李虎

斯蒂文·霍尔建筑师事务所，建筑与城市设计

斯蒂文·霍尔先生于1976年在纽约创立斯蒂文·霍尔建筑师事务所，并于1989年正式注册。斯蒂文·霍尔建筑师事务所以设计为导向，分别在纽约和北京设有办公室，员工约50人。该事务所在国际享有盛誉，以设计的高质量而多次获奖，包括九次美国建筑师协会荣誉大奖，以及北京当代MOMA荣获由国际高层建筑与城市住宅协会（CTBUH）所颁发的“2009年世界最佳高层建筑”大奖以及2009年西班牙对外银行（BBVA）基金会知识前沿奖。在许多事务所获取的美国建筑师协会荣誉奖项中，包括了北京当代MOMA项目获得的2008年美国建筑师协会纽约分会可持续设计大奖和凭尼尔森-阿特金斯美术馆获得的2008年美国建筑师协会纽约分会建筑荣誉大奖。

斯蒂文·霍尔建筑师事务所擅长有关艺术和高等教育类型的建筑设计，包括赫尔辛基当代美术馆、纽约普拉特学院设计学院楼、爱荷华大学艺术与艺术史学院楼、西雅图圣伊格内修斯小教堂。最近建成的项目有中国深圳的万科中心、北京当代MOMA、丹麦海宁艺术博物馆、挪威哈姆生中心和在国际上普遍获得好评的堪萨斯市尼尔森-阿特金斯美术馆——《纽约人》杂志称该美术馆为“近一代美术馆中的佼佼者”。正在设计中的项目包括法国比亚里茨滑浪与海洋中心、纽约哥伦比亚大学体育中心以及普林斯顿大学创意表演艺术中心。

被《时代》杂志誉为美国最优秀的建筑师，斯蒂文•霍尔对“同时满足心灵和眼睛的建筑”具有独一无二的设计敏感。

Steven Holl Architects, (SHA)Architecture and Urban Design

Steven Holl founded Steven Holl Architects (SHA) in New York in 1976, and incorporated in 1989. SHA is a design-oriented office with locations in New York and Beijing and a total staff of 50. The firm has been recognized internationally with numerous awards, publications and exhibitions for quality and excellence in design, including nine National AIA Honor Awards and, most recently, the Council on Tall Buildings and Urban Habitat's "2009 Best Tall Building Overall" award for Linked Hybrid and the 2009 BBVA Foundation Frontiers of Knowledge Award. SHA has also won numerous AIA awards, including the 2008 AIA New York Chapter Sustainable Design Award for Linked Hybrid and the 2008 AIA New York Chapter Architecture Honor Award for the Nelson-Atkins Museum of Art.

SHA specializes in works for the arts and higher education that include Kiasma: The Museum of Contemporary Art, Helsinki; Higgins Hall Center Section at Pratt Institute; School of Art and Art History, University of Iowa; and the Chapel of St. Ignatius, Seattle, Washington. Recent completions include the Horizontal Skyscraper in Shenzhen, China, Linked Hybrid in Beijing, the Herning Museum of Contemporary Art, Denmark, the Knut Hamsun Center in Norway, the highly acclaimed, internationally recognized Nelson-Atkins Museum of Art, Kansas City, MO, which The New Yorker calls "one of the best museums of the last generation." The firm's current work includes: the Cite de L'Ocean et du Surf, with Solange Fabião in Biarritz, France; the Campbell Sport Center at Baker Field, Columbia University, New York; and the Creative and Performing Arts Center at Princeton University.

Named by Time Magazine as America's Best Architect, Steven Holl has a unique design sensibility for "buildings that satisfy the spirit as well as the eye."

公司合伙人及主要设计师

- 斯蒂文•霍尔　Steven Holl　事务所主持人及首席设计师

斯蒂文•霍尔于1947年生于美国华盛顿州的布雷默顿市。他毕业于华盛顿大学，并在1970年到罗马对建筑学研究进行深造。1976年他加入了伦敦的建筑协会，并设立纽约市的斯蒂文•霍尔建筑事务所。作为美国当代建筑师中的代表人物之一，斯蒂文•霍尔是公认能将空间与光线细腻巧妙地融入到所处场所和意境中，和运用每个项目的独特性质来创造以概念为驱动的建筑设计。他还擅长通过设计使现代的建筑项目与其特定的历史文化环境浑然成一体。

霍尔先生在学术、文化、公民和住宅领域的建筑作品遍布美国本土及海外。他被公认的代表作为密苏里州堪萨斯市的尼尔森-阿特金斯美术馆（2007年）和芬兰赫尔辛基的奇亚斯玛当代美术馆（1998年）。其他优秀的设计作品包括中国北京当代MOMA（2009年）以及位于华盛顿州西雅图的圣伊格内修斯小教堂（1997年）。斯蒂文•霍尔事务所近期完成的项目包括丹麦海宁当代博物馆（2009年）和挪威哈姆生中心（2009年）。

- 李虎　Li Hu　合伙人

李虎是斯蒂文霍尔建筑师事务所的合伙人和其北京工作室的负责人。他和斯蒂文霍尔一起设计了事务所在亚洲一些最重要的作品，包括北京当代moma和深圳的万科中心项目。李虎还是实验设计工作室开放建筑的创始人。他是美国哥伦比亚大学建筑学院的兼职助理教授及其Studio-X北京建筑中心的主任。

当代 MOMA— 联接复合体

LINKED HYBRID

项目地点：中国 · 北京　建筑面积：221 462 m^2
建筑设计：斯蒂文 · 霍尔建筑师事务所
建筑师：斯蒂文 · 霍尔，李虎
摄影：Iwan Baan，Shu He

LOCATION: Beijing, China　BUILDING AREA: 221,462 m^2
DESIGN CORPORATION: Steven Holl Architects
ARCHITECTS: Steven Holl, Li Hu
PHOTOGRAPHERS: Iwan Baan, Shu He

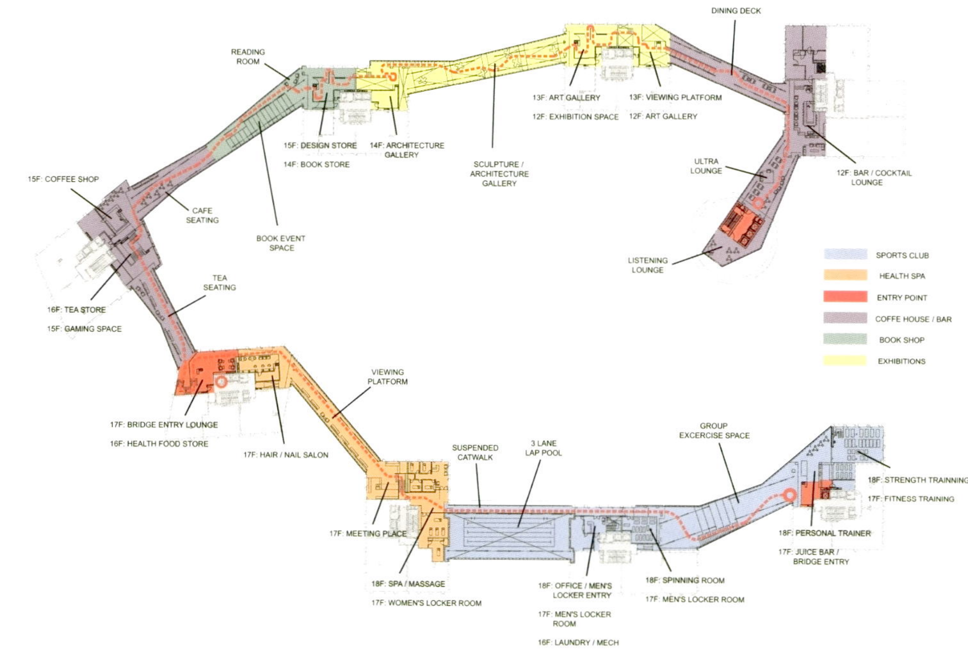

220 000 m^2 的北京当代 MOMA—联接复合体，旨在面对现今中国的城市发展创造一个 21 世纪的渗透型城市空间，自各方向引入并向公众开放。类似电影的城市经验的空间：环绕、跨越、通过多层次的空间，以及穿过该项目的许多通道，使“联接复合体”成为“城市中的开放城市”。此项目不仅增进互动关系并鼓励商业、住宅、教育至休闲的公共场所交错，是一个生动的立体公共城市空间。

地面层有很多可供人们（住户及访客）通行穿梭的开放式通道。这些通道包括由小规模商店形成的“微型都市”，它们也激发环绕于中心映水池的城市空间。建筑的中间层有公共屋顶花园，提供了静谧的绿地空间；八座塔楼住宅的顶端有私人屋顶花园连接至阁楼；首层的所有公共设施，包括餐厅、酒店、蒙特梭利学校、幼儿园和电影院，都与项目中无所不在的绿色空间相连。电梯的移动如电影“跳跃剪辑”至另一系列更高楼层的通道。12~18 层附有游泳池、健身房、咖啡厅、画廊等一系列多功能的连廊，连结 8 座住宅塔楼和酒店楼，并提供这座绵延城市的完整视野。此环状设计上为半格栅状，而不是过于简单的直线形。建筑师希望公共的空中环廊与基地网可以不断衍生出随机关系，作为一个社交凝结器运行，为住户与访客带来特殊的都市生活体验。

项目中心的大型城市空间因中水回收池中的水莲和水草焕发出生机，电影院与酒店看起来似乎漂浮在其中。在冬季，水池结冰形成一个溜冰场。电影院不仅仅是一个聚集场所，也是这片区域的视觉中心。这座电影院建筑底面上投影着正在上映的影片片段，影像同时倒映在水面上。建筑的第一层拥有整个景观视野，从左面向社区开放。中国佛教建筑彩饰启发了色彩方面的灵感。悬臂部分的底面为夜晚发光的彩色薄膜材料，门窗边框则是依据《易经》中阴阳五行的五种元素，用古代庙宇建筑中的色彩为之上色。

可持续性设计策略

“当代 MOMA”在空间性与通道的预期经验在它的可持续性上有极大的冲击和影响力。它是一个公共与私人空间的结合，居民和来访者可徒步穿行其间，鼓励资源共享，减少了不必要的运输需求。它同时也将最尖端的可持续性技术组合到一起。据估算所有公寓单元每天有 22 万公升（220 m^3）中水被回收，并被再利用于景观与绿色屋顶的灌溉、洗手间冲水以及再平衡水池用水，使得饮用水的使用量减少了 41%。“当代 MOMA”的地源热泵系统是在住宅建筑中最大的地源热泵系统之一，它最具有开创性的创新。该系统由位于地基之下 100 m 深的 655 面地热墙组成，肩负着这座复合建筑每年制冷与采暖负载的 70%。此外，这些位于地下的墙面取代了地上空间原本通常需要的制冷塔，增加了可用的绿色区域，将噪声污染降到最低，并且显著减少了由传统采暖 / 制冷方式产生的二氧化碳排放。该项目引以为傲的还有它外部窗户用于太阳能采集和热量控制的遮光格栅与中空镀膜 Low-e 涂层玻璃，以及高性能的建筑聚合板采暖与制冷系统。另外，当代 MOMA 使用一项称做置换通风的技术，将比预期略低温度的空气从地板中释放出来。较凉空气将较暖空气从房间中置换出去，创造了凉爽空间和空气清新的环境。它是一方城市中的绿洲，在北京这样一个人口爆炸的大都市中提供了一个平和恬静的绿色空间。

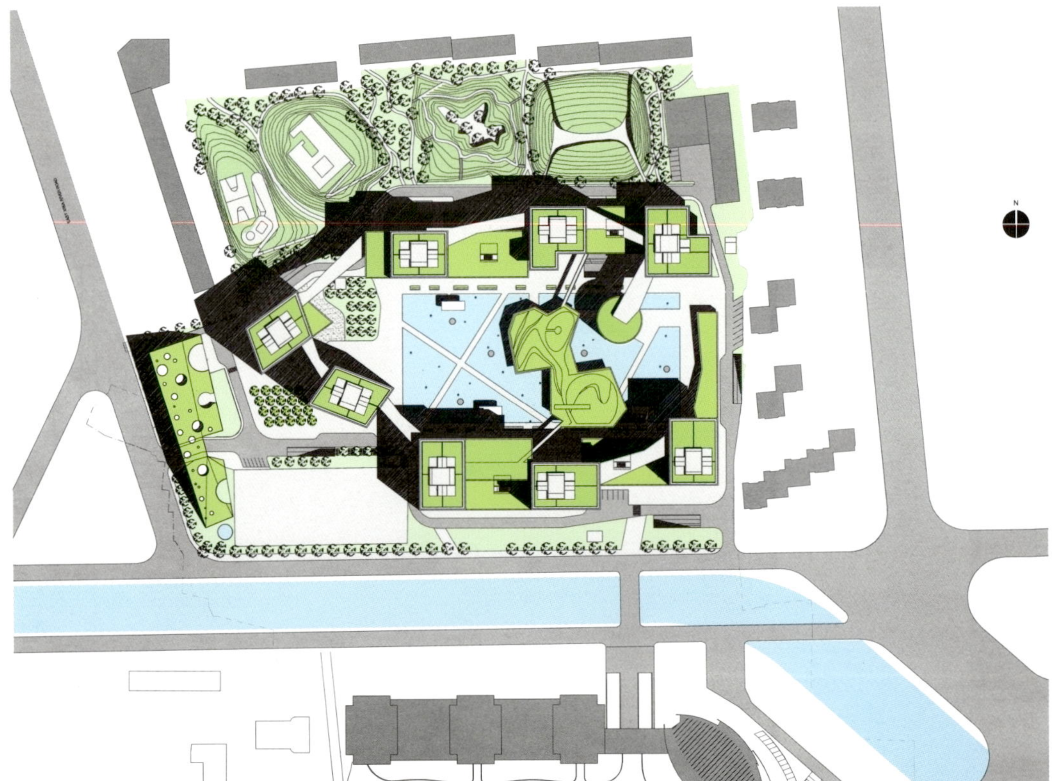

The 220,000 square-meter Linked Hybrid complex in Beijing, aims to counter the current privatized urban developments in China by creating a twenty-first century porous urban space, inviting and open to the public from every side. A filmic urban experience of space; around, over and through multifaceted spatial layers, as well as the many passages through the project, make the Linked Hybrid an "open city within a city." The project promotes interactive relations and encourages encounters in the public spaces that vary from commercial, residential, and educational to recreational; a three-dimensional public urban space.

The ground level offers a number of open passages for all people (residents and visitors) to walk through. These passages include "micro-urbanisms" of small scale shops which also activate the urban space surrounding the large central reflecting pond. On the intermediate level of the lower buildings, public roof gardens offer tranquil green spaces, and at the top of the eight residential towers private roof gardens are connected to the penthouses. All public functions on the ground level, including a restaurant, hotel, Montessori school, kindergarten, and cinema have connections with the green spaces surrounding and penetrating the project. Elevators displace like a "jump cut" to another series of passages on higher levels. From the 12 to the18th floor a multi-functional series of skybridges with a swimming pool, a fitness room, a café, a gallery, etcetera connects the eight residential towers and the hotel tower, and offers views over the unfolding city. Programmatically this loop aspires to be semilattice-like rather than simplistically linear. We hope the public sky-loop and the base-loop will constantly generate random relationships; functioning as social condensers in a special experience of city life to both residents and visitors.

The large urban space in the center of the project is activated by a greywater recycling pond with water lilies and grasses in which the cinematheque and the hotel appear to float. In the winter the pool freezes to become an ice-skating rink. The cinematheque is not only a gathering venue but also a visual focus to the area. The cinematheque architecture floats on its reflection in the shallow pond, and projections on its facades indicate films playing within. The first floor of the building, with views over the landscape, is left open to the community. The polychrome of Chinese Buddhist architecture inspires a chromatic dimension. The undersides of the bridges and cantilevered portions are colored membranes that glow with projected nightlight and the window jambs have been colored by chance operations based on the "Book of Changes" with colors found in ancient temples.

Sustainable design strategies

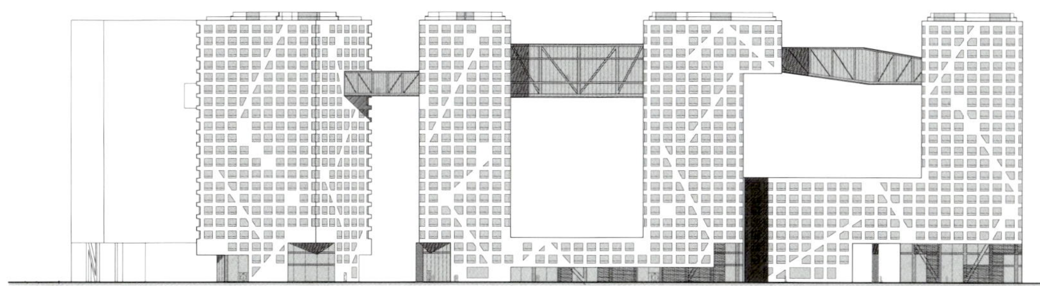

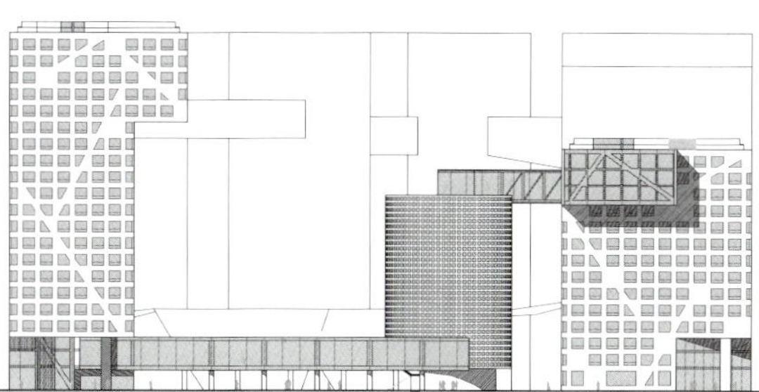

Linked Hybrid's intended experience of spatiality and passage has a tremendous impact on its sustainability. It is a pedestrian-oriented combination of public and private space that encourages the use of shared resources and reduces the need for wasteful modes of transit. It also brings cutting-edge sustainable technologies together. An estimated 220,000 liters of gray water from all apartment units will be recycled each day and reused for landscape and green roof irrigation, toilet flushing, and rebalancing pond water, resulting in a 41% decrease in potable water usage. Linked Hybrid's ground source heat pump system, one of the largest in residential construction, is its most groundbreaking innovation. Shouldering 70% of the complex's yearly heating and cooling load, the system is comprised of 655 geothermal wells, 100 meters below the basement foundation. Additionally, the underground wells have taken the place of above-ground space normally needed for cooling towers, increasing available green areas, minimizing noise pollution and significantly reducing the CO_2 emissions created by traditional heating/cooling methods. The project also boasts exterior window louvers and low-e coated glass for solar gain and heat control, as well as a high-performance building envelope and integrated slab heating and cooling system. In addition, Linked Hybrid makes use of a technique called displacement ventilation, in which air that is slightly below desired temperature in a room is released from the floor. The cooler air displaces the warmer air, causing it to be released from the room and resulting in a cooler overall space and a fresh breathing environment. It is an urban oasis, proving that peaceful, green spaces can exist in an populous metropolis such as Beijing.

来福士广场

SLICED POROSITY BLOCK

项目地点：中国 · 成都　占地面积：17 500 m^2　建筑面积：310 000 m^2

建筑设计：斯蒂文 · 霍尔建筑师事务所

建筑师：斯蒂文 · 霍尔，李虎

摄影：Iwan Baan，Steven Holl Architects

LOCATION: Chengdu, China

SITE AREA: 17,500 m^2　BUILDING AREA: 310,000 m^2

DESIGN CORPORATION: Steven Holl Architects

ARCHITECTS: Steven Holl, Li Hu

PHOTOGRAPHERS: Iwan Baan, Steven Holl Architects

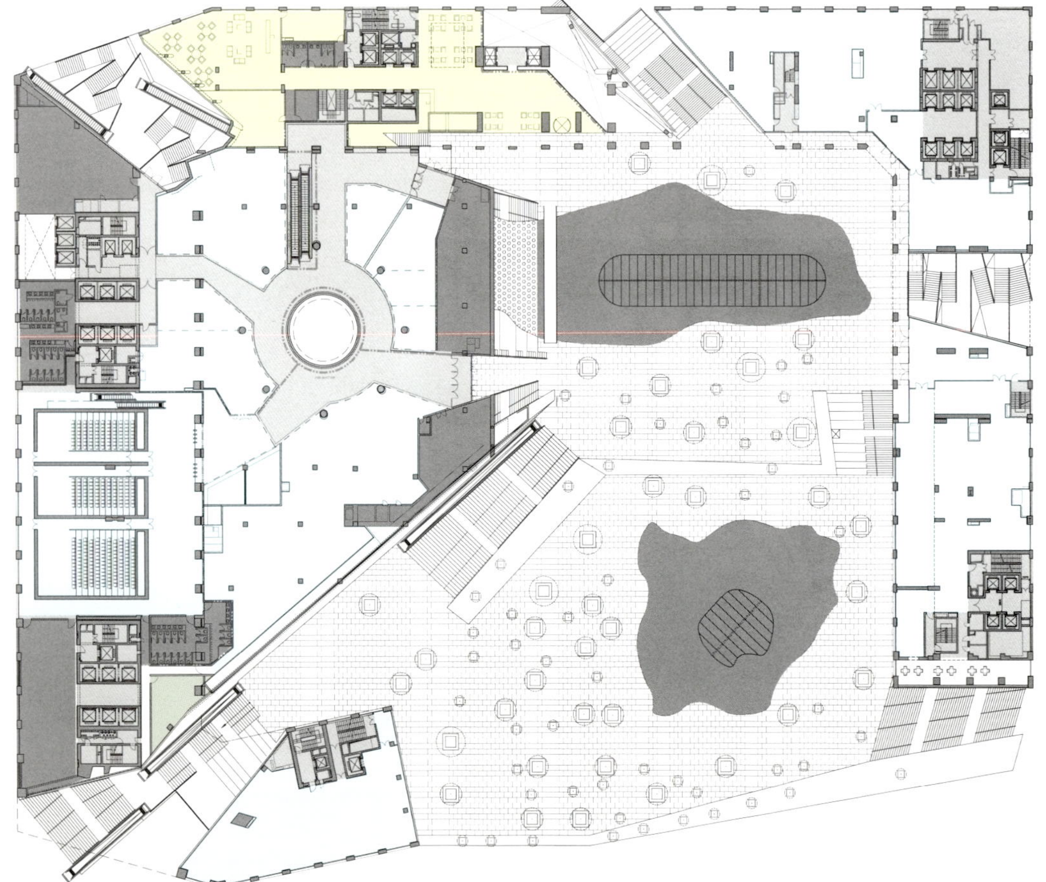

凯德置地（中国）投资所建的成都来福士广场，是多功能混合建筑群体，形如都市中的巨型块。此建筑位于成都市一环路与人民南路交口南侧。日照切割的几何体依据建筑规范要求对周边建筑最小日照值，经精确几何日照角度计算切割而成。

大型公共空间由建筑体块围合而出，受杜甫（713—770）的诗灵感启发的“三峡”，在一些多孔开口处插入了不同建筑体块。我们的微观城市策略将创造一个公共空间的新地平，其规模与洛克菲勒中心的城市平台相近。这个新城市地平是由石阶、坡道和水池蔓延至阶梯式喷泉、咖啡馆旁连接着树木、植被和长凳。屋顶花园通过各自连结可至酒店咖啡厅。商铺门脸将由亮色、霓虹灯，以及背部打光的色块组成，犹如安德列 · 塔可夫斯基黑白电影中那突然呈现的一抹亮色。成都来福士广场着重于创造新公共空间并实现成都绿化建筑新标准。本项目将由 400 个地热井提供冷热调节。裙房屋顶上的大型水池收集雨水并循环将其用于草坪的浇灌，而池中的百合花可以达到对裙房降温的作用。

The Sliced Porosity Block, Capitaland China's new Raffles City in Chengdu, is a hybrid of different functions like a giant chunk of a metropolis. It will be located just south of the intersection of the First Ring Road and Ren Min Nan Road. Its sun-sliced geometry results from required minimum daylight exposures to the surrounding urban fabric prescribed by code and calculated by the precise geometry of sun angles.

The large public space framed by the block is formed into three valleys inspired by a poem by Du Fu (713-770). In some of the porous openings, chunks of different buildings are inserted. Our micro urban strategy will create a new terrain of public space, an urban terrace on the metropolitan scale of Rockefeller Center. This new terrain is sculpted by stone steps and ramps with large pools that spill into stepped fountains. Trees, plantings and benches are flanked with cafes. Roof gardens are cultivated through their individual connections to hotel cafes. At the shop fronts, there will be luminous color, neon, backlit color block, like the wash of color that suddenly appears in the great black and white films by Andrei Tarkovsky. The aim for the Sliced Porosity Block is to form new public space and to realize new levels of green construction in Chengdu. The complex is heated and cooled by 400 geothermal wells. The large podium ponds harvest recycled rainwater with natural grasses and lily pads creating a cooling effect.

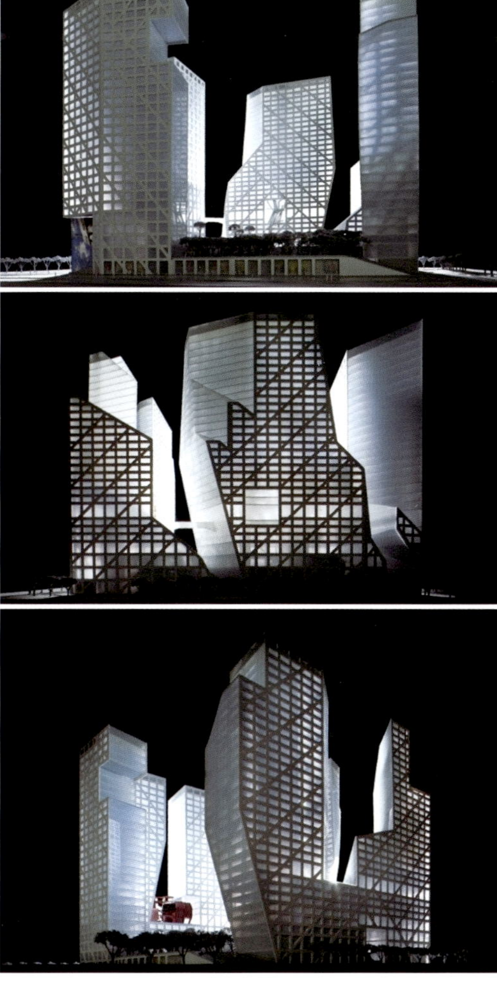

南京艺术与建筑博物馆

NANJING MUSEUM OF ART AND ARCHITECTURE

项目地点：中国 · 南京　建筑面积：3000 m²
建筑设计：斯蒂文 · 霍尔建筑师事务所
建筑师：斯蒂文 · 霍尔，李虎
摄影：Iwan Baan，Steven Holl Architects

LOCATION: Nanjing, China　BUILDING AREA: 3000 m²
DESIGN CORPORATION: Steven Holl Architects
ARCHITECTS: Steven Holl, Li Hu
PHOTOGRAPHERS: Iwan Baan, Steven Holl Architects

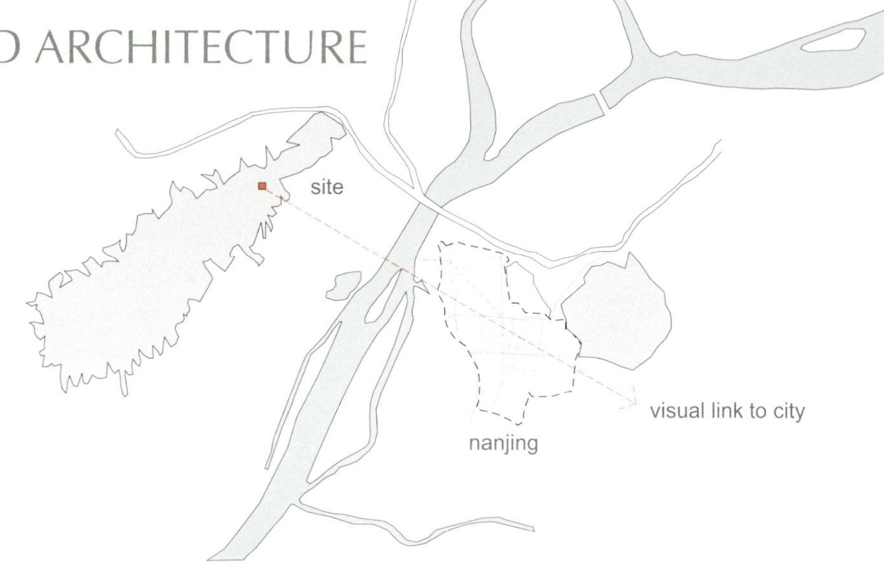

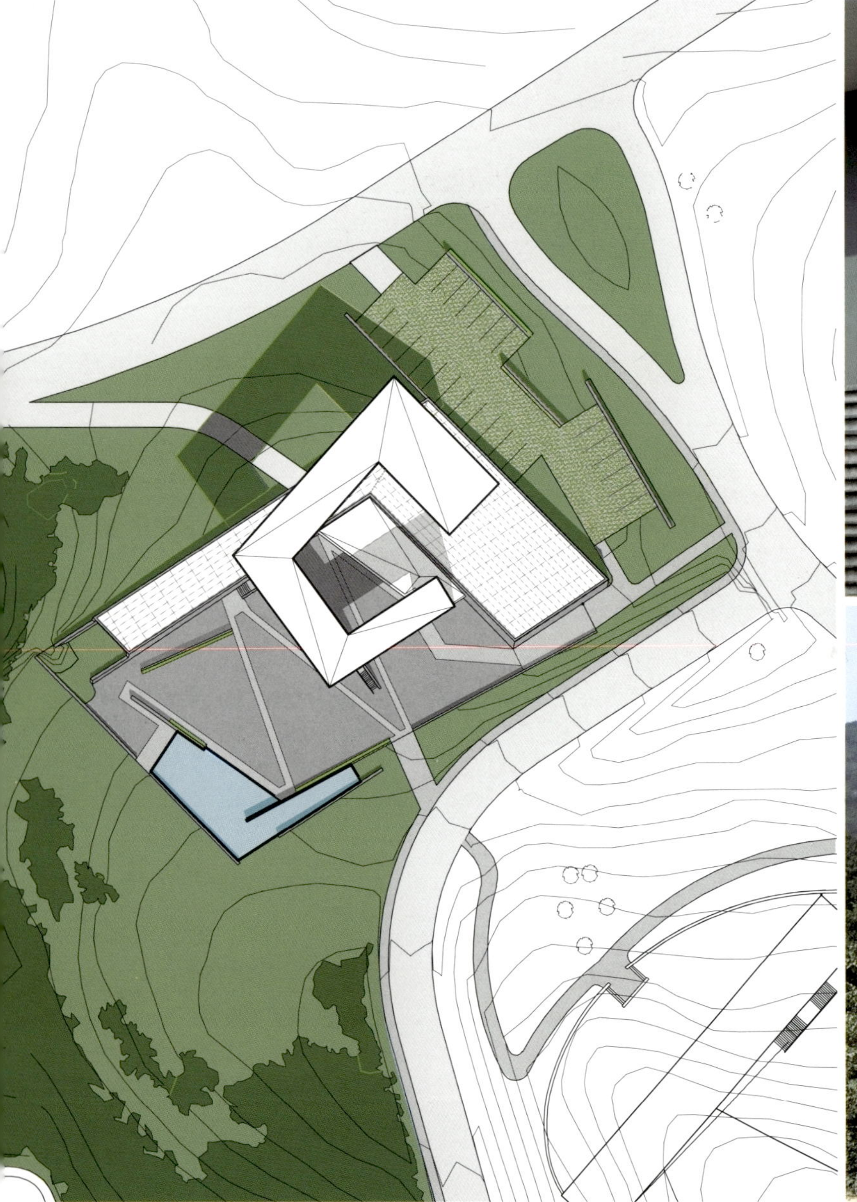

博物馆的新址在南京珍珠泉茂盛碧绿的草木景观环绕间，中国国际实践建筑艺术展览区入口处。博物馆力图通过多角度的理论观点、空间层次、雾气与水的扩展等几个主题，来探索中国古代绘画深层次变化的空间奥秘。博物馆项目由平行的透视空间"场域"、黑竹混凝土花园墙壁以及其上方悬空的轻盈"形体"组成。地平面笔直的道路渐渐变成了悬空形体内的弯曲通道。上层的画廊被悬挂在高空，按顺时针顺序逐渐暴露，最后聚集于"观景位"的顶点眺望远方的南京城。这块乡村场地由于建立了与明朝古城南京相连的视觉轴线而被赋予了新的城市意义。

"悬空形体"半透明墙体与地板结构的截面是其下方半透明天花板建筑截面的倒置。此天花板则由室内透光率30%的太阳能收集装置组成。为博物馆3000 m²的灵活通透的展览空间旁增光添色的是旁边的茶馆和策展人公寓，两者享受着南面的阳光，并面对着一座循环流水池塘。

馆庭院的地面由南京城中心已经埋没的古老胡同旧砖石铺成。设计中将整个项目的色彩限制为黑白两色，不但使设计与中国古代绘画艺术联系起来，并且赋予未来展出作品一个宏伟的观赏背景，有利于展览建筑与艺术作品的色彩及纹理。原本在新建筑附近生长的竹子被重新利用作为黑竹混凝土。地热循环加热／冷却系统和低耗能源材料的使用等方案使项目的节能环保主题的得到了充分的体现。

The new museum is situated at the gateway to the Contemporary International Practical Exhibition of Architecture (CIPEA) in the lush green landscape of the Pearl Spring near Nanjing. The museum explores the shifting viewpoints, layers of space, and expanses of mist and water, which characterize the deep alternating spatial mysteries of early Chinese painting. The museum is formed by a "field" of parallel perspective spaces and garden walls in black bamboo-formed concrete over which a light "figure" hovers. The straight passages on the ground level gradually turn into the winding passage of the figure above. The upper gallery, suspended high in the air, unwraps in a clockwise turning sequence and culminates at "in-position" viewing of the city of Nanjing in the distance. The meaning of this rural site becomes urban through this visual axis to the great Ming Dynasty capital city, Nanjing.

The section of the "figure" with translucent walls and floor is an inversion of the section of the building below with a translucent ceiling. This ceiling is made of solar collectors which permit 30 % light passage to the interior. The 3,000 sq. m. museum's flexible exhibition spaces are complimented by a tea house and curator's residence facing the south light and re-circulated water of the pond.

The courtyard is paved in recycled Old Hutong bricks from the destroyed courtyards in the center of Nanjing. Limiting the colors of the museum to black and white connects it to the ancient paintings, also gives a background to feature the colors and textures of the artworks and architecture to be exhibited within. Bamboo, previously growing on the site, has been used in bamboo-formed concrete, with a black penetrating stain. Geothermal cooling and heating, recycled storm water, and low embodied energy materials are part of the green building aims of the project.

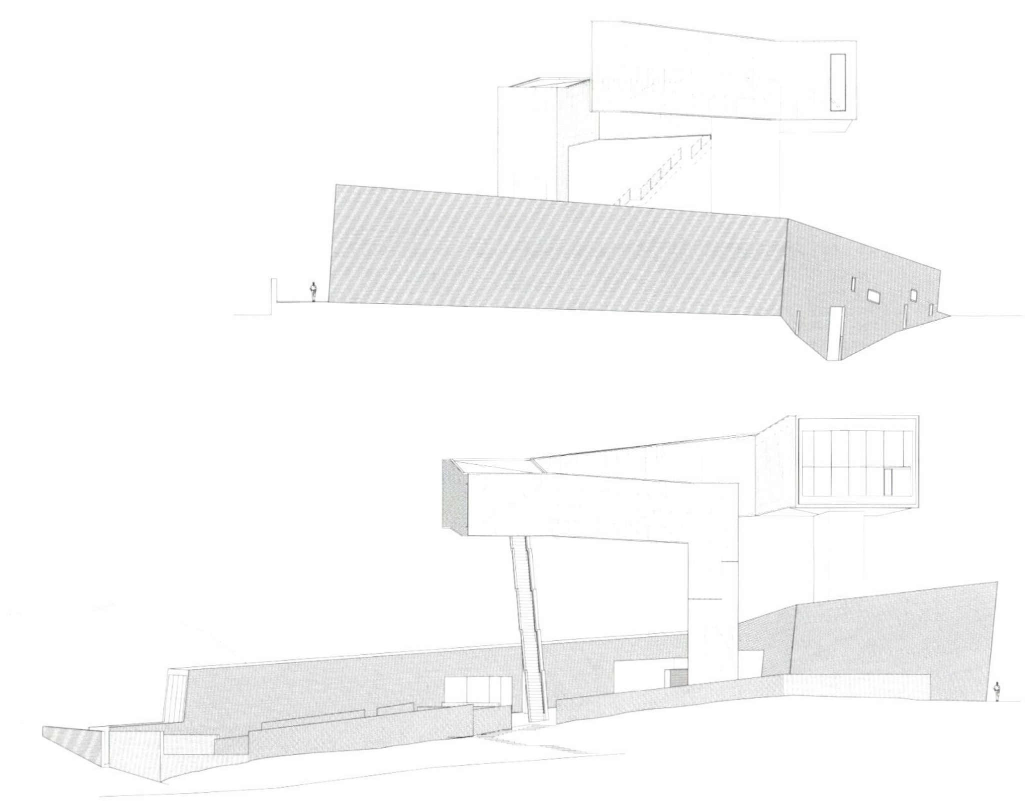

万科中心

VANKE CENTER

项目地点：中国 · 北京　建筑面积：120 445 m^2
建筑设计：斯蒂文 · 霍尔建筑师事务所
建筑师：斯蒂文 · 霍尔，李虎
摄影：Iwan Baan，Steven Holl Architects

LOCATION: Beijing, China　BUILDING AREA: 120,445 m^2
DESIGN CORPORATION: Steven Holl Architects
ARCHITECTS: Steven Holl, Li Hu
PHOTOGRAPHERS: Iwan Baan, Steven Holl Architects

盘旋在公共热带花园上空的“水平摩天楼”是一个综合建筑群，其功能包括公寓、酒店，以及万科集团总部办公室。会议中心、水疗中心和停车场都位于热带花园绿地下方，特色景观包括土丘下的餐厅和五百人座的报告厅。

此建筑看似曾经一度漂浮在较高海面上，如今海面已经退去，留下结构高高屹立在八个支撑腿上。在 35 m 限高下抬起一个整体的结构以取代数个小结构体分别满足特定功能，使绿化空间最大限度地敞向地面层公众。

悬浮结构的下方成了它的主立面——第六个立面——深圳之窗在此提供了 360° 的全景视野，可鸟瞰下方苍翠繁茂的热带景观。一个起始于“龙头”的公共通道连接酒店，并由公寓楼划分出办公楼翼。

由于是热带方案，建筑和景观融合了几个新的可持续发展方向：通过中水系统运作将水池温度冷却而形成微观气候环境。建筑屋面是绿化花园和太阳能板，材料为当地材料和可再生的竹材。大楼的玻璃幕墙能透过外在多孔百叶设计以阻挡强光及风力。该建筑是一个海啸防悬停架构，创建了一个可渗透的微气候公共休憩景观。

可持续性设计策略

万科中心将会是中国第一个荣获绿色建筑评估体系白金认证的建筑。此项目利用了光伏电板、中水回收、雨水采集、绿色屋顶以及多功能可控制电动百叶来最大化地增加自然光，并提供被动式太阳能冷却效能。安装在建筑屋面的 1400 m² 光伏电板可提供整个万科总部 12.5% 的电能要求。

此建筑的二十六个面向的每一面都是依全年太阳能吸取量为根据来计算，它的百叶可随太阳光向调整。除此之外，百叶建立了双层外观，其中间隙的中空部分可以产生对流的烟囱效应，冷空气由建筑的下方流入，再通过建筑顶部靠近屋顶的地方排放出热空气。多孔百叶在闭合状况下提供了主要的遮阳效果，它们在最大日晒下能减低 70% 的太阳能热量采集，而通过洞孔仍然能提供 15% 的透光率。由于拥有热带强烈的日照，现场测量计算出当百叶在闭合模式下，15% 的透光度已可以提供充足的自然光线来满足办公室日常照明需求，大多数空间（75%）无需辅以人工照明。笔直的玻璃幕墙允许日光渗入室内空间，最新锐的高性能玻璃涂层（双层银色中空镀膜 Low-e 玻璃）被使用于整个项目中，通过降低冷却负载进而达到节能效果。

在项目的办公室区，室外百叶、室内遮阳板、空气调节和照明系统的运行都由室内外感应器调度操作，从而平衡周围环境的照明度、太阳能热量采集和周围温度以达到最大程度的节能。多数办公室都有照明和遮光的独立控制。个别的任务／聚光照明是用于下班时段或其他使用。

两米宽的特殊大型可操控窗户可在一年较凉爽的期间为室内提供自然通风和大量的和煦微风。机械通风装置系统据估算在此季节中，至少 60% 的时间可以是关闭的，这样每年将减少 5 kWh/m² 的电能消耗。

此项目既是建筑也是景观，是复杂工程与自然环境精美的交织。通过将建筑由地面升起，一个开放、可公共进出的公园在一个其他方面都封闭和私有的社区内创造了新的社交空间。项目占地面积约有 60 000 m²，其中 45 000 m² 万平方米被植被覆盖，加上主建筑植被绿化屋顶（约 15 000 m²），总绿化面积已与这片区域未开发前的面积大致相等。

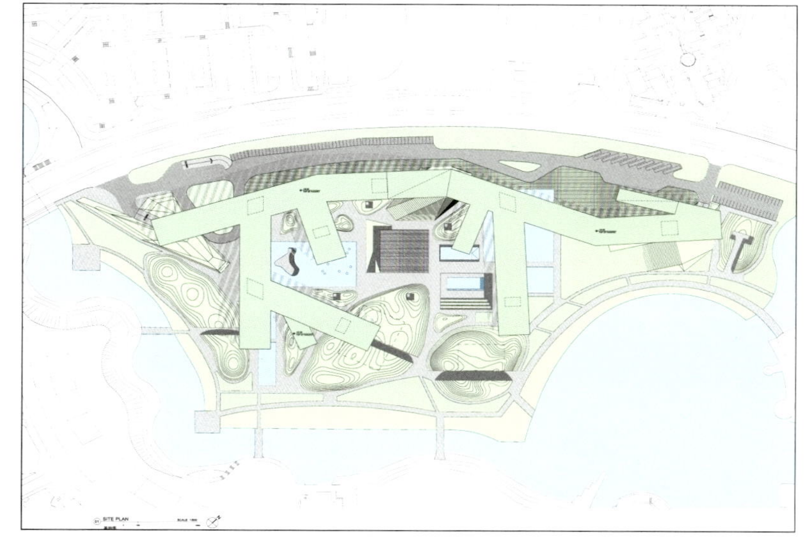

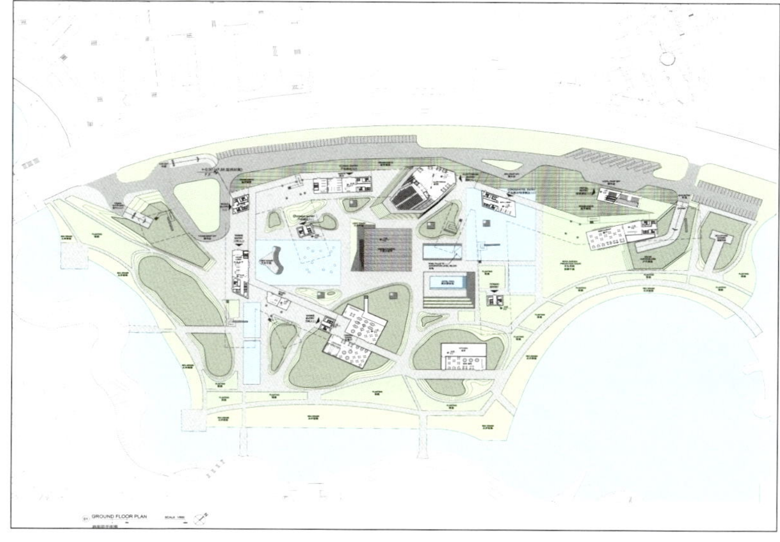

Hovering over a public tropical garden, this "horizontal skyscraper" with the length the same as the Empire State Building's height is tall - is a hybrid building including apartments, a hotel, and offices for the headquarters for Vanke Co. Ltd. A conference center, spa and parking are located under the large green, tropical landscape which is characterized by mounds containing restaurants and a 500 seat auditorium.

The building appears as if it were once floating on a higher sea that has now subsided; leaving the structure ropped up high on eight legs. The decision to float one large structure right under the 35-meter height limit, instead of several smaller structures each catering to a specific program, generates the largest possible green space open to the public on the ground level.

The underside of the floating structure becomes its main elevation -- the sixth elevation -- from which Shenzhen Window offers 360-degree views over the lush tropical landscape below. A public path beginning in the "dragon's head" connects through the hotel, and the apartment zones up to the office wings.

As a tropical strategy, the building and the landscape integrate several new sustainable aspects: a micro-climate is created by cooling ponds fed by a recycled water system. The building has a green roof with solar panels and uses local materials such as bamboo The glass façade of the building will be protected against the sun and wind by perforated louvers. The building is a tsunami-proof hovering architecture that creates a porous micro-climate of public open landscape.

Sustainable design strategies

The Vanke Center will be one of the first LEED Platinum certified buildings in China. The project utilizes photovoltaics, recycled water recycling, rain water harvesting, green roofs, and dynamically controlled operable louvers which maximize natural light and provides solar passive cooling. 1400 square meters photovoltaic panels installed on the roof of the building provide 12.5% of the total electric energy demand for Vanke Headquarters.

Each face of the 26 faces of the building has been calculated based on solar heat gain throughout the year and its louvers are fine-tuned to the orientation of the sun. In addition, the louvers create a double skinned facade where the interstitial cavity creates a convective stack-effect, drawing cool air in through the underside of the building and hot air out at the top of the structure near the roof. The perforated louvers provide extensive primary sun protection in closed condition. They reduce up to 70% of solar heat gain at its peak load, yet still provide 15% of light transmittance through the perforations. Given the intensity of the tropical sunlight, field measurements have calculated that this 15% light transmittance in closed mode is sufficient natural lighting to perform routine office functions without the need for secondary artificial lighting in most (75%) of spaces. The full height glass curtain wall brings daylight deep into all interiors spaces, and the latest high-performance glass coatings (double silver low-e) are used throughout the project. This saves energy by reducing cooling loads.

In the office portion of the project the operation of the exterior louvers, interior shades, air conditioning and lighting systems are coordinated by a series of interior and exterior sensors which balance ambient light levels, solar heat gain and ambient temperatures for maximum energy efficiency. There are individual controls for lighting and shade operation in most offices. Individual task/spot lights are provided for off hour, additional use.

Exceptionally large operable windows of two meters wide provide natural ventilation and generous cross breezes for the interiors during the cooler months of the year. It is estimated that during this season mechanical ventilation systems can be switched off for at least 60% of the time. This will reduce electric energy consumption annually by 5 kWh per square meter.

The project is both a building and a landscape, a delicate intertwining of sophisticated engineering and the natural environment. By raising the building off of the ground plane, an open, publicly accessible park creates new social space in an otherwise closed and privatized community. The site area is approximately 60,000 square meters, of which 45,000 square meters is planted. With the addition of the planted roof area of the main building (approximately 15,000 square meters) - the total planted area of the project is roughly equal to the site before development.

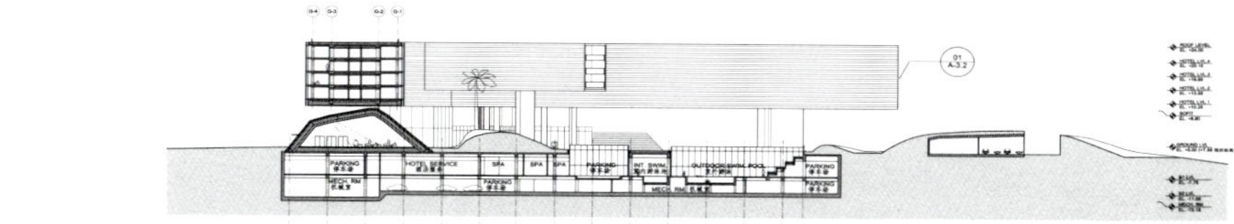

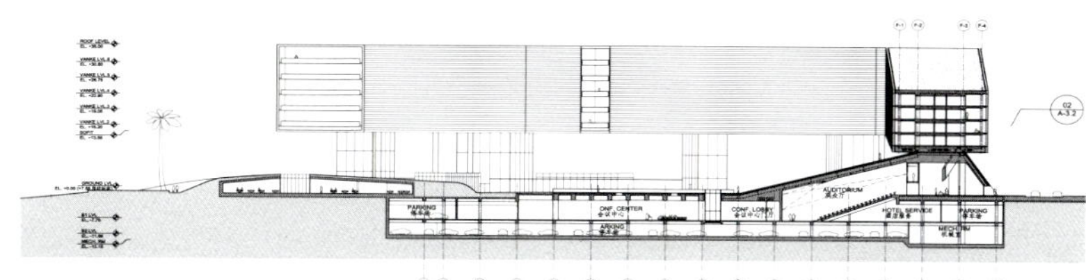

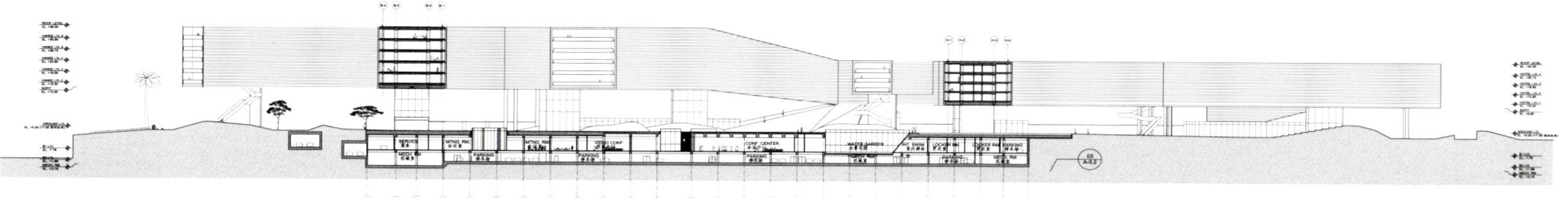

STI STUDIO

STI思图意象工作室

杭州STI思图意象建筑设计咨询有限公司成立于2004年。公司负责人秦洛峰博士曾留学于德国斯图加特大学建筑与城市规划学院，获得工学博士学位。杭州STI思图意象公司的设计项目涵盖了公建、住宅以及城市设计等领域，项目主要分布在浙江及周边地区。

公司员工由资深建筑师、结构工程师、德国建筑师及实习生组成。设计理念具有创新性和地域特色，得到了业主和专家的广泛好评。作品多次获得国内各项建筑设计竞赛奖，在浙江及周边地区享有声誉。

STI-studio was established in 2004. Dr. Qin Luofeng, the executive of whom studied at the University of Stuttgart Institute majoring in Architecture and Urban Planning, and received Ph.D. degree of science. The design projects of STI-studio in Hangzhou cover public construction, residential and urban design and other fields, mainly distributed in Zhejiang and the surrounding areas.

The team of the studio is constituted by the senior architects, structural engineers,German architects and interns. Its design concepts combine innovation and local characteristics, and it has been widely acclaimed by the owners and experts. Its works has won national design competition awards and good reputations in the surrounding areas of Zhejiang

杭州经济开发区城建文化馆

URBAN CULTURAL CENTER IN HANGZHOU ECONOMIC DEVELOPMENT AREA

项目地点：中国 · 杭州 **用地面积**：12 000 m^2 **建筑面积**：15 000 m^2

建筑设计：STI 思图意象工作室

建筑师：秦洛峰，Anna Birk，魏薇，黄蓉，俞淳流等

LOCATION: Hangzhou, China SITE AREA: 12,000 m^2

BUILDING AREA: 15,000 m^2

DESIGN CORPORATION: STI Studio

ARCHITECTS: Qin Luofeng, Anna Birk, Wei Wei, Huang Rong, Yu Chunliu

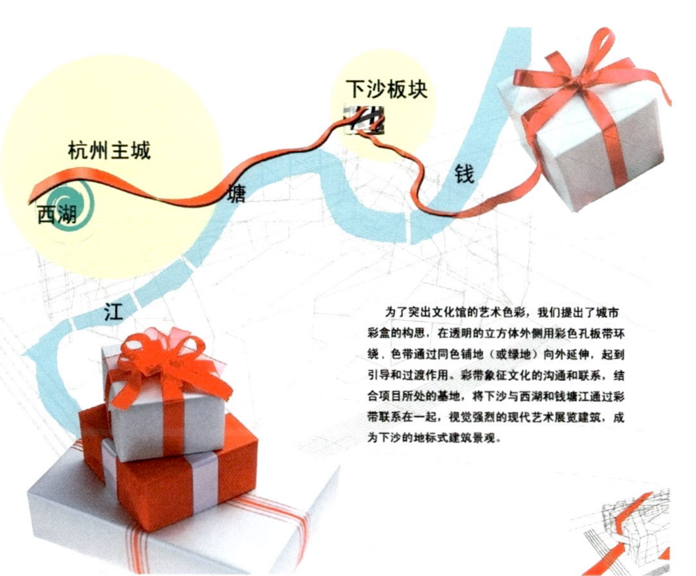

作为下沙新的地标建筑，文化性是设计中至关重要的，而作为这个城市的新建筑，其文化内涵必然要与这个城市所积淀的文化接轨，以延续这个城市的文化特质，只有这样，才能更好地挖掘一个建筑潜在的意义，使其起到地标建筑应有的作用。

设计理念：为了突出文化馆的城市文化，我们提出了"城市彩盒"的构思，城市彩盒的"彩"字象征了多元文化、多元社会的和谐，具有丰富的社会内涵与意义。我们在透明的立方体外侧用彩色孔板带环绕，色带通过同色铺地（或绿地）向外延伸，起到引导和过渡作用。彩带象征了文化的沟通和联系，以塑造出富有特色的、视觉强烈的现代艺术展览建筑，从而成为下沙的地标式建筑景观。同时，分别从文化性、技术性、社会性和经济性来表达我们的设计理念。

文化性：设计从"天圆地方"的"方"字出发，用最简单的形体塑造最具魅力的形象，作为文化建筑如何体现出文化性是设计的关键。从中国传统文化的四合院，希腊文化中雅典的神庙，到现代巴塞罗纳展厅及卢浮宫地下展馆，方形的建筑穿越时空成为永恒的主题。城市彩盒的"盒"以方为体，形成外盒（外形）与内盒（内庭与屋顶内院），延续了传统（内庭内院），融汇了中西方文化。方形的建筑体现了力量的凝聚，代表了阳刚之美，建筑外侧用彩色孔板带环绕，通过同色铺地（或绿地）向外延伸，让红色孔板带得以延续，远远望去正像一条飘逸的丝带，代表了阴柔之美，体现了刚柔相济的建筑品质，突出了建筑的文化价值。

技术性：文化馆采用先进的工程技术手段，把高科技与功能、艺术有机地结合起来，在设计中从材料的特性及表现力出发，以植物园的樟树为原型，创造性地以大小不同的圆孔排列出树影，塑造建筑的"第二层皮"，同时也充分考虑了现代文化展馆的采光、通风、保温、节能的技术要求，使技术性不仅仅成为表现及应用手段，而成为设计的出发点和目标。在空中花园里，彩色孔板带缠绕形成的富有肌理的构架在不同时段太阳光的照射下，屋顶的光影如同富有节奏的音符轻轻跳动。

社会性：下沙城建文化馆位于下沙城中心地块，不仅要体现积极的社会价值和精神，还要体现社会凝聚力与包容性，能够成为下沙人的社区活动交往中心，因此，我们拟设计出这样的实用空间：竖向空间的渗透和横向空间的延伸。通过将一层墙面向内收缩，在建筑外围形成一圈室外围廊，一层到三层中庭通高，用于模型展览，这样的空间方便了不同范围人群的交流，表现了人与人、人与社会、人与自然相互联系与共生共荣的多样统一，并具有丰富的社会内涵与意义。

经济性：建筑格局方整，内部功能分区明确，房间规整，组织紧凑，内部与外部空间统一，以方形柱网为模数，便于空间的灵活划分，有利于根据使用要求变化进行调整，建筑形体的方整提高了建筑的经济性和功能性，适用于展览参观、会议接待及办公活动等多种功能，符合文化馆的多层次、多方位、多样性的社会要求。

建筑建成后，从外形到内部空间，都充分体现了社会凝聚力与包容性，并逐渐成为下沙人的社区活动交往中心，给人留下了深刻的印象，象征了多元文化、多元社会的和谐，真正起到了地标建筑举足轻重的作用。

In terms of a new landmark in Xiasha, culture is essential to the design, but as the city's new architecture, its cultural connotation has to be bound to the cultural accumulation of this city to extend the city's cultural identity, which is the only way to better grub the potential significance of a building so that it can play the role as a landmark building.

Design Concept: To highlight urban culture of the cultural center, we propose the concept of "city foldingcarton" which covers the meaning of a harmonious multicultural and pluralistic society. The name has a rich social meaning and significance. We made it a transparent cube wrapped with "color ribbons" that extends out through the same color floor (or green floor), which plays a guiding and transition role. The ribbon symbolizes the cultural communication and contact. The distinctive contemporary art exhibitions building with strong visual impact makes itself a landmark landscape architecture in Xiasha. Meanwhile it expresses our different design concepts from the cultural, artistic, technical, social and economic perspectives.

Cultural: The design starts from "fang" (which means square or rectagular in Chinese idiom "Tian Yuan Di Fang")to show the most charming image with the simplest shape. As a cultural construction the most important is to show its cultural characteristics. Since the ancient

time the square building has become the eternal theme through time and space ranging from the Chinese traditional courtyard to the Greek Athens. Temple, from the modern Barcelona Gallery to the Louvre underground exhibition hall. The body of the foldingcarton is square to form the outer box(shape) and the inner box (the inner courtyard and roof). It continues the traditional(the inner chambers), fused the Chinese and Western styles. Square of the building reflects the strength condensation, representing masculine beauty. The outside of the building is surrounded by colorful "ribbons" which extends out through the same color paved floor so that seen from afar it's just like an elegant ribbon, representing the feminine beauty. So it totally reflects alternate strength and tenderness, which highlights the cultural value of architecture.

Technical: The museum uses advanced engineering techniques to integrate high-tech, art and functions. Taking the features of construction materials into consideration and based on the architype of the camphortrees in the botanical garden, the design creatively arranged holes of different sizes to compose the shadows of the trees and built "another skin" of the construction. Meanwhile it takes full account of the lighting, ventilation, insulation and energy-saving technical requirements of the modern culture exhibition hall to make the technology not only a practical means but also the starting point and objective of the design. In the air garden the framework of rich texture exposed to the sun the roof light and shadow gently and rhythmically beat at different times.

Social: Xiasha Urban Cultural Center is located in the central area of the city. It's required to reflect not only positive social values and spirit but also the social cohesion and inclusiveness to be the exchange center of Xiasha community. Therefore we planned such a practical space as infiltration of the vertical space and extension of horizontal space. By contracting the bottom wall inward it forms a surrounding corridor outside the building. The square hollow shape from the first to the third floor makes it easy for model displays, and is convinient for people to exchange in different areas, which shows the coordinated relationship of interdependence and coexistence between people, between man and society nd between human and nature, and has a wealth of social meaning and significance.

Economical: The building pattern is regularly squre. The internal functional areas are clearly separated. The rooms are neat and compactly organized that has reached the unity of internal and external spaces. It makes the square column as the module to facilitate the flexible division of space, and is beneficial to adjust according to changes in use requirements. The regularly square structure strengthes the economical and functional characters of the building, and is feasible for exhibition,conference reception and office activities and other functions, to meet the multi-level, multi-faceted and diversified social demands of the cultural center.

The completion of construction fully embodies the social cohesion and inclusiveness from the appearance to the interior space, and gradually becomes the communication and exchange center of Xiasha community,which has left a deep impression on people symbolizing harmony of multicultural and pluralistic society. It plays a pivotal role as a real landmark building.

安吉县龙山体育中心

LONGSHAN SPORTS CENTER IN ANJI COUNTY

项目地点：中国 · 浙江　用地面积：79 856 m^2　建筑面积：19 890 m^2
建筑设计：STI 思图意象工作室、浙江大学建筑设计研究院
建筑师：秦洛峰，Neol Shardt，沈晓鸣，魏薇，张晓静，黄蓉，俞淳流等

LOCATION: Zhejiang, China　SITE AREA: 79,856 m^2　BUILDING AREA: 19,890 m^2
DESIGN CORPORATION: STI Studio　Architectural Design and Research Institute of Zhejiang University
ARCHITECTS: Qin Luo Feng, Neol Shardt, Shen Xiaoming, Wei Wei, Zhang Xiaojing, Huang Rong, Yu Chun-flow

1. 城市环境

建设用地处于安吉将来城市发展轴的附近，龙山森林公园山麓，体育场馆作为大型的公共建筑，既要考虑到建成后带动社会、经济、文化齐步发展，又要赋予城市形象以新的时代特征，同时还要考虑到“竹乡”安吉的生态性，这就是生态型城市发展的需求。

“石开竹韵”体育场馆设计，我们的落脚点在于通过体育场馆极具动势的建筑形象、富于变化的立面设计，塑造具有艺术特色、视觉效果，展现人文精神和现代建筑技术的特色场馆。场馆结合自然山势延展开来，保持自然地形，几乎对原有山体进行一种没有破坏性的土建施工，通过与原有场地结合的处理手法，整体布局有别于现代常规的体育场馆处理手法，场馆的选址和建筑空间都从一个新的视角阐释了现代体育运动场馆建筑设计的发展方向和演变趋势。

2. 功能布局

体育中心建筑规模包括：3500 座体育馆主场馆、300 座网球馆、室外标准游泳池一个、室外网球场六片以及户外拓展运动场地若干。

因为体育场馆选址的特殊性，不可能像常规场馆一样进行周边疏散模式，所以主场馆一层为接待和相关的辅助用房，通过二层平台的覆盖，自成一个比较完整隐蔽的功能区块，提供辅助服务和运动员、贵宾记者流线；主场馆二层以上的空间主要为观众席区块，所有的观众流线可以完全通过二层平台直接疏散到场地北侧和东北侧的出入口，二层平台构成建筑造型空间的同时，主要功能是作为观众室外疏散平台，平台向北侧和向东侧延展的大坡道，与城市主要干道连接，很明确地引导人群的疏散方向，这种空间的功能组织在竖直方向上分区明确独立，即使在山地这样特殊的建设场地，也能很好地达到体育场馆疏散的功能要求。

3. 构思理念

原创性：我们采用了“石开竹韵”建筑设计构思理念，体育场馆组团的主体是两个不规则体块。形体大小一主一次，体态形象一反一正，两个体块的走势向着两个不同的方向延展，形成一种生长的动态，而架空平台的整合使整个建筑组团达到了一种平衡性和统一性。两个建筑体块之间的平台进行留白处理，形成一个开放式庭院，这种处理形成了一定的张力感，同时又使自然山体融入到了建筑当中。

建筑材料以石材和玻璃为主，两个建筑主体的二层立面运用了大片的玻璃幕墙，通过材质的变化产生视觉的收缩感，从立面层次上突出了两个主体的体块造型，犹如两个简洁的体块从竹山中生长出来，宛如两块玉石，净润温婉，平衡中蕴含动感，变化中充满活力，塑造出视觉动感极强、独

具个性的现代体育建筑。

文化性：从中国山水文化中提取“山之体，石为骨”的脉络，用最简洁的形体塑造出山骨嶙峋最具魅力的建筑形象，是一种体现现代人文和文化精神的建筑艺术表达手法。“石开竹韵”体育场馆设计作为自然山水画意境的写意和抒情，洗练的建筑轮廓线表达了一种建筑的山水文化。

地域性：考虑到龙山体育中心场馆建筑的功能性与建设地点的特殊性，我们充分利用了项目建设用地的自然环境条件，结合“竹乡”的地域性特色，适当地运用了“竹”这一元素丰富建筑空间和立面，设计将周边竹林的光影之美寓意于建筑立面和屋面的开窗细部设计之中。体育馆的立面和屋面开设了许多不规则的“竹叶形”直接采光窗户，进行一系列光影的规划与设计。室内空间通过天光的投射，不仅能够节能和充分利用自然光，而且在场馆内部空间和疏散休息空间勾勒出自然竹林光影，丰富了建筑立面。

生态性：龙山体育中心作为一个可持续发展的绿色建筑，提出了自己的生态性建筑理念。建筑方案采用先进的工程技术手段，把高科技与功能、艺术有机地结合起来，充分考虑了现代体育建筑的采光、通风、保温、节能的技术要求，使体育馆与自然融为一体。可持续发展的建筑设计应关注自然环境与建筑的平衡关系，在“石开竹韵”设计中我们利用了现有的自然山体形态和天然绿化条件，保护现状环境，最大限度地使建筑与山体结合，做到“建筑与山体合一”，“体育、文化与自然合一”，既保护绿色的山体，又减少对山体的开挖，同时达成了建筑的经济性。

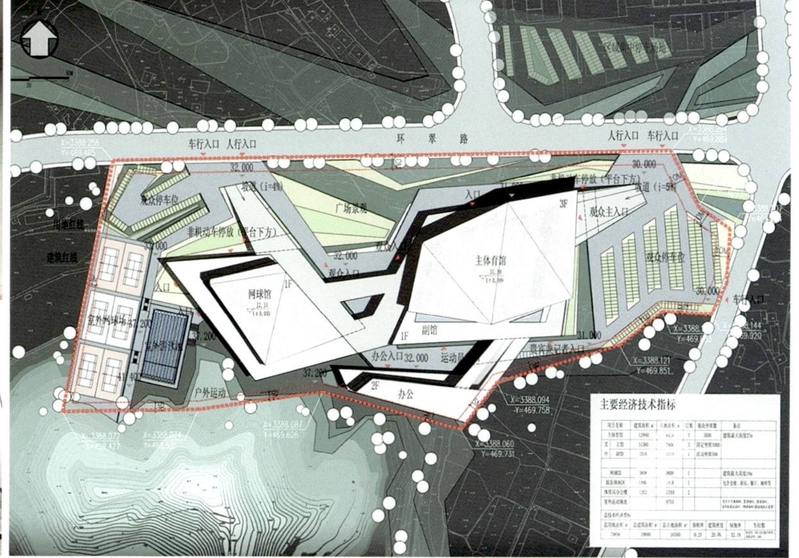

1.City environment

The construction land is in the future urban development axis of Anji County, near the foothills of Longshan Forest Park. To design it as a large public building we should not only take into account its leading position to the synchronous social, economic and cultural development, but also endow the urben image with characteristics of a new era and take into account the ecology of the "Bamboo Town", which is the needs of eco-city development.

As for the design of bamboo and forest stadium, our ultimate goal is to create artistic character and visual effects and show humanism and modern construction technical characteristics by displaying its dynamic architectual image and richly varied facade. The venue extends itself by conforming to the natural mountain topography. It's a civil engineering construction almost without distroying the original mountain. Combined with the handling of the original site, the overall layout is differentiated from the conventional modern stadium design approach. Both the venue site and construction space reinterpret construction and the design orientation of the modern stadium.

2.Functional layout

The building size of the Sports Center include: 3500 main arenas, 300 tennis halls,a standard outdoor swimming pool, 6 outdoor tennis courts and some outward bound venues.

Due to the specificity of the stadium site it's impossible to use the same all-around evacuation mode as the regular venues,therefore the first floor of the main arena is the ancillary space for reception, and covered by the platform of the second floor it forms a more complete self-contained hidden space providing related services and a flow line for athletes. Guest and reporters; the second and other spaces of the main arena mainly auditorium space , all the spectators can be completely evacuated to the north and the northeast entrances through the second-floor platform. The platform constitutes architectural modeling space and in the meantime functions as an outside audience evacuation platform. The large ramps extending from the platform to the east and the north are connected to the main roads of the city clearly showing the direction to audience evacuation. The functional organization of this space is clear and independent on the upper partitioned area in the vertical direction, so that even in such a special mountain it can meet the functional needs of the evacuation.

3.The idea of concept

Original: We adopted the bamboo and stone design concept. The main stadium group are two irregular blocks, with one size bigger and the other smaller, one positive and the other negative in the posture image. The two blocks extend to two different directions, forming a dynamic growing trend While the integrated body of the superterranean platforms realizes a balance and unity of the whole building group. The platform between the two building blocks were left blank to form an open courtyard. This architectural treatment provide a certain sense of tension and at the same time integrates the the natural mountain scenery into the whole construction.

Stones and glass are the main construction materials. The two-storey facades of the two main bodies use large glass curtain walls to produce the visual sense of contraction by material changes, which highlights the modeling of the two main blocks in facade, as if two simple pieces grow out of bamboo mountains,just like two pieces of jade, net and gentle to show the interactions betwen balance and mobility as well as diversity and energy. It just creates a modern sports building with highly dynamic visual effect and unique personality.

Cultural:We extract "mountain body and stone bone " concept from Chinese landscpe cultural context,and create the most fascinating architectural image through the most simple shape, which is an architectural expression of modern humanity and cultural spirit. The "bamboo and stone"stadium design, as an impressionistic and lyrical mood of natural landscape, expresses a architectural landscape culture by simple but refined building outline.

Regional: Considering the functionality and specificity of the construction site,we took full advantage of the natural conditions, combining the local characteristics of "bamboo town" ,and appropriately used the bamboo element to enrich architectural space

and the facade. The design displays the light and shadow magnificence of the bamboo forest in the detailed design of the fenestration of the facade and the roof. The facade and roof of the stadium were designed with many lighting windows of irregular bamboo shape. The interior space can not only save energy and make full use of natural light through the projection of daylight, but also outlines natural charm of bamboo shade against the light in the venues and the evacuation space to enrich the building facade.

Ecological: As a sustainable green building it put forward its own concept of ecological construction. The construction program used advanced engineering techniques to organically integrate the high-tech and the functional, artistic skills,fully considering the technical needs of lighting, ventilation, insulation and energy-saving of the modern sport construction, and thus integrate sports into nature. Sustainable building design and construction should be concerned about the balance between the natural environment and the architecture. In the bamboo and stone design we used the natural shape and green mountain conditions, to achieve ultimate unity of the building and the mountain and maximum integration of sport, culture and nature without destroying the status of the environment, both to protect the green of the mountain, and to reduce excavation, and meanwhile increase economy of the building.

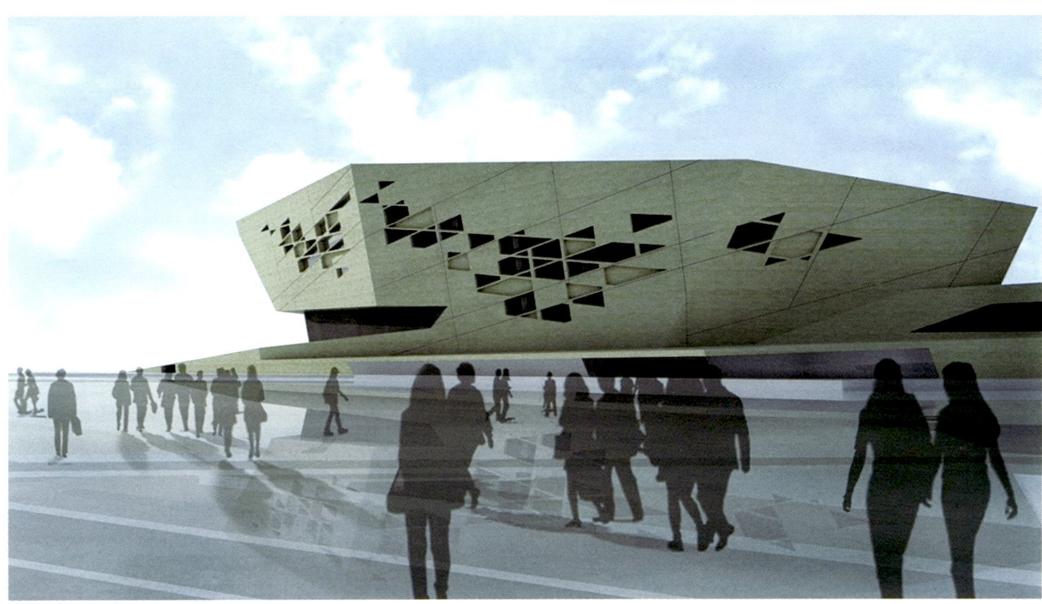

杭州火车东站综合体广场概念设计

COMPLEX SQUARE CONCEPTUAL DESIGN OF HANGZHOU EAST RAILWAY STATION

项目地点：中国 · 杭州　用地面积：45 ha

建筑设计：STI 思图意象工作室

建筑师：秦洛峰，K. Bodelle，俞淳流，黄蓉，陈蓉蓉等

LOCATION: Hangzhou, China　SITE AREA: 45 ha

DESIGN CORPORATION: STI Studio

ARCHITECTS: Qin Luofeng, K. Bodelle, Yu Chunliu, Huang Rong, Chen Rongrong

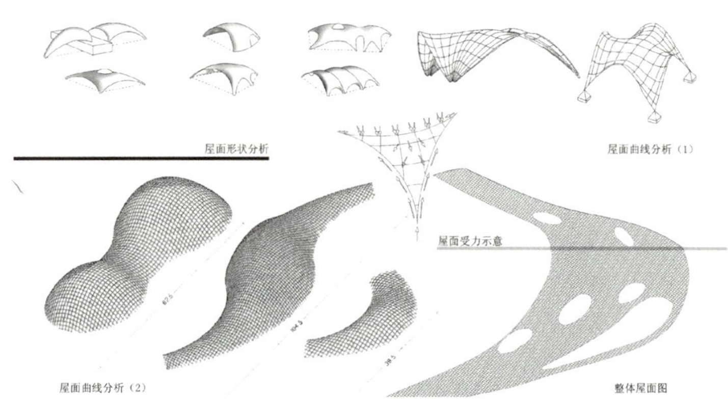

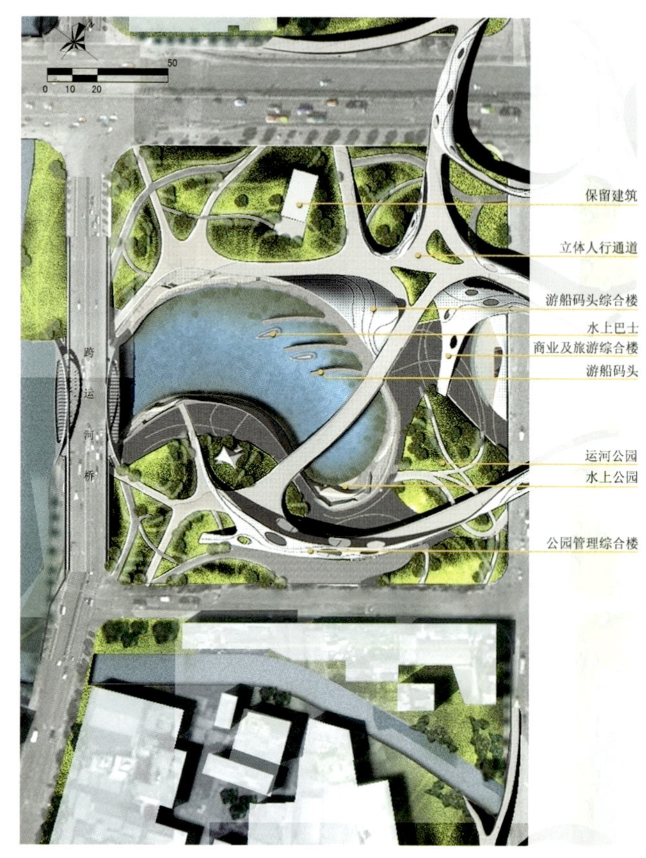

设计构思——蝴蝶·水域文化

杭州东站是衔接长三角的重要客运枢纽，综合体广场的设计将会和东站一起给长三角地区带来巨大的“蝴蝶效应”。在设计中我们将东站广场的形象构思定位在“蝴蝶”之上。在主体站房对称布局的设计理念上，将站房立面延续投影到东西两广场中，形成蝴蝶振翅之势。

规划布局：对于一个大型交通类建筑的站前广场来讲，处理好繁杂的人流、车流关系，构筑良好的交通、景观和建筑之间体系关系至关重要。在设计中，我们将各个结构分置于不同的空间层面之上，分别是建筑体系、桥面步行体系、地面广场步行体系、景观体系，使不同的体系能够发挥各自最大的效能。

西部广场南北两块区域主要以车流交通疏散为主，区域的北边布置旅游集散中心及社会停车，南边是公交汽车车站。站房主体西部入口的前场区域以人流疏散为主，布置西站前广场，同时引入“运河”这一杭州水域文化主题，将水上巴士码头和公园结合，在担负水上交通职责的同时成为广场景观的一部分。东部广场区域布置格局大致延续西部广场的布置，但在空间尺度处理上较之西广场有所扩大。由北往南依次布置短途汽车客运站、东站前广场、公交车站和邮政处理中心，加强并补充了广场的配套设施。

建筑设计：建筑单体的布置顺应整个广场肌理，形体运用多重曲面，表现自然的生长之势。如运河水上巴士码头公园，建筑造型自然扭曲，犹如一叶小舟浮于水面，既体现了其功能特点，又与东站站房建筑群体之间有良好的协调关系，构筑了变化有序的空间和丰富的滨水天际线。区块内建筑方案设计尽量结合并利用现状地形，避免破坏环境，强调建筑与环境的亲水性。东广场处的邮政支局的布置也是如此，其顶部和人行天桥相连接，整体建筑运用弧线造型蜿蜒和广场周围环境融合。同时根据邮政支局的流线，布置了邮运地道、接发、分拣、管理、停车等设施和场地。邮运地道设置在邮政支局地块的南侧，从站台通过 9 m 的地道到达地下一层。地面层则设置接发、分拣用房及邮包临时存放处。发送用房还配备了室外平台和车辆停放场地以满足功能需要。

道路桥梁构造设计的规划理念是以道路桥梁为主要覆盖面，建筑则是隐藏其中。为此我们将各功能区块布置在各个人行桥面之下，桥面在担负疏散人流的同时，也成为各个建筑的屋顶。在这里，道路桥梁的构造可行性成为了我们设计的重点。我们对不同的道路桥梁进行了一系列力学分析，得到了各自合适扭曲的形式。

建筑不仅仅是建造，它应延伸到更广意义上的层面，赋予一个地块乃至一座城市建造形式的思想。作为杭州实施“城市国际化”战略之一的东站城市综合体及其综合体广场将引领未来杭州东站枢纽成为“长三角”重要的现代化综合交通枢纽中心，展现杭州“精致和谐、大气开放”的城市形象。

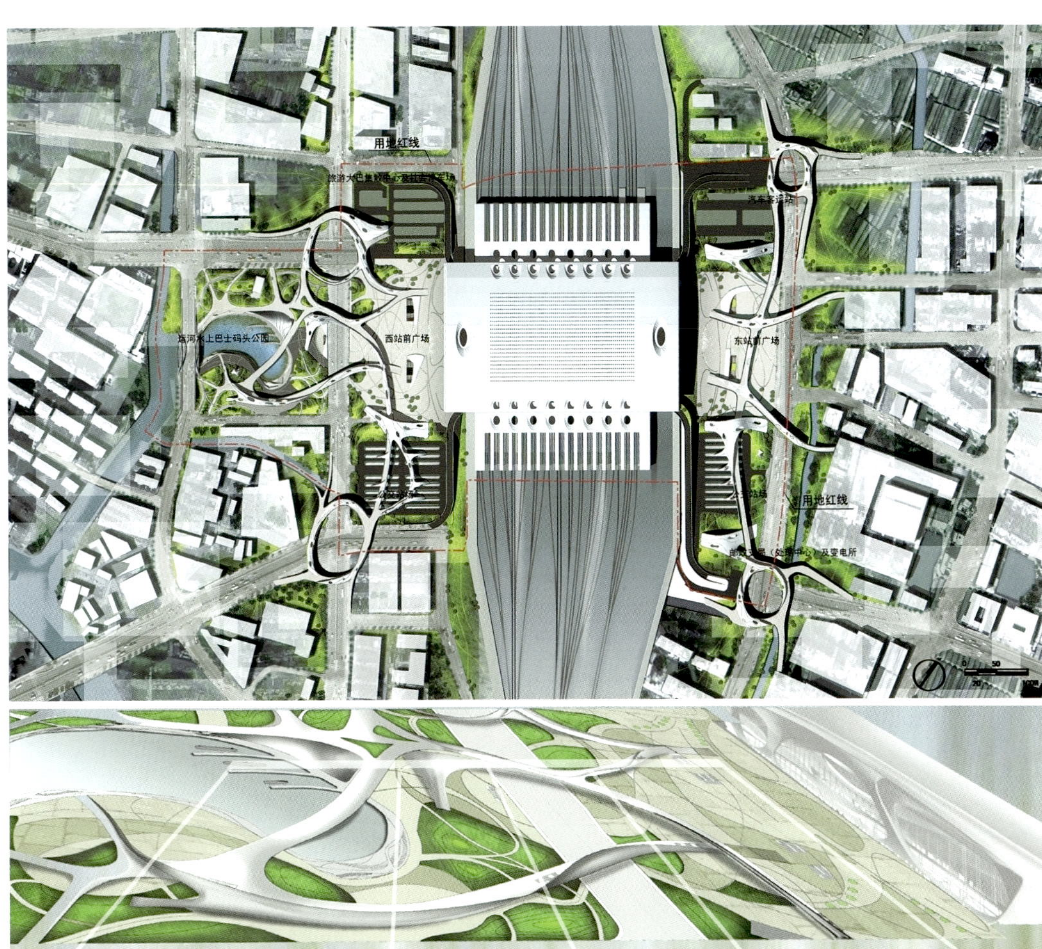

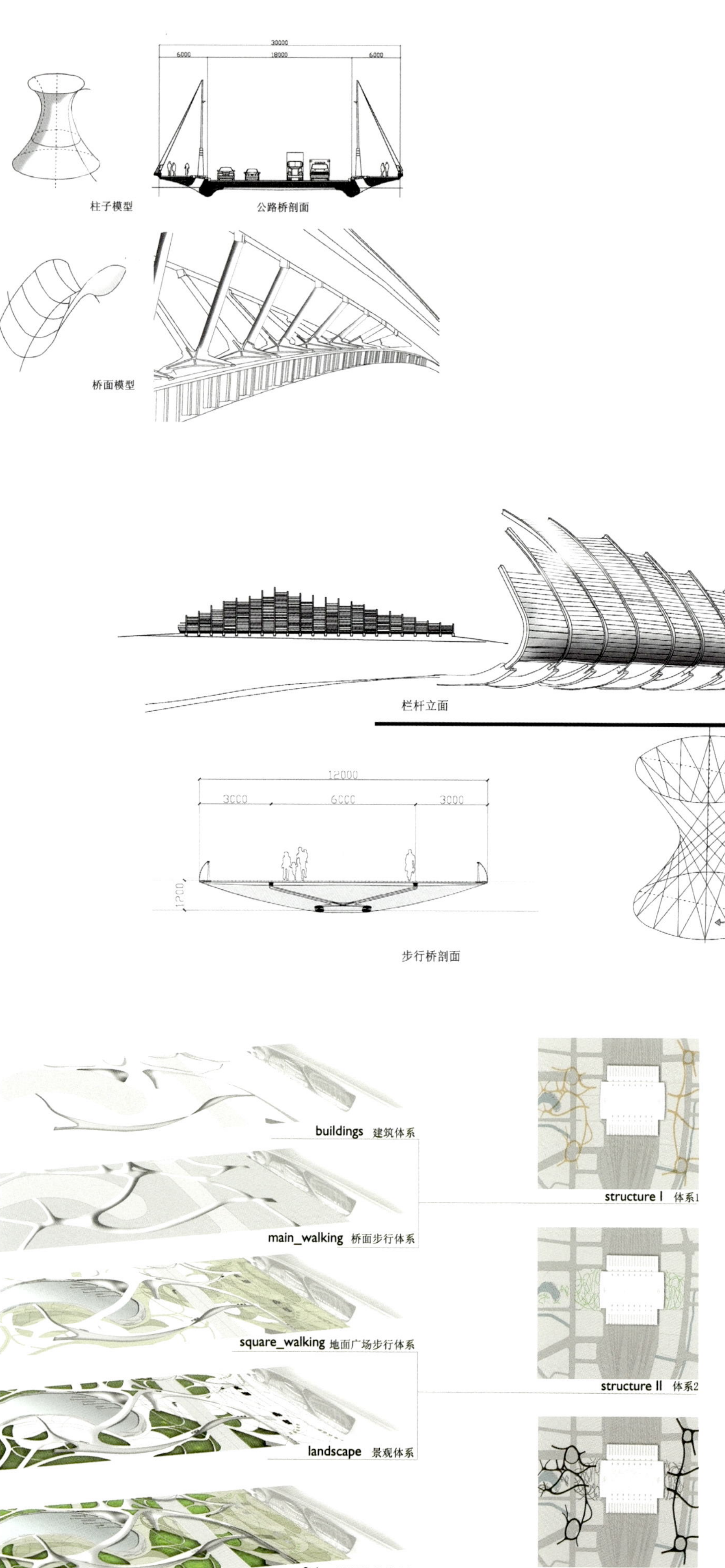

Design concept - Butterfly • Water Culture

Hangzhou East Railway Station is a major convergence to the Yangtze Delta for passenger transport. The complex square design against the background of the East Square station will bring conspicuous "butterfly effect" to the Yangtze Delta. In the design we fixed the concept of"butterfly"to the image of East Point Plaza. In the design concept of symmetrical layout of the main house in the station,it's continually projected onto the east and west squares forming the scene of a fluttering butterfly.

Layout:In terms of a station square of a large traffic building it is essential to well handle the complicated population transportation and construct a coordinated system of traffic, landscape and the buildings.In the design, each structure will be divided into the different dimensions of space above the building system. In the design we put structures on different space systems so that they can play their roles to maximum efficiency.They are building system, bridge walking system, square ground walking system and the landscape system.

North and south areas of the west square are mainly used for transportation evacuation. The north of the square is the collecting and distributing travel center and the parking area. while the south is the bus station. The front area of the west entrance to the main house of the station is mainly used for people's evacuation. The layout of the west station square introduces the "canal"theme to combine the water bus terminal with the park,which becomes an integral part of the landscape apart from its function as a transportation area. The layout of the east square is almost a maximized copy of the west square. The short distance passenger station, the east station square,the bus station and the post processing center are sequently arranged from the north to the south, which strengthen and complement the square facilities.

Architectural design: The arrangement of single building conforms to the square texture. The shape uses multiple surfaces, to express natural growth trend. For example, the water bus terminal park of the canal was built naturally tortuous, just like a boat floating on the water, which not only embodies its functional characteristics,but also is well coordinated with the east station house groups to build a orderly diversified space and a wealth of waterfront skyline. The architectural design within the area makes full use of existing topography to avoid damage to the environment, emphasizing hydropathy of the construction and environment. Layout of the branch post office is just the same. The top is connected with the pedestrian bridge, and the overall construction use fusion of the arc modeling and surrounding environment of the square. Meanwhile the mail road, receiving, sorting, management, parking and other facilities and venues are arranged according to the flow line of the branch post office. The mail road was arranged in the south of the branch post office, reaching the basement from the site through the 9-meter tunnel, while the ground floor is occupied as the sending and receiving and sorting office space and the

temporary packet storage. The sending space is also equipped with outdoor platform and mobile parking facilities to meet the functional needs.

The design and planning of roads and bridges is based on the concept of making the roads and bridges the main coverage which the building hides itself in. Therefore we arranged all functional blocks under each footbridge deck which functions not only as passenger evacuation but also as the roof of each building. Here, the feasibility of the construction of roads and bridges has become the focus of our design. We conducted a series of mechanical analysis to different roads and bridges to get the diferent suitable contorted forms .

Architecture is not just about construction but should be extended to a broader sense, to endow the land or even the city with the ideological construction form. The east station complex and its complex square, as one of the strategies of " city internationalization" of Hangzhou will lead Hangzhou East Station to the most important modern integrated transport hub of Yangtz Delta to display an exquisite, harmonious and open image of Hangzhou.

杭州市西湖小学教育集团维修校舍——外立面整治方案

HANGZHOU WEST LAKE PRIMARY SCHOOL EDUCATION GROUP MAINTENANCE—FACADE RENOVATION PROGRAM

项目地点：中国 · 杭州　用地面积：6700 m²　建筑面积：4857 m²
建筑设计：STI 思图意象工作室
建筑师：秦洛峰，Sascha Ellenberg，Felix Fritz

LOCATION: Hangzhou, China　SITE AREA: 6700 m²　BUILDING AREA: 4857 m²
DESIGN CORPORATION: STI Studio
ARCHITECTS: Qin Luofeng, Sascha Ellenberg, Felix Fritz

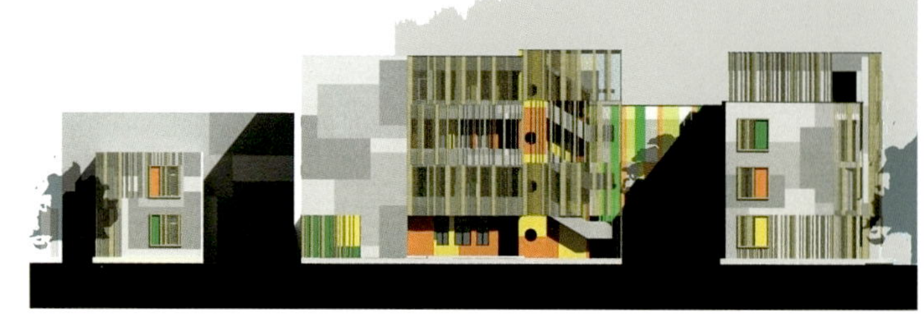

西湖小学始创于 1933 年，有着深厚的人文历史积淀。学校坐落在西子湖畔风景区内，校园倚山临湖，四周树木葱茏，环境清幽宜人。以下三点是针对现状，此次方案的重点和首要解决的问题：

· 体现该校特有的文化氛围，尊重历史，又不失现代气息；
· 体现其位于西湖风景区的环境特色，融于西湖，升华环境；
· 体现出小学所应有的个性、活泼、生动的教学氛围；

色彩构成：2006 年，市规划局讨论决定将灰色定为杭州城区的主色调。西湖小学处于主城区与西湖风景区交接点的重要位置上，考虑对此规划政策的贯彻，方案采用不同浓淡、冷暖、明暗的灰色涂料对立面进行构图，丰富的墨色相交错，与西湖的碧水、绿树相映成趣，加以烟雨朦胧，为建筑铺就一层江南水墨画的底。

蒙德里安的抽象构成在方案中进行了运用，形成了建筑与艺术的有机结合。立面采用不同浓淡、冷暖、明暗的灰色涂料以及水平线与垂直线的相交来组织其结构关系，使立面给人以画面形式的均衡感，传达出视觉形态的简约、洁净与动态的节奏等，在现代时尚文化内反映出水墨江南建筑所具有的秀丽、细腻、端庄。

木格栅干挂立面：在局部设置竖向干挂木格栅，利用其在建筑立面上放置位置和角度的变化，形成不同的光影效果，以及虚实律动，充满趣味和韵律的建筑空间。同时在立面上，木格栅是一种暖色调的东西，表现出一种东方建筑的气质。灰色涂料墙面与木格栅共同构成的建筑外表皮，赋予建筑历史的基调与和环境相融的优雅氛围。

亮色点缀：橙色是欢快活泼的色彩，使人联想到金色的秋天，丰硕的果实，是一种富足、快乐而幸福的颜色。

黄色是灿烂、辉煌的色彩，它有着太阳般的光辉，象征着照亮黑暗的智慧之光。

绿色是美丽、优雅的颜色，它使人联想到草、树等植物，生机勃勃，象征着生命。

方案在灰色调的基础上，在走廊内墙面上用这三种颜色对建筑进行点缀，给学校创造了一种欢乐与愉悦、单纯与年轻、生态与健康的氛围，符合少年儿童趣味心理，符合学校的办学理念。

生态、环保理念：尊重自然，整个方案力求使小学的建筑风格与周边环境、西湖风景区这个大背景、大自然环境和谐融洽。整个建筑的灰白色调加上竖向的木格栅，点缀在青山绿水间显得清新干练、韵味无穷，有机地融入自然山色中间。

在材质的选用上，木材为环保和自然的建材，构造上，采用预制干挂体系，简单易施工，工期短，占用的地少，施工现场干净，造价相对便宜，尽其可能减少对环境的破坏。

West lake Primary School founded in 1933, has deep cultural and historical accumulation. It is located in West Lake Scenic Area. It is close to the mountains and the lake and surrounded by verdant trees,which is quiet and pleasant.

The following three points are the primary focuses and problems of this program against the current status:

· To reflect the unique culture of the school with respect for history but without losing the modern flavor;

·To reflect its environmental characteristics in the West Lake Scenic Area and fuse it into the West Lake landscape to reach the sublimation of the environment;

· To reflect the individuality and the lively and vivid teaching atmosphere that a primary school should have.

Color composition: In 2006, the City Planning Board decided to make gray as the main color of Hangzhou City. The West Lake Primary School is located in the important position of the intersection between the main city zone and the West Lake Scenic Area. Taking this policy into account the program takes gray paint of various degrees, warm or cold, light or dark, brgiht and vague, to coat the facades. Rich and varoius inks stagger with each other and cast beautiful reflections by the West Lake's clear water and trees, and the misty and the rainy scene cast a south ink painting background for the building.

Mondrian's abstract form was applied to the program, forming a combination of architecture and art. The facades take gray paint of various degrees, warm or cold, light or dark, brgiht and vague, and the intersecting between horizontal and vertical lines to organize structural relationship to give a balanced sense of picture form and deliver visual simplicity, neatness and dynamic rhythm, which mirrors the beauty, delicacy and dignity of the Southern Chinese buildings in the modern fashion culture of picturesque south.

Hanging wooden grille facade: The various positions of the vertical wooden grille partially hanging in the facade form indistinct and rhythmical lighting effects, which is full of fun and melody. Moreover the wooden barrier is warm-colored,which shows an oriental architecture temperament. The gray walls and the wooden grille which compose the facade endow the building with historical tone and elegant atmosphere that are coordinated with the environment.

Interspersed bright colors: Orange is a cheerful and lively color, reminiscent of the golden autumn, rich fruit.

Yellow is a brilliant and splendid color, and it's brilliant like the sun, symbolizing the light of wisdom in the darkness.

Green is a beautiful, elegant color, reminiscent of grass, trees and other plants which are all vibrant and symbols of life.

The corridor wall is decorated with these three colors against the gray background, which created a happy and dilightful, young and pure, ecological and healthy atmosphere that is of psychological interest of the children and meets educational philosophy of school.

The ecological and environmental philosophy :It is the respect for nature. The program seeks to integrate the architectural style into nature and the background of the surroundings and the West Lake Scenic Area. With the gray underpainting and vertical wooden grille, it dotted in the mountains and the water is harmoniously blended with nature, which is fresh and clear and full of lingering charm.

In the material, the wood is natural and environment friendly. In construction, the vertical hanging grille system can not only be easy to build and reduce the construction period but also occupy the less area and produce less pollution to the construction site with relatively lower cost and environmental damage.

TIANJIN TIANZITUOWEI ARCHITECTURAL DESIGN CO., LTD.

天津天咨拓维建筑设计有限公司

公司合伙人及主要设计师

- 宝建华　Bao Jianhua

1984-1988年天津城建学院建筑学学士学位

1999-2004年天津市建工设计院建筑二所所长

2004-2007年至今天津天咨拓维建筑设计有限公司合伙人建筑师

2006年入选《城市环境设计》青年建筑师栏目

2006年设计作品天津红桥区民族中学获“天津市优秀设计二等奖”

2008年设计作品天津津南艺术文化中心获“天津市优秀设计三等奖”

2007-2009年中国建筑设计作品年鉴特约编委

- 方轶　Fang Yi

1996-2001年天津大学建筑学学士学位

2001-2004年天津大学建筑学硕士学位

2004年至今天津天咨拓维建筑设计有限公司合伙人建筑师

2006年入选《城市环境设计》青年建筑师栏目

2008年设计作品天津津南艺术文化中心获“天津市优秀设计三等奖”

2007-2009年中国建筑设计作品年鉴特约编委

- 郑盟　Zheng Meng

公司OPEN工作室主持建筑师

2001-2004年天津大学建筑学院城市规划专业 硕士学位

2004-2008年中国建筑设计研究院崔愷工作室

2008年至今 公司OPEN工作室主持建筑师

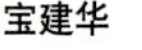

宝建华

方轶

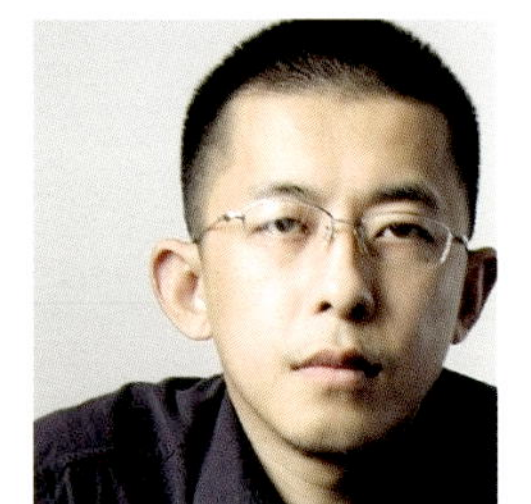

郑盟

天津天咨拓维建筑设计有限公司成立于1999年。在建筑设计日益专业化、开放化、市场化的背景下，公司于2004年进行了改革，并以现代企业制度的要求做到了运作规范化、管理先进化、模式专业化。公司拥有自成体系的设计理念，充满创意的工作方法，融洽的人际关系，能最大限度地激发设计人员的创作激情。公司中拥有一批高水平专家，具有很强的设计实力和极其丰富的设计经验，在方案设计和施工图设计方面也均有较强的实力。公司的主要业务范围为规划、建筑设计、景观设计和室内设计，8年间已完成各类工程70多项。

团队精神

方案创作强调以设计主创人为中心，设计组随时研讨，根据分工，同步协调展开设计工作，形成互动式的工作状态。工程设计强调以工程主持人为中心，主创人对整体空间进行处理，细部效果则根据工程设计进度同步把握。

创新意识

从机械的工作中跳出来，去体验活生生的、创造性的现实生活。主张多视角、多思维、见效快、收益高的消费文化，推崇具有个性的原创文化。

服务意识

在建筑创作活动中，设计师们学会了换位思考，更多地去倾听、分析和满足业主的需求。建立诚信的合作关系。公司强调行为自律和职业道德。在创造精品与追求经济利益之间，强调道德与良好习惯的平衡作用。

Tianjin Tianzituowei Architectural Design Co., Ltd. was set up in 1999. Under the background that the architectural design becomes increasingly specialized, open and market-oriented, the company undertook enterprise innovation in 2004, and realized the standardized operation, advanced management and specialized mode according to the requirements of modern enterprise system. The company has the design concept which forms its own system, the working methods that are full of originality and the harmonious interpersonal relationship, and can maximize the creation enthusiasm of design staff. The company owns a group of high-level experts who have great design abilities and rich design experience, and are expert in conceptual design and construction drawing design. The main business scope of the company covers planning, architectural design, landscape design and interior design, and has fulfilled more than 70 projects of various kinds during the past eight years.

The Group Spirit

For the schematic creation, the main designers are centered with coordination of the project designers. For the project design, the project designer in charge is to control the space design and detailed impression according to the design process.

Creation Consciousness

We expect to extricate ourselves from the mechanical working and to experience so active life. We maintain the consumption culture with multi-views, multi-thinking and with quick result and high profit. and we praise the original creative culture with special characteristics.

Service Consciousness

During the architectural creation ,we have understood how to listen ,analyze and meet the requirements of the clients based on their conditions and to establish honest cooperation. We emphasize in-corruptive behavior and professional moral. Both of the moral and the better habit should be balanced during creating the excellent works and concentrating on the economic benefit.

汽车研究中心

AUTOMOTIVE RESEARCH CENTER

项目地点：中国·天津　建筑面积：150 000 m^2
建筑设计：天津天咨拓维建筑设计有限公司、开放建筑工作室
建筑师：郑萌，方轶

LOCATION: Tianjin, China　BUILDING AREA: 150,000 m^2
DESIGN CORPORATION: Tianjin Tianzituowei Architectural Design Co., Ltd. , Open Architecture Studio
ARCHITECTS: Zheng Meng, Fang Yi

我们所处的时代无疑是与汽车相关的，汽车在城市中的流动，无论从微观或是宏观的层面上都建构了新的秩序和景观。所以，着手于中国汽车技术研究中心这个项目时，对于由汽车能带来的新的城市要素（如屋顶停车、汽车专用道、立交系统等），我们给予了特殊的关注。如果说当代城市为汽车而产生了新的形象，那么汽车研究中心应该是这种形象的强化和功能的集合。

在宽阔的用地，我们用车行道串联起用地的各个部分。车行道本身又是自循环和立交的，它保证了汽车在其上的行驶质量，用最精简的方式描述出理想的汽车交通的组织形态。道路上下穿梭于屋顶停车场及各个试验区，无疑将车辆作为场地的主角和主人。

将用地根据滋扰和适宜的程度，分为绿区、蓝区和灰区。这是针对于人的感受进行设定的，并通过这种区分来建立功能的组织关系。厂房和试验建筑的外墙采用被动式的太阳能墙体，让环保的概念能落实到高效和低成本的层面上。建筑体造型目标在于反映出新的建造技术和精细加工的当代特征，采用不易污染变旧的材料。

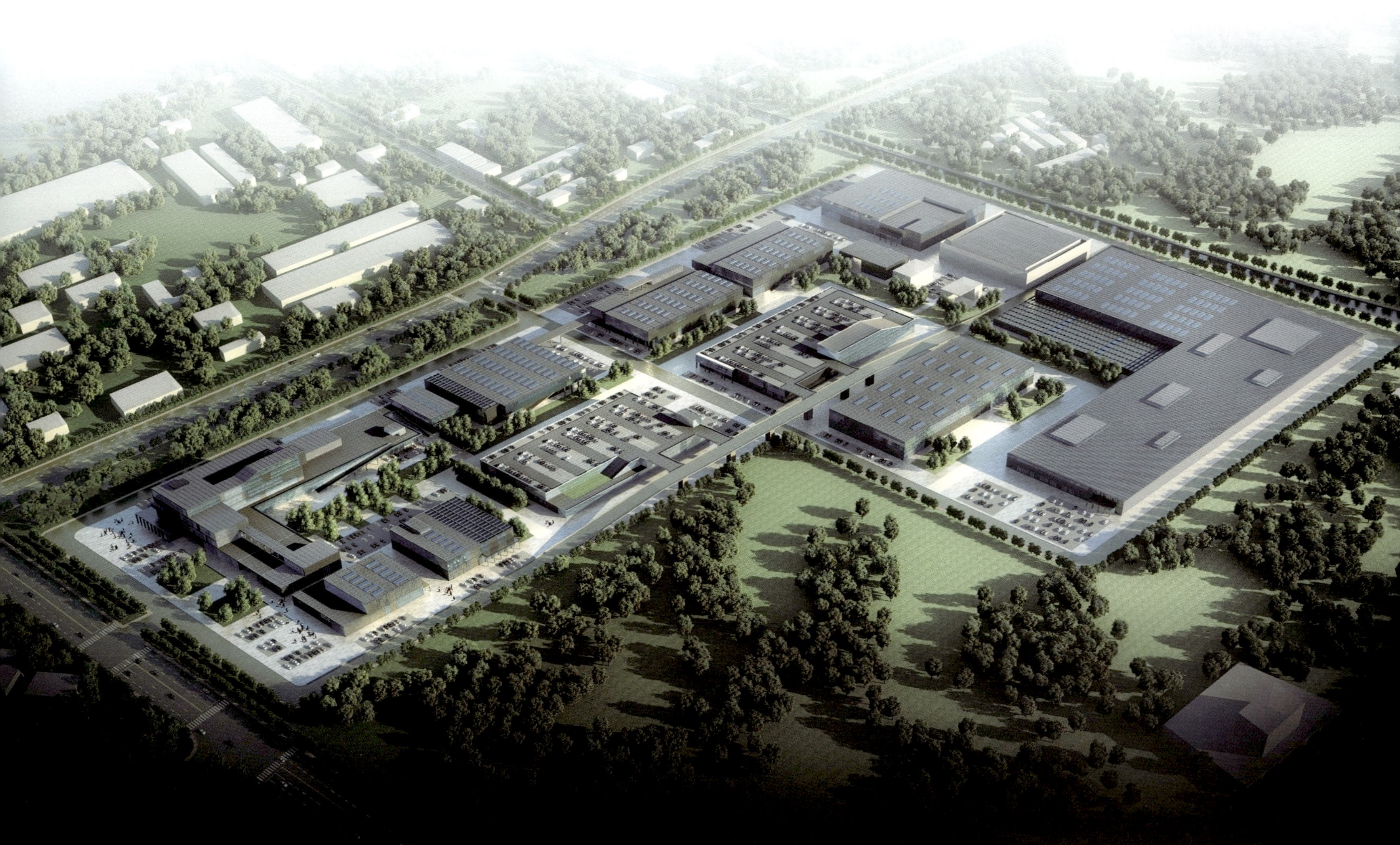

Our time is undoubtedly associated with the vehicle, whose flowing in the city creates new order and the landscape whether in micro or macro sense. Therefore, at the beginning of the China Automotive Technology Research Center design,we paid special attention to new urban elements brought by the mobiles(such as rooftop parking, driveway, flyover systems, etc.), we have given special attention. If the modern city created a new image for its cars, the Automotive Research Center should be the strengthened condensation of this image.

In the wide land, we use the driveway to completely get through parts of the site. Driveway itself is a loop and an interchange to guarantee that the travel quality of the car, with the most concise way to describe the ideal organization of automobile traffic patterns.The car shuttles back and force through the roof park and the pilot site undoubtedly dominating the whole site.

We divided the site into different areas of green, blue and grey according to the degree of disturbance and appropriateness. It serves to be in line with huamn feelings. Through this distinction we establish organizational relationship between functions.The factories and the test buildings use the outer passive solar power wall to ahieve environment conservation on a basis of low costs and high efficiency.

The physical modeling of the building is to reflect contemporary features of new construction techniques and fine processing with pollution-free and endurable materials.

天津市优联集团第六城商业中心区

THE CENTRAL BUSINESS DISTRICT OF THE SIXTH CITY OF YOULIAN GROUP TIANJIN

项目地点：中国 · 天津　建筑规模：330 000 m²
建筑设计：天津天咨拓维建筑设计有限公司
建筑师：宝建华，张杰，徐珩

LOCATION: Tianjin, China　SCALE: 330,000 m²
DESIGN CORPORATION: Tianjin Tianzituowei Architectural Design Co., Ltd.
ARCHITECTS: Bao Jianhua, Zhang Jie, Xu Heng

项目设计试图以大型滑雪馆为商业核心，合理组织大型百货商场与特色步行街、购物与餐饮休闲、酒店与公寓、地下与地上等诸多关系，将商业功能，社会生活和城市设计三者有机地结合起来。

建筑布局点、线、面相结合，综合商业以及大型室内滑雪馆摆放在东南角的交叉路口处，发挥主力百货和大规模体育娱乐的龙头作用，配以特色风情街、商业步行街、商务酒店，打造创造高品质的大型综合商业区。

通过对人群的步行习惯、流动线路的研究，综合考虑走道、店铺、竖向交通核，使得购物路线顺畅自由，四通八达，避免了综合商业和步行街中商业死角的出现。同时在步行街的设计中还考虑到楼层对商业价值的影响，试图削弱步行习惯对楼层的敏感反应。用大平台、室外楼梯、内街共享等手法，使得各个楼层与顾客的关系更为亲切，商业氛围更加自由、愉悦。

在步行街中设置下沉广场，结合台阶设置的舞台，形成地下购物广场，使之成为融购物、活动、景观、交通的局部核心，并在设计中融入了“二首层”的设计概念，更好地引导人流走向，弱化地下层与首层的感官区别，使地下购物广场的商业价值，接近于首层。

在建筑单体的处理上，迎合商业建筑的特殊要求，最大限度地满足未来商业区内的每个商家的利益。创造尽可能多的沿街、沿广场的商业界面，提高商业价值。同时在建筑形体和建筑立面的塑造时，将建筑品质感和商业气氛感有机地融为一体。建筑群塑造连续的城市界面和丰富的天际线，统筹设计广告位置，维护沿街形象完整，为消费者创造高品质的购物环境，营造良好的购物气氛，激发消费者的购物欲望。

在设计中对于业态的分布上，也有一定的考虑。在人流的主要界面上作一些建筑节点的处理，所谓主力店面。一方面丰富了建筑立面，给购物人流带来不同的新鲜感。另一方面，在业态上使主力点分布较为均匀，更好地带动整个商业区的商业价值。

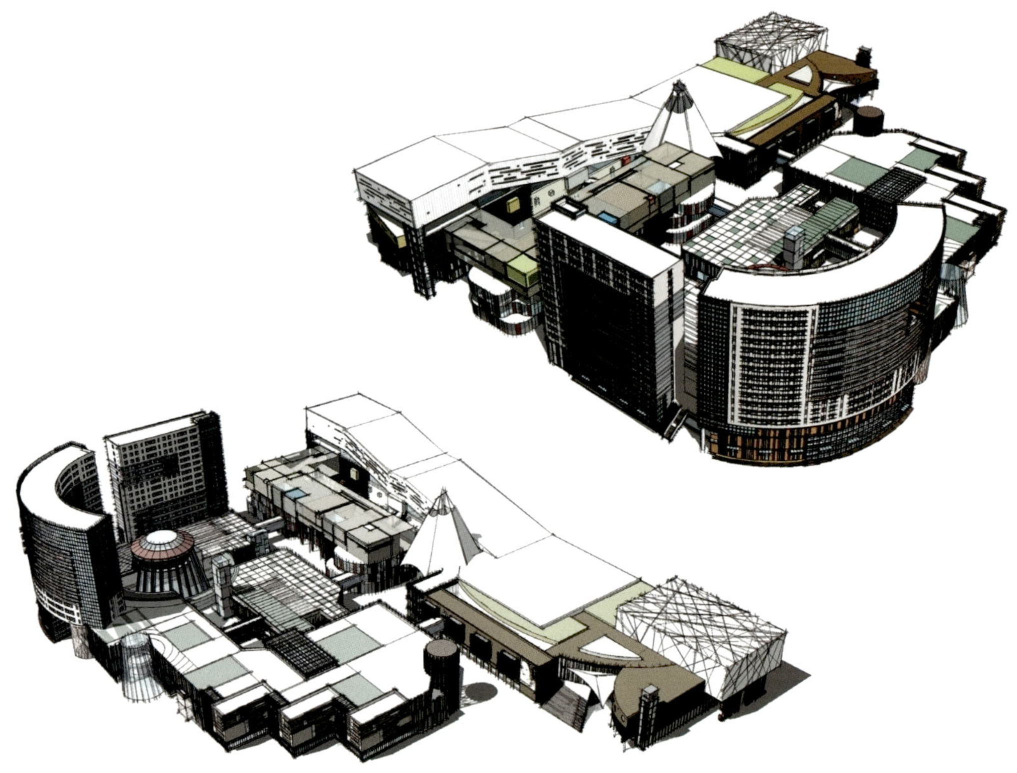

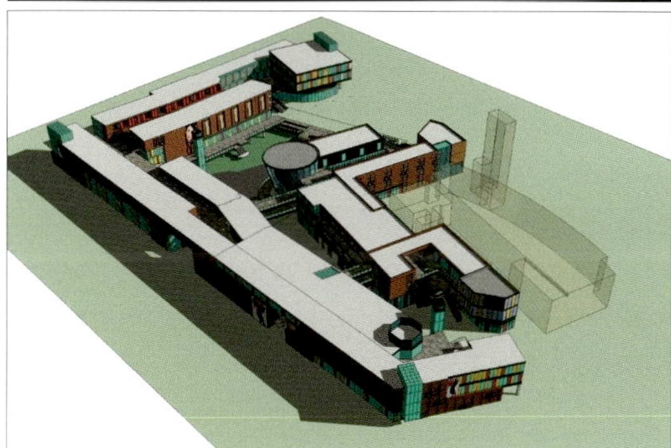

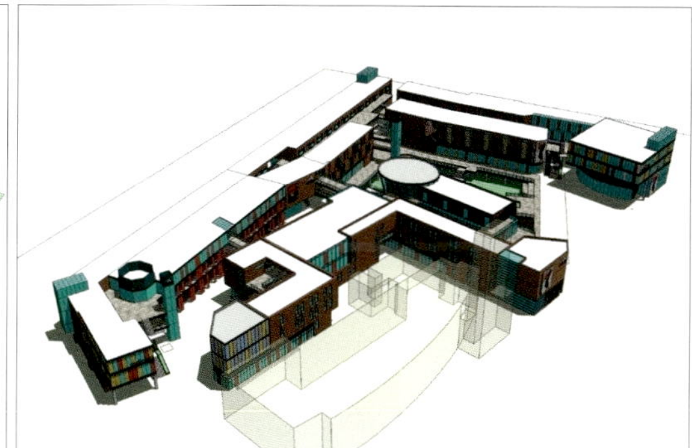

Project design attempts to make the large ski hall as the commercial center and properly dealt with the relation between large department stores and characteristic pedestrians, shopping,dining and recreation, hotel and apartments, the underground and the overground to achieve rational organization of the business function, social life and urban design.

Construction layout integrates points, lines and planes. The whole commercial area and large indoor ski hall at the intersection in the southeast corner function as the major department stores and large-scale sports entertainment areas. While the characteristic style street, commercial walking street, business hotel together create high-quality and large-scale comprehensive commercial district.

Taking full consideration of the walking habit and the flow route of the crowd and the walkways, shops and the vertical transport core, the design attempts to make the shopping lines smooth and free extending in all directions. It also avoids dead commercial corners in the comprehensive business and pedestrian streets. The design of pedestrian street also takes into account the impact on commercial value brought by building floors to reduce the disagreement between the walking habit and the floors. The techniques of using large platforms, outdoor stairs, shared inner streets to make it possible to build intimate relationship between floors and customers and casual and pleasant business climate.

The sunken plaza set in the pedestrian street, combined with steps of the stage forms the underground shopping plaza which is local core of the shopping, activing, landscape, and transport area. It fused into "second head floor" design concept,taking advantage of the downward population flow to better guide the crowd moving trend and weaken the sensory difference between the basement and first floor, so that the commercial value of underground shopping mall can approach the first floor level.

The handling of construction monomers meets the special demands of commercial buildings, and ultimately fulfills the interests of the merchants of the future business district. We created as many commercial interfaces as possible along the streets and the plaza to increase business value. Meanwhile the building form and the facade fully integrate the quality sense of architecture into business atmosphere. The buildings create continuous interfaces and rich city skyline to organize the ad positions as a whole and maintain integration of the street image, providing high-quality shopping environment, joyful shopping atmosphere for consumers,and stimulate their consuming desires.

We also give consideration to the industry distribution, that is, to use construction nodes as the main store,which on one hand enriches the building facade to bring different senses to the crowd, and on the other hand equalizes the main points distribution to better promote the commercial value of the business district.

天津普育学校

TIANJIN PUYU SCHOOL

项目地点：中国 · 天津　建筑面积：39 000 m²
建筑设计：天津天咨拓维建筑设计有限公司
建筑师：宝建华，方轶

LOCATION: Tianjin, China　BUILDING AREA: 39,000 m²
DESIGN CORPORATION: Tianjin Tianzituowei Architectural Design Co., Ltd.
ARCHITECTS: Bao Jianhua, Fang Yi

设计的构思来源于对旧学堂的呼应，对传统四合院的空间分析——重视礼法，中轴分明，整体空间格局整齐，由建筑实体围护形成院落，对外封闭，对内开敞；整体布局层层递进，每进空间以垂花门或月亮门相连，彼此渗透。在分析了基地的现状后，设计者采用了入口局部运用中轴线、体块顺应地形转折的方法。建筑整体布局方正有序，北侧小学区建筑体块则顺应地形做转折；在学校的主入口，围绕复原的旧学堂设置礼仪广场，两侧建筑形体、立面处理手法基本一致，广场尽端的大门体形简练对称，形成了中轴对称、严谨大气的入口空间。小学区和中学区各自形成两进院落，院落使建筑形成内敛向心的形态，而院落与院落之间以月亮门和玻璃连廊相隔，或开或闭，有效地丰富了空间层次，使相邻空间之间既相互连通又彼此隔断，成为一个互相渗透的有机体。

Design ideas come from the correspondence to the old school and spatial analysis of the traditional courtyard-emphasis on etiquette, the clear axis and the neat spatial layout; building entities enclose to form courtyard internally open and externally close; the overall arrangement go froward layer by layer, spaces are connected with each other by festoon gate or moon gate. In the analysis of the base status the designer takes use of axis and body mass at the entrance to comply with the terrain. The entire layout of buildings is regular and in order, and the north side of school district building turns to adapt to the terrain; in the main entrance of the school is the etiquette square

around the restored old school. The building forms and the facade are basically the same. The gate form at the terminal of the plaza is concise and symmetrical, forming a rigorous and dignified entrance space. The primary zone and secondary zone are both two-entrance courtyards which are formed the inward-looking shape enclosed by the buildings within it. While the courtyards are partitioned by open moon gates and glass corridors,which are opened or closed effectively enriching the space levels, and making the adjacent spaces connected or disconnected to form a mutually-penetrated organism.

天津市友谊路 35 号地块

BLOCK 35 OF TIANJIN FRIENDSHIP ROAD

项目地点：中国·天津 建筑面积：90 000 m^2
建筑设计：天津天咨拓维建筑设计有限公司
建筑师：方轶，宝建华

LOCATION: Tianjin, China BUILDING AREA: 90,000 m^2
DESIGN CORPORATION: Tianjin Tianzituowei Architectural Design Co., Ltd.
ARCHITECTS: Fang Yi, Bao Jianhua

· 延续友谊路的城市空间结构，丰富街道空间形态，基地的空间形态对于区域空间的完整起到丰富和加强的作用。
· 满足和解决地盘发展与区域功能的需求关系。
· 项目形象的整体性和标志性。
· 考虑建立现代的人文空间环境。
· 考虑建立功能完备的公众服务和商务服务。

• To maintain the urban spatial structure of the Friendship Road and enrich street space form which plays an important role in the completeness of the regional space.
• To handle the relationship between the regional development and the functional needs.
• The integrity and representability of the project.
• To establish a modern human space environment.
• To establish complete functional public services and business services.

WSP ARCHITECTS

维思平建筑设计

张瑛　克劳德·罗森　吴钢　陈凌

WSP是一个国际化的建筑设计事务所，1996年在德国慕尼黑成立，经过15年的发展，已成长为在德国慕尼黑和中国北京、中国杭州三地拥有70多名规划师、建筑师、环境设计师和室内设计师，具有丰富的大型项目运作经验和众多成功合作客户的著名设计企业和行业先锋。

WSP在四位拥有欧洲教育背景和国际化的设计实践经验的建筑师和环境设计师：吴钢、张瑛、陈凌、Knud Rossen的领导下，其在城市设计、城市综合体和办公建筑、住宅区设计、商业与休闲项目、大型公园和环境项目和室内工程项目的规划和设计中所展示的创造性能力、先锋的设计理念和不懈的探索精神得到了公众和学术界广泛的认可。

WSP主持和参与的项目遍及欧洲和中国40多个城市和地区，其合作的客户包括德国西门子公司、中国华润集团、中国万通集团、中国金地集团、万科集团、中国用友软件公司和新加坡凯德置地等著名企业和许多城市的政府机构。WSP最重要的项目有奥地利INFINEON总部、新加坡西门子总部、南京长发中心、北京新首都机场酒店、北京龙山教堂、北京中信国安会议中心、北京渡上别墅区、深圳金地梅陇镇等。

WSP的核心优势来源于其独特的多专业、多文化和国际型的主设计师群和设计团队。WSP云集了一批不同年龄，来自不同背景经历的设计师，他们带来了创新的设计概念和手法、先进的设计管理经验和各种文化的精华。基于这一核心优势，WSP建立了一个由多学科专家、工程师、生产厂家和客户共同积极参与的多领域综合设计的工作模式和团队协同合作的工作流程。这一独特的工作模式和流程保证了每一个项目中必需的思想的活跃、设计的创新和最终项目的成功。

WSP的核心优势还来源于其对卓越设计的不懈追求和探索。不论是在南京长发中心150 m的超高层项目中，高技术要求的西门子工业厂房项目中，还是在只有三个一层农家小院改造的北京会议中心项目中，WSP都给予同样的关心和重视，充分分析设计任务中每个要素的重要性和整体性，自始至终将创造性、和谐性和超越性完美统一作为一个完整的卓越的设计提供给客户。

WSP的作品在大量的专业杂志和公共媒体上出版、发表，并在2008年“荷兰设计周”、2008年“西班牙建筑节”、2004年“中国国际建筑艺术双年展”、2003年德国杜塞尔多夫市“中国当代建筑展”、2002年中国北京“WA2002建筑奖”展等展览上展出。

WSP的作品至今已获得了广泛的关注，并赢得了40余项国际竞赛和建筑奖，其中包括英国AR AWARDS奖、LEAF AWARD全球华人青年建筑师奖、CNBC建筑奖、中国建筑学会建筑创作大奖、美国芝加哥国际建筑奖、德国BAUWELT建筑奖、WA中国建筑奖等。WSP主设计师入选中国100位最具影响力的建筑大师、中国十大新锐建筑师。主设计师们不仅在专业领域和房地产界具有广泛的学术影响，同时还执教于著名学府南京大学及香港中文大学等。

WSP Architects is an international comprehensive design firm. WSP was founded in Munich, Germany in 1996. Due to the steady development in last 15 years, WSP has established offices in Munich, Beijing and Hangzhou, with over 70 planners, architects, landscape architects and interior designers. It has become one of the leading design institutes in the industry, holding rich management experiences for large-scale projects and keeping successful cooperative relations with many clients.

All the four principal architects and landscape designers, Mr. Wu Gang, Ms. Zhang Ying, Mr. Chen Ling and Mr. Knud Rossen, boast their international designing practice and European educational background. Now the public and the professionals have well recognized its creativity, its advanced design concept and its explore spirit which are well displayed in their planning and designing works of urban comprehensive developments, public buildings, large-scale industry and enterprise headquarters, social housing projects, commercial and recreational facilities, main theme parks and landscaped environments, as well as the interior engineering projects.

WSP takes charge or participates in projects in over 40 cities or regions across Europe and China. Its clients include famous enterprises such as the Siemens, the China Resource Group, Beijing Vantone Industry Co.Ltd, Vanke Group, Gedmale Group, UFsoft and the government in many cities. The featuring projects of WSP are the Infineon Headquarter in Austria, the Siemens Headquarter in Singapore, Nanjing Changfa Center, New Capital International Hotel Beijing, Longshan Church Beijing, Beijing Citic Guoan Meeting Center, Beijing Other Villa, Gedmale Meilong Town etc.

The core advantage of WSP comes from its multi-discipline, multicultural and multi-national group of principle designers and its staff team. WSP attracts a group of designers of various ages and with individual background and experience. They bring about the innovative design concepts and techniques, the advanced design management and different cultural soul. Based upon this core advantage, WSP establishes a work method of interdisciplinary design with the active participation of multi-disciplined experts, engineers, manufacturers and clients. The collaborative working process is achieved by the coordination and cooperation of all team members. This unique work method and process ensure the active ideas, the innovative designs and the final success of every project.

The core advantage of WSP also comes from its persistent pursuit and its tough exploration for design excellence. Whether it is a highrise project such as the 150-metre-high Changfa Center in Nanjing, or it is a project with hightech requirement such as the industrial workshops for Siemens, or the Beijing Convention Center which is of the renovation of 3 courtyard houses, WSP always pay equal attention and regards to them. WSP will thoroughly analyze the importance and the integrity of every element in the design, combine the creativity, the harmony and the transcendentalism together from the beginning to the end and provide the client an integrated and excellent design work.

The works of WSP is published in many professional journals and public medias. They are also exhibited at 2008 "Netherlands Design Week", 2008 "Spanish World Architecture Festival" ,2004 "China Contemporary Architecture exhibition" in Germany in 2003 and at the Exhibition of WA 2002 architecture award in Beijing. Thus far, the works of WSP has attained broad attention and won over 40 international competitions and architectural awards, including AR Award in England. Leaf Award, World Young Chinese Architects Award, CNBC Asia Pacific Property Awards, the Chicago Athenaeum International Architecture Award, etc. The principal designers of WSP are elected as the Ten Greatest Architects in China in 2005 and the 100 Most Influential Chinese Architects in 2004. They not only have wide academic influences in the professional field and the real estate industry, but also deliver lectures in many famous schools and universities.

苏州金墅商业街坊

JINSHU COMMERCIAL CENTER SUZHOU

项目地点：中国 · 苏州　用地面积：41 000 m^2　建筑面积：60 000 m^2
建筑设计：维思平建筑设计

LOCATION: Suzhou, China　SITE AREA: 41,000 m^2　BUILDING AREA: 60,000 m^2
DESIGN CORPORATION: WSP Architects

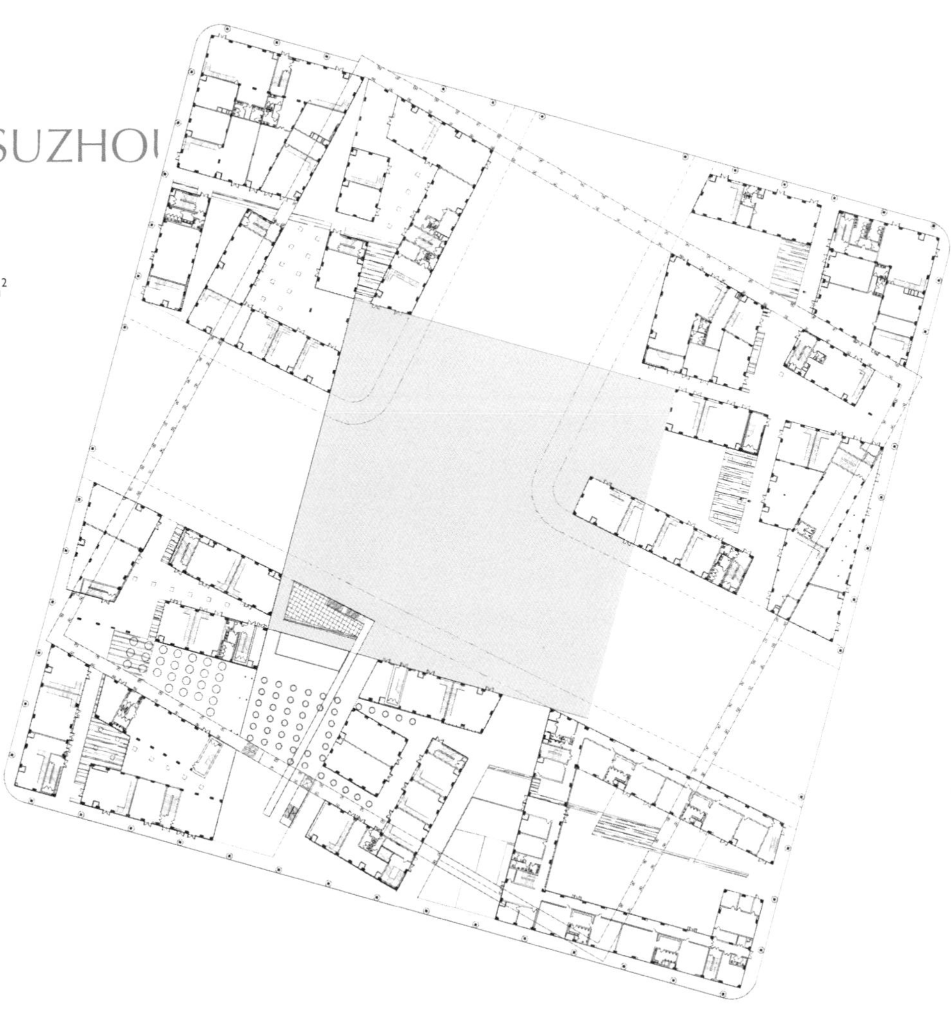

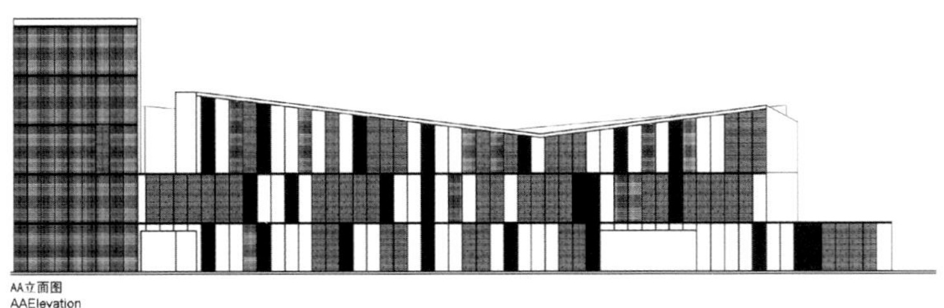

AA立面图
AAElevation

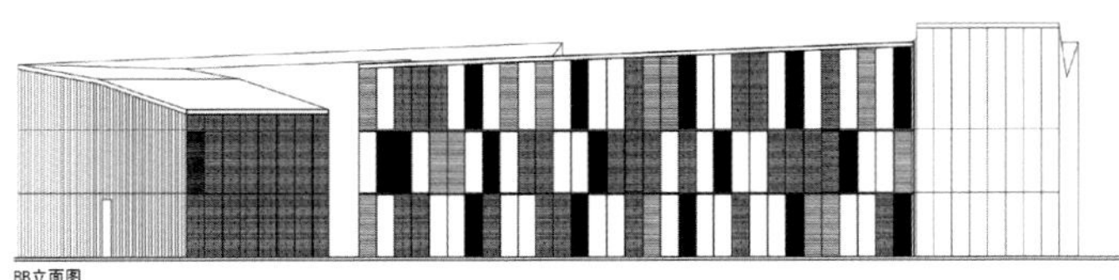

BB立面图
BBElevation

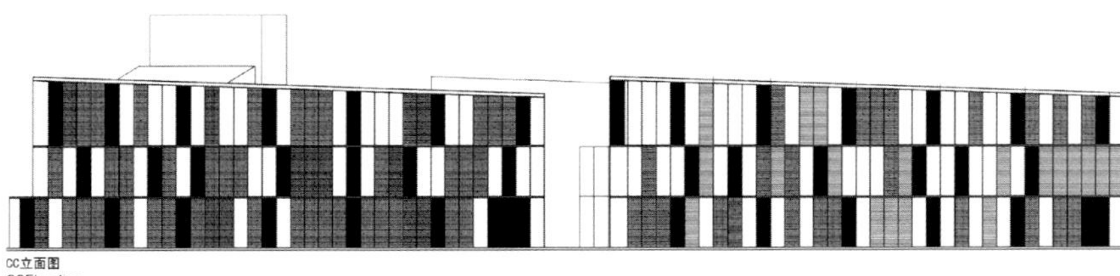

CC立面图
CCElevation

院落、天井和街巷的空间结合

项目位于苏州工业新区内部。设计师希望在苏州固有城市肌理和传统传承的基础上创造一座新型的商业场所。

构思来源于对苏州街坊式建筑形态的分析，结合院落、天井和通道的空间，这些空间界定街坊，形成公共集合场所，由此编织出特定的空间形态，展示苏州传统居住和商业文化特质。

设计师在设计手法上运用了化整为零、多重轴线的引入、体块的转折等方法。因地块接近于方形，三个向心缩小的方形构成形体的骨架，沿此骨架布置的商业单元形成了流动的街，每一个单元都是沿街的。利用中央T形现状规划路构成的方形中央广场，辅助景观灯阵，界面上建筑体量局部拔高的手法形成整个商业街区的中心。

方案对商铺、餐饮酒吧、会所娱乐、幼儿园等四类不同使用功能的研究从类型学出发，将建筑最小单元确定为 9 mX7.2 m，由此规定了基本的构成方式并留有供多种不同意愿使用而发展的可能性。外墙系统的双层外皮使商业体更具有特质性和整体性。外围的外层皮采用有遮光作用的金属网，部分位于空调机位前起遮挡作用，对整个外立面也表达了内敛、闭合的特征。和外围不同，面对内庭院的立面表达的是生动、更加开放的主题。内外立面之间的并置契合了传统文化里内与外、张扬与内敛的辨证关系。

庭院在方案里不仅承担建筑内活动拓展的容器，也是充满感性的墙面背景，通过今后使用者对院墙的开闭及不同的围合材料和方式的应用，庭院甚至可以变成公共小广场，也可以成为完全封闭的服务性内院，以满足使用者今后可能改变的实际需要。

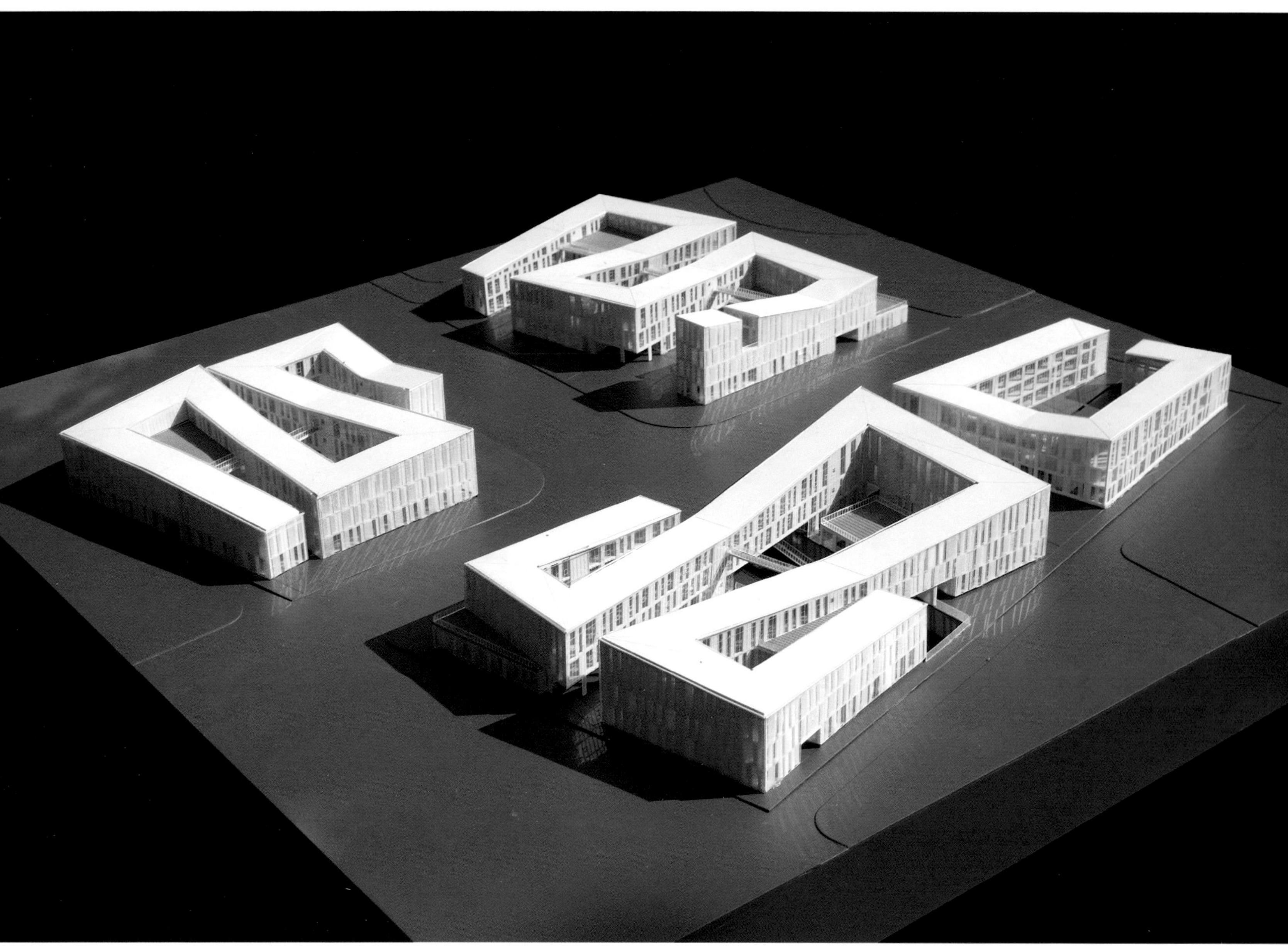

Negative spaces of yards, patios and alleyways

Our initial design idea came from analysis of the existing city layout of the Suzhou lanes that are composed of residences in a way similar to cell structure with different functional forms. The outside protective structures form the negative spaces of yards, patios and alleyways.

After analyzing the situations of the existing plot, we adopt the methods in our design of disassembling integrity into pieces, multifold-axe introduction and mass turning. Typology is used in the designing of the four different functions of shops, restaurants/bars, clubs/entertainments and kindergarten, with a minimum cell of 9 m × 7.2 m, which defines their basic formation manner while reserves different development possibilities for different uses.

Our finalized design has different sizes and scales in space, while the cells composed of different plots have the same origin typologically. Thus, the dialectic relations of interiority and exteriority of the common cells of Suzhou's traditional blocks are inherited here, forming the starting point of the reconstruction, instead of the simplified imitation of traditional manners of white walls and gray tiles. As far as the exterior wall system is concerned, the double-exterior coating design will have a clear logic division functionally and ecologically. The differences of commercial masses set in a large surrounding of residences will make them more characterized and integral. Being a vessel containing activities and a wall-background full of sensibilities, the design of courtyards in the plans can be boasted as the most and diversified task for us.

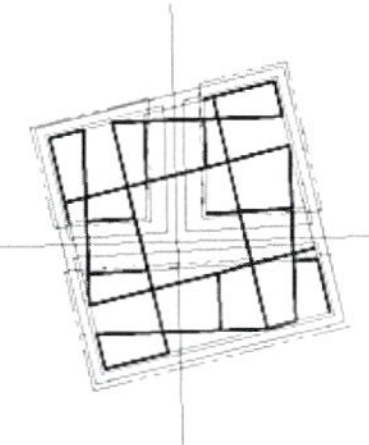

北京龙山教堂

LONGSHAN CHURCH BEIJING

项目地点：中国 · 北京　用地面积：3813 m^2　建筑面积：1380 m^2
建筑设计：维思平建筑设计
摄影：舒赫，姚力

LOCATION: Beijing, China　SITE AREA: 3813 m^2　BUILDING AREA: 1380 m^2
DESIGN CORPORATION: WSP Architects
PHOTOGRAPHERS: Shu He ,Yao Li

盛满光的容器

富有西方宗教文化内蕴又独具圣灵气息的教堂建筑，诞生在北京怀柔旅游风景区大规模低密度别墅区——龙山新新小镇中，在中国式高速度、大批量的开发模式中营造出人文色彩的文化氛围。

两个体量前后相接组成教堂的主体：入口处的附院呈现出 4 m × 4 m 的内院空间，这是通向主厅的前院，又是一处净化、沉淀心灵的空间。院内平铺的灰色石子，墙面裸露的混凝土恰到好处将沉淀的意境在厚重与质感中蔓延。与前院并置的是同样平面大小的主厅，一主一附的屋面都呈 45° 斜面，却表现为两个一大一小，一凹一凸，一敛一扬的相反空间，于稳重的宗教空间里，融入了视觉上的变化。

建筑的立面材料采用了极具质感的蓝灰色玄武岩，厚重而肃穆。一系列 20 cm 宽的竖向窄条开窗看似随意却又暗藏规律地排列其上，按一定比例系数向高处递增。阳光透过高高低低的窗口洒进教堂里，给肃穆的教堂平添些许温暖，又让一道道的光束引导心灵净化。

与教堂所处的小镇优美而不失现代气息的风景相对应，教堂周围的景观设计简练而概括，挺拔的水杉树、毛石围墙，以及广场上的自然条石坐凳，都塑造出与周边居住环境不同的气氛。钟塔简洁而具标志性，每天为居住在小镇 400 000 m^2 建筑面积的居民送去悦耳的钟声。

The light vessel

A religious building, rich in both the contents of western religious culture and the atmosphere of Holy Spirit, was created at Longshan Town, a large scale low density housing area in the scenery zone in Huairou County, Beijing, China. Under the Chinese style of high speed and large-scale development, the church produces a culture atmosphere with human culture character. The church consists of two linking-up masses: The first space to enter is a space of 4 m × 4 m, a place of both the forecourt toward the main hall and the space for meditation and soul purification. The front hall is paved with raw wood floor and the sand stone uncovered on the walls spreads the deposited artistic conception among the thickness and mass. Behind the front hall and the forecourt is the main hall with the same size as the front hall. The roofs of the two halls are all declined at 45°, giving an image of two contrary spaces that one is big and the other is small; one is cupped and the other is raised. Massive and solemn, the exterior elevation of the construction is made of blue-gray basalt stone. A series of vertically narrowed widows of 20 cm wide are rich in rhythm. The elevations of the windows are raised at a given slope. The sunlight spreads into the church through the fluctuated windows and makes the solemn church warm and peaceful.

Corresponding to the elegancy and scenery of the small town where the church is located, the design of the landscapes surrounding the church is simple and recapitulatory. With the tall cedar trees, raw stone walls and the natural stone benches in the square, it delivers an atmosphere different from its neighboring residential surroundings. The clock tower delivers pleasant bell sound to the citizens who live this small town.

北京中信国安会议中心庭院式客房

THE COURTYARD SUITES OF SPRING VALLEY RESORT BEIJING

项目地点：中国 · 北京　建筑面积：8172 m^2
建筑设计：维思平建筑设计

LOCATION: Beijing, China　BUILDING AREA: 8172 m^2
DESIGN CORPORATION: WSP Architects

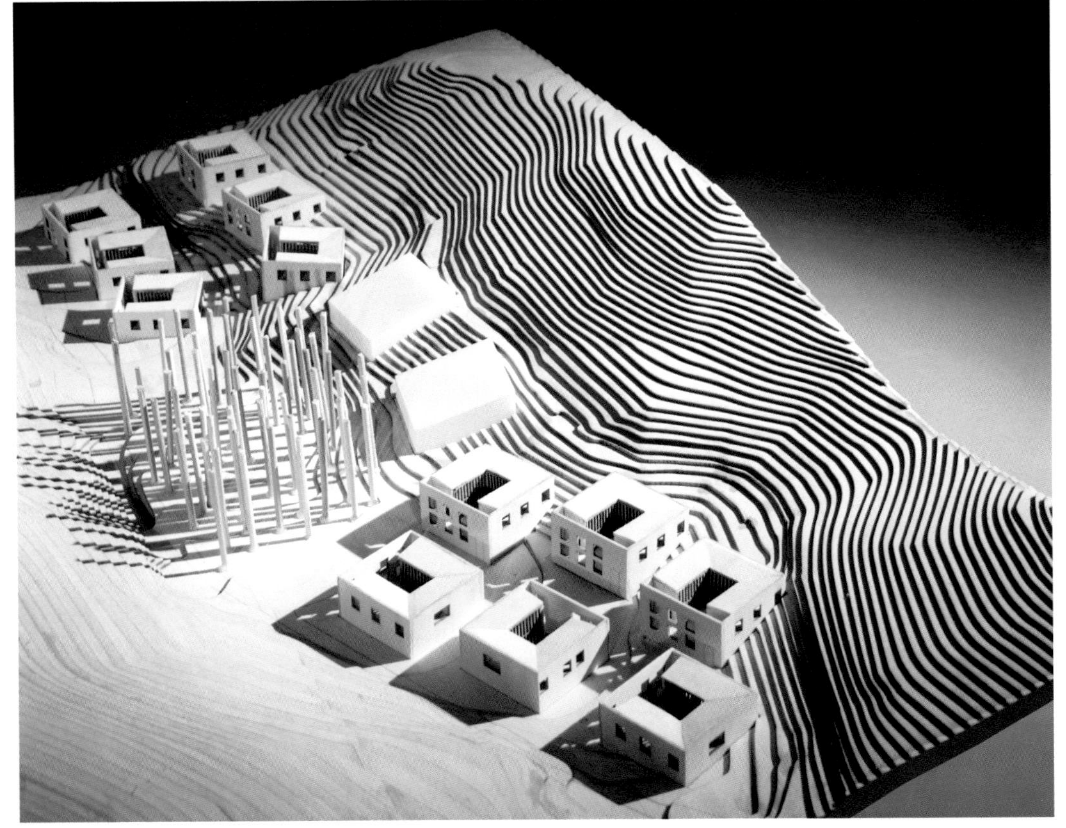

北京中信国安会议中心庭院式客房坐落在一个从东南向西北方向跌落的谷地里，谷地的中部是一片优美的杨树林。设计保留了杨树林。十二幢点式的庭院式客房分成两组布置在杨树林的东西两侧，杨树林的北侧预留了两幢独立式别墅的基地，南侧为区内主要道路。每幢庭院式客房的位置和方向依据山地的坡度和道路的走向自由安排，形成了丰富的室外空间。

庭院式客房的内部围绕着一个内庭院成 U 形布置；由于地势不同，庭院的入口有时在地下层，有时在一层。建筑共分三层，其中一层为门厅、起居室、餐厅和两间标准客房；二层为两套从一层由两个不同的楼梯进入的完全独立的套房；地下室则是娱乐空间——桌球室、酒吧和一个家庭影院、设备用房也布置在地下室。在平面布置上，主要空间——门厅，起居室，餐厅，客房，全部面向内庭院；辅助空间——楼梯，阳台，走廊，卫生间，厨房，全部面向山谷。

建筑为混凝土剪力墙结构。面向山谷的外墙面和屋顶均外挂毛面花岗岩，使得十二幢建筑像一群散落在山谷里的“石头盒子”；面向内庭院的外墙采用了木格珊和涂料。

建筑使用了地源热泵低温热水地板辐射采暖；外墙在混凝土墙和外挂花岗岩之间是保温层和空气层；门窗也采用了断桥铝合金框材和中空玻璃，以保证项目的节能效果。

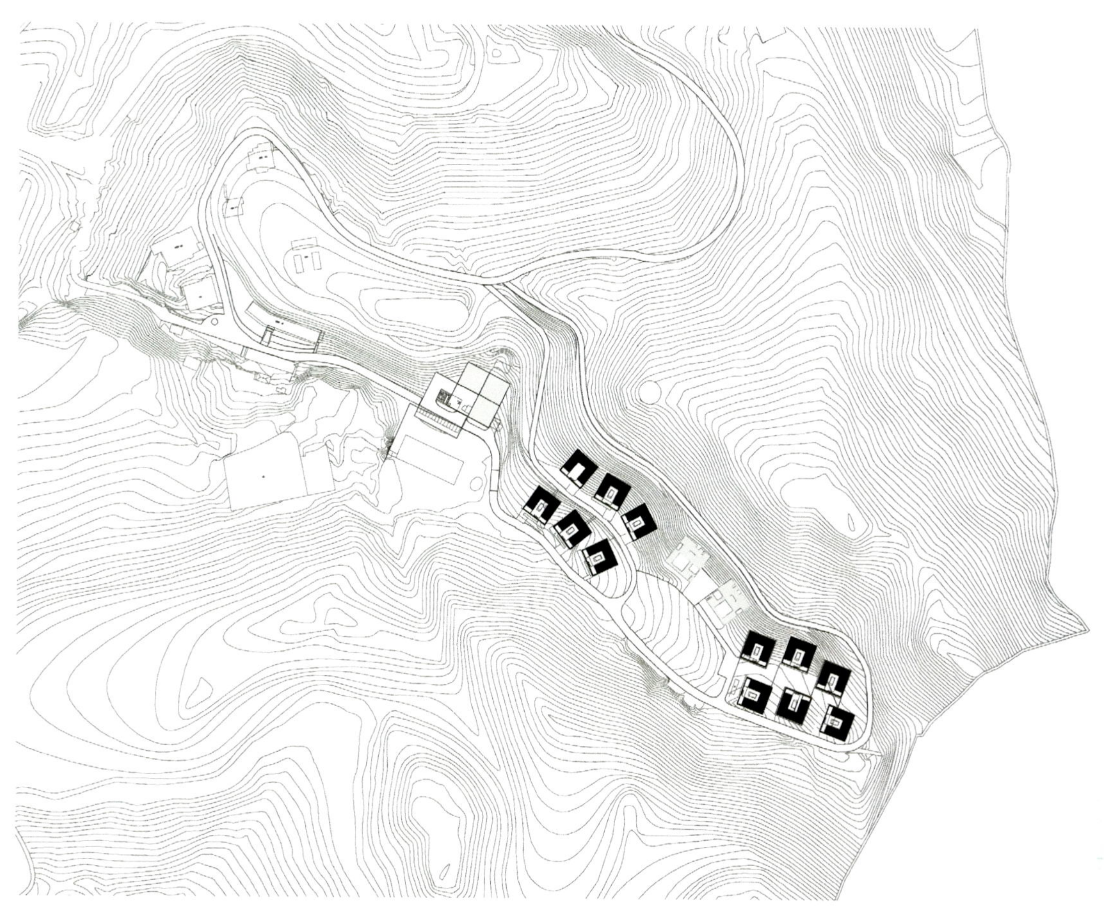

The courtyard suites of spring valley resort are located inside a valley descending from the southeast to the northwest. There is a beautiful poplar grove in the middle, which has been preserved in the design. Twelve courtyard suites are arranged by the east and west sides of the poplar grove in two groups. Sites for two independent villas have been reserved on the north side of the poplar grove. The main road within the district is on the south side. The location and direction of each courtyard suite are freely arranged according to the gradient of the hill and the direction of the road and form rich outdoor spaces between the courtyard suites.

There is an inner courtyard for each courtyard suite. The building is arranged in U shape around the inner courtyard. Due to difference of the gradient of the hill, the entrance of the courtyard is sometimes on the underground floor, and sometimes on the first floor. Each building is of three stories. On the first floor is a lobby, a living room, a dining room and two standard guestrooms; on the second floor, there are two completely independent suites, each of which has an independent stair; the basement is an entertainment space including a billiards room, a bar and a cinema, in addition to equipment rooms. All main rooms like lobby, living room, dining room and guestrooms are facing the inner courtyard; and all secondary rooms like corridor, staircase, balcony, bathroom and kitchen are facing the valley.

The building is reinforced concrete shear wall construction. Outside the external walls facing the valley and the rooftop are hanged rough granite tiles, which make the buildings look like a group of "stone boxes" dispersed in the valley. And the external walls facing the inner courtyard are wooden louvres and painted.

GSHP (Ground Source Heat Pump System) is used to provide energy for the hot water floor radiant heating of the building. Insulation and an air cavity are set between the concrete wall (inside) and granite (outside). These four layers together constitute the outer wall. In order to improve the energy-saving effect, we have chosen thermally separated aluminum window frame and insulating glass in the window components.

SHANGHAI INSTITUTE OF ARCHITECTURAL DESIGN & RESEARCH CO.,LTD.

上海建筑设计研究院有限公司

赵晨

于鹏

- 赵 晨 Zhao Chen

1986.9-1990.7 天津大学建筑系，学士

1990.9-1993.7 东南大学建筑研究所，建筑学硕士

1993.9-1996.12 东南大学建筑研究所，建筑学博士

1995.5-1995.10 香港理工大学建筑及房地产学系交换学者

1996.12至今 上海建筑设计研究院有限公司（原名上海市民用建筑设计院），高级建筑师，国家一级注册建筑师,历任副主任建筑师，主任建筑师，总建筑师助理，第二综合设计所副所长，所长

2001.1高级建筑师，国家一级注册建筑师

2002.11-2003.3 入选“100名建筑师在法国”,在巴黎机场公司ADP工作/实习

2004.09 国际项目管理专业资质认证IPMP项目管理资格

2005年上海第二届建筑师新秀入围奖

2005年上海市重点工程建设功臣荣誉称号

2007.1 教授级高级建筑师

2007.11 上海现代建筑设计集团一级项目经理资格

2004-2006年上海市劳动模范荣誉称号

2009年度全国五一劳动奖章

- 于 鹏 Yu Peng

1997.9-2002.7 哈尔滨建筑大学，建筑学学士

2001.4-2001.12 上海现代集团魏敦山体育建筑研究所 实习工作

2002.9-2005.7 哈尔滨工业大学，建筑学硕士

2005.7至今 上海建筑设计研究院有限公司（原名上海市民用建筑设计院），主创建筑师;

主要奖项

2002年全国经济适用住宅设计竞赛 佳作奖

2009年第三届上海建筑学会创作优秀奖大奖获得者

主要完成作品

2009年杭州奥特莱斯（建成）（上海院优秀方案）

阿尔及利亚体育中心（上海院优秀方案，现代集团优秀方案）

广西南宁体育中心（上海院优秀方案）

2008年沈阳奥体中心 游泳馆及网球中心（建成）（上海院优秀方案，上海院优秀工程大奖，现代集团优秀方案，现代集团优秀工程大奖，第三届上海建筑学会创作优秀大奖）

乌克兰国家体育场设计（上海院优秀方案）

2007年山东潍坊体育中心（建成）（上海院优秀方案）

2006年福建大学生体育中心（建成）

2005年上海金山体育中心（建成）

阿尔及利亚体育中心

ALGERIA SPORTS CENTER

项目地点：阿尔及利亚　用地面积：75 ha　建筑面积：450 000 m^2
建筑设计：上海建筑设计研究院有限公司
设计指导：魏敦山，赵晨
建筑师：于鹏，潘海迅，雷峻，潘迪

LOCATION: Algeria　SITE AREA: 75 ha　BUILDING AREA: 450,000 m^2
DESIGN CORPORATION: Shanghai Institute of Architectural Design & Research Co., Ltd.
DESIGN DIRECTORS: Wei Dunshan, Zhao Chen
ARCHITECTS: Yu Peng, Pan Haixun, Lei Jun, Pan Di

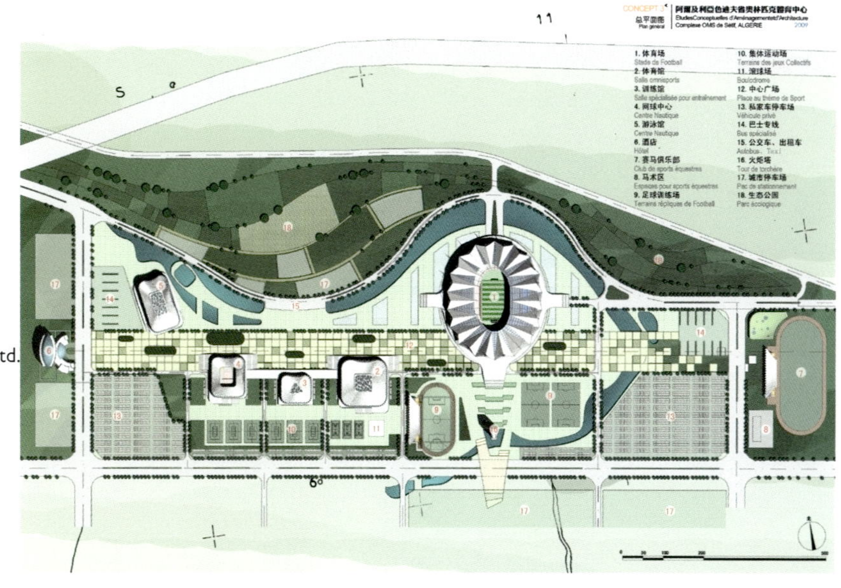

方案将以“绿色、科技、人文”三大建设理念为指导，在阿尔及利亚色迪夫省建设一个环境优雅、造型独特、绿色节能、生态环保、设施先进、功能完备、具备能够承办国内外综合性体育赛事的体育中心。项目建设将促进色迪夫省体育产业和全民健身运动水平的提高，将完善提高城市功能和文化品位，建成后的综合体育中心平时又可为广大市民的健身、娱乐、观光旅游提供良好的场所，并将起到拉动区域经济发展的重要功用。

体育中心建设力求实现以下目标：

(1) 功能完备：建设一流的奥林匹克训练基地，能够承办国内外综合性体育赛事的场地；培养和输送高水平体育人才的后备人才基地。

(2) 高效运营：既要满足大型运动会及国际比赛的要求，又要成为群众健身、休闲、娱乐的载体。

(3) 形象现代：展示现代体育建筑形象，融合地域性特色，尊重传统审美品位，使之成为城市新时代的标志。

(4) 技术先进：研究中水利用、地热采暖、光纤通信、电视传输、信息平台、智能化管理系统等科技手段，打造高科技含量的体育中心。

设计充分体现与功能和谐，与地域和谐，与时代和谐的整体设计思想。

整体方案以“蓝天、大地、白云”为主题，希望通过自由形态与秩序网格的叠加，强化规划布局的序列感，营造一个城市空间与自然空间渗透交融，自然与人，与城市，与奥林匹克精神和谐共存的新城市公共空间。

体育场位于整个基地的东侧，其最显著的部分是由雪白的膜结构构成的建筑整体形象。薄薄的屋顶尽可能地伸展，表现本身的动感，层层叠叠的形式，如白云，如风帆，如花朵。体育场的膜结构顶部既能遮阳，又能避雨，同时获得了柔和的光线，其突出的形态与观者心理产生共鸣。体育场震撼观众的心灵，屋顶的形式隐喻运动员的情感与精神，向他们的技能、决心、自信致意，这些品质使他们不断打破生理极限而创造新的纪录。

The program will be guided by the three contructive concepts-"green, technology, humanity" to build a sports center in Setif province of Algerian which is elegant and uniquely shaped , environmental friendly and energy-saving with advanced facilities and full functions and capable of undertaking domestic and international sports events. The project construction will promote the Sports Industry and public fitness exercises level of Setif and improve the urban functions and cultural tastes. The completed comprehensive sports center will provide a good place for the general public's fitness, recreation and sight seeing. It will play an important role in stimulating the regional economic development.

The project seeks to achieve the following objectives:

(1) Complete functions: to build the first-class Olympic training base;to be capabale of undertaking a comprehensive domestic and international sports events: to train and provide high-level back-up sports.

(2) Efficient operation: both to meet the demands of large sports meetings and international competitions, and carry activities of public fitnesses, leisure,recreation.

(3) Modern image: to show the image of modern sports construction, integrate local features, respect for traditional aesthetic taste and make it a sign of in the new era of the city.

(4) Advanced technology: to explore the use of reclaimed water, geothermal heating, optical fiber communications, television transmission, information platforms, intelligent management systems and other scientific and technological means to create high-tech sports center.

The design fully reflects the concepts of functional harmony,regional harmony and time harmony.

The overall program makes "blue sky, earth and clouds" as its theme, hoping to strengthen the sense of the sequencial layout through the overlay of free forms and order grids and create a new urban public space that is an infiltration of urban space and natural space and the harmonious co-existence between nature,human, the city and the Olympic spirit.

The stadium is located in the east of the base. The most significant part is the overall building image constituted by the snow white membrane structure. Thin roof extending freely and the overlay forms like clouds,sails or flowers just show their dynamic senses. The top of the membrane structure at the stadium can not only shade and be rainproof but also takes soft light. Its prominent form produces psychological resonance with the audience. The stadium shock the audience with their minds and hearts. The forms the roof indicate athletes' emotions and spirits as if paying tribute to their skills, determination as well as self-confidence which lead them to continuously break the physical limits and create new records.

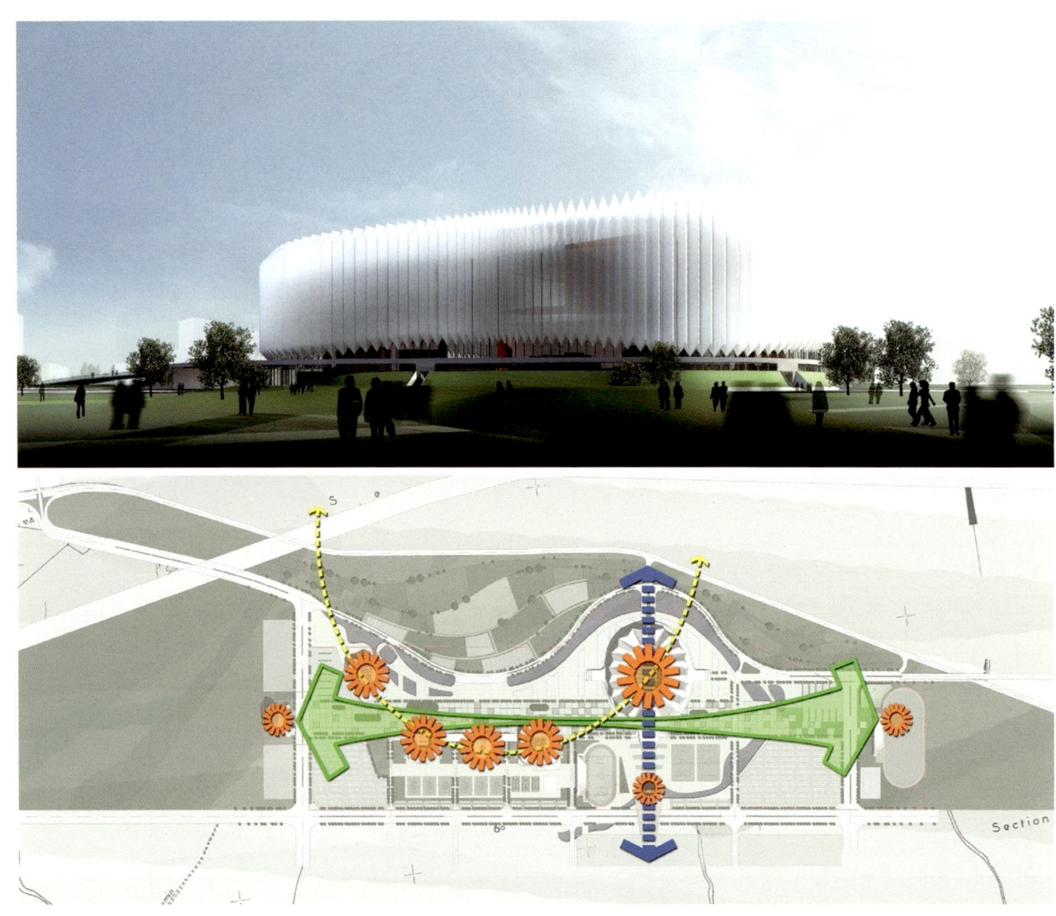
Section

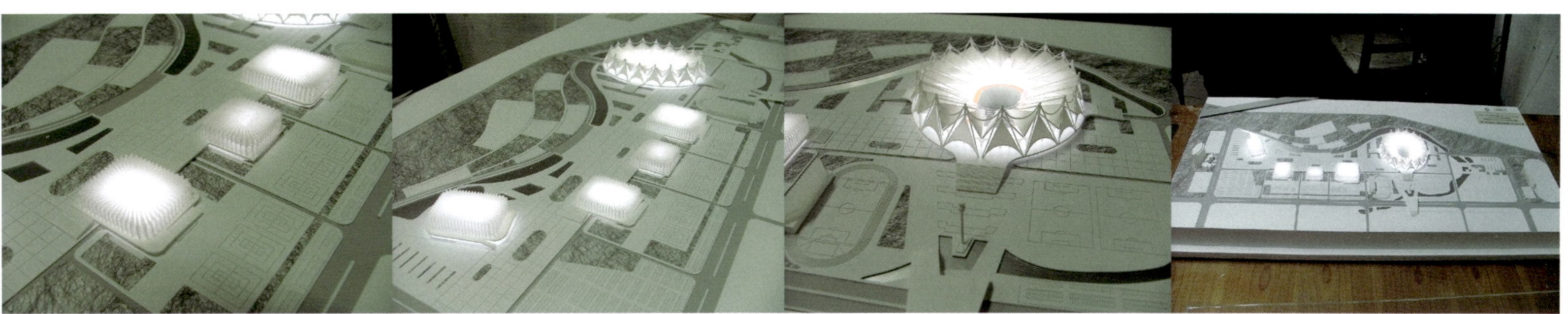

SHENZHEN ZHUBO ARCHITECTURAL & AMP, ENGINEERING DESIGN CO.,LTD.

深圳筑博工程设计有限公司工作室

筑博建筑工作室是筑博设计下属的研发型设计团队。本工作室的工作集中于研究建筑与日常生活的关系，关心如何塑造人与空间的关系以及如何通过建筑影响人与人的关系，避免陷入纯形式操作和美学化的窠臼。我们积极投身于中国的城市化进程，以建筑为媒介参与并影响社会。我们的工作总是从现实情况入手，处理海量信息，通过图表逐渐显影空间中的社会结构，并基于现实提出构想，最终推导出实体空间。随着这种现实主义的思路日益成熟，我们在深圳等地的多个重要的国际竞赛中获奖，如深圳南方科技大学（规划建筑第一名）、深圳华强北立体街道改造国际（第一名）、深圳水晶岛（第二名）、南宁规划展示馆（第一名）。

Zhubo Architect Studio is a researching and developing design team subordinate to Zhubo Design. It focuses on the study of relationship between architecture and daily life, and concerns about how to shape the relationship between people and space and how architecture affects human relations, so that architecture reaches wider issues beyond pure form of operation and aesthetics. We are actively engaged in the process of urbanization in China, and participate and influence society with the media of construction. Our work is always to start from the reality to deal with massive information, and gradually make clear the social structure of space by chart where we propose ideas based on reality and ultimately get the physical space. With the maturity of this realistic idea, we have won a number of important prizes in international competitions in Shenzhen and other places, such as the South University of Technology (the first place for building Plan), Shenzhen Huaqiang North D Street Reconstruction International (the first place), Shenzhen Crystal Island (the second place), Nanning Planning Exhibition Hall (the first place).

顺德水街

DESIGN SCHEME OF CRYSTAL ISLAND IN

项目地点：中国 · 广东　用地面积：76 510 m^2　建筑面积：38 329 m^2
建筑设计：深圳筑博工程设计有限公司工作室
建筑师：钟乔，邢果，龚晓文

LOCATION: Guangdong, China　SITE AREA: 76,510 m^2　BUILDING AREA: 38,329 m^2
DESIGN CORPORATION: Shenzhen Zhubo Architectural & amp , Engineering Design Co., Ltd.
ARCHITECTS: Zhong Qiao, Xing Guo, Gong Xiaowen

中国水文化的集中体现，一个是沿水路交通发展来的商业文明，一个是江南园林的人工引水造景。前者适用于市井繁荣，河道成为隐形的地标，“上河”成为约定俗成的逛街活动。后者则是偏重人文，努力将建筑和自然有机结合，并进行艺术加工和浓缩。

园林艺术里包含建筑艺术，建筑艺术里包含了书法、雕刻、彩绘等多门类艺术。

造园的主旨也是利用园林式景观艺术对于身处其中的艺术创造进行熏陶和影响。

本项目位于北滘新城中心区的文化中心和海琴水岸高层住宅区之间，北面是建设中的君兰高尔夫球场，南连北滘公园。现有水系可南北贯穿地块。

中心区现代规划和配套逐步完善虽然显示了小城的发展进步，但是小城自我认知感并不强。

一方面，需要满足当地急需的时尚休闲的需要，一方面，水文化街应当打造成为一个向外宣传的品牌形象，以加强北滘在广佛地区的认同感和吸引力。

整个地块被市政路划分为南北两个区，功能开发定位也有所不同，北区地块滨临美的大道，建筑与城市关系不应过于生

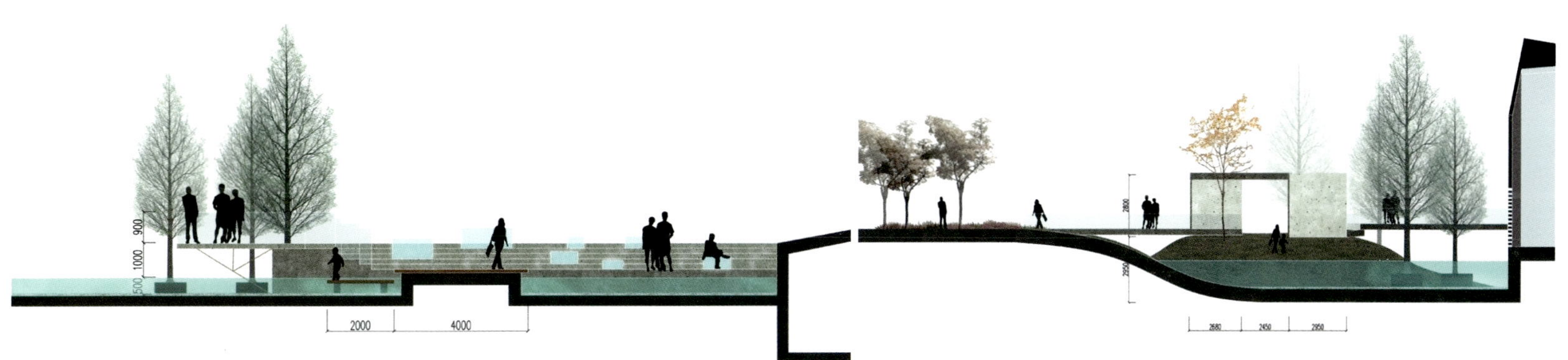

硬。因此，以水环绕建筑，既突出水景形象，又可形成双层的水面边界，一面邻接城市，一面包围水城，水城形成水面中岛城的景观意象。北区以漫游园林式意象为主，景观宜移步换景，因此在各个主要入口都布置有小桥、门廊和花树等景观层次，通过这些进入公共庭院，再进入各自的围院组团。

南区地块狭长，顺应水街布局。一侧是正常街铺边界厚度，另一侧建筑形成水院式、空中院落式、半包围等自由式布局；朝向中心绿轴一侧更加开放，中心形成主要水景活动区，带动两端的商业、休闲及餐饮区域。南区以休闲餐饮商业业态为主，沿街面主要是街铺，形成良好的城市界面，院落以大型集中式餐饮为主，庭院结合公共活动空间，如戏台、泳池等以刺激更多行为的可能性。

将首层和二层的交通系统通过水上廊桥和屋顶露台联系起来，既丰富了景观视线，又加大了公共活动的区域，为民间的艺术活动和传统娱乐提供了演艺和展示的平台。

清河坊

STARBUCKS

The condensed embodiment of Chinese water culture, one is the commercial civilization developed along the waterway transport, the other is the artificial water garden landscaping of southern gardens. The former applies in marketplace prosperity, in which river way becomes an invisible landmark, and "Shang He" has become the conventional shopping activities. The latter focuses more on human culture, attempting to combine architecture and nature and make them artistic and condensed. Landscape art includes garden architecture art, while architectural art includes calligraphy, sculpture, painting and many other art categories. The aim of the garden design is to nurture and impact the art creation within it with garden type landscape.

The project is located between the cultural center of Beijiao Town Center Area and high-rise residential area of Haiqin waterfront, the north to the Royal Orchid Golf Course in construction, the south connected to Beijiao Park. The existing water can run through the block from north to south. The gradual improvement modern planning and supporting services of the center area show the development and progress of the town, but the self-awareness of the small town is not as advanced. On the one hand, it is to meet the needs of local fashion and leisure activities; on the other, the water culture street should be created as a brand image publicizing the area, so that the regional identity and attractiveness of Beijiao Town will be strengthened in Guangfuo area.

The entire area was divided into two areas of north and south by municipal road with different functional orientations. The north block is close to Meidi Road and the relation between the city and architecture should not be too stiff. Therefore, the water around the building both

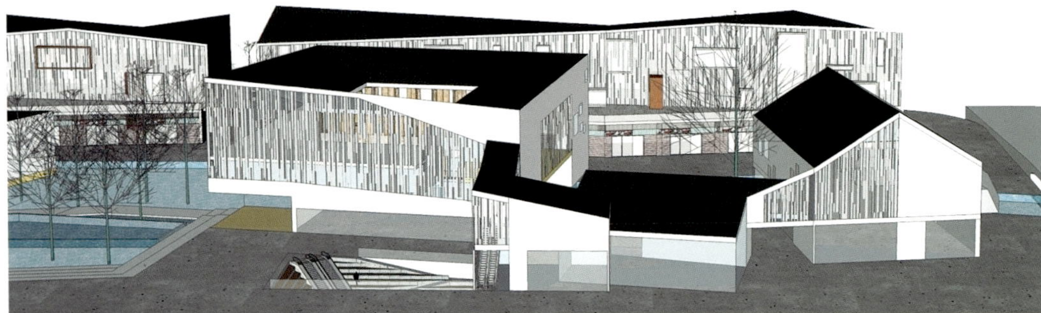

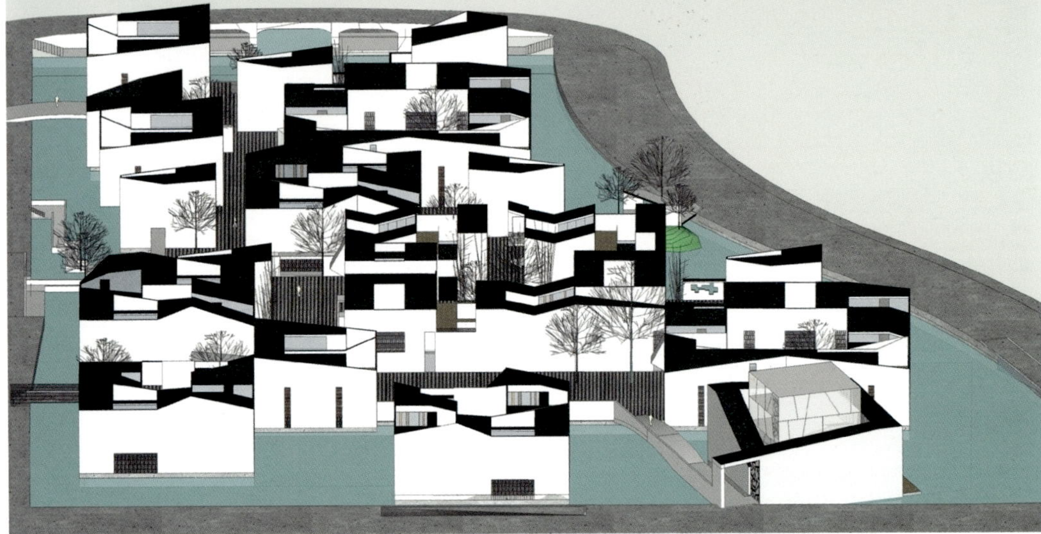

highlights its water feature image and forms double boundary layer, with one adjacent to the city and the other surround the water city which forms island city landscape image. The north area mainly takes the roam garden image, and the landscape should be changed by viewing position changing, so all the main entrances are set with small bridges, porches, trees and flowers and other landscape decorations, through which the public gardens are accessible, and finally to the yard enclosed by building groups.

The south block is long and narrow following the Water Street layout. One side is thickness of normal Street, and the other buildings form free style layouts, such as the water courtyard, air courtyard, semi-surrounded style; It's more open toward the green axis with the central area forming the main water activity area driving the business, leisure and catering on both sides. The south zone focuses on leisure and catering commercial activities, and the street stalls from good urban interface. The courtyard mainly holds the large-scale centralized dining area, combined with public garden spaces, such as the stage, swimming pool, etc. to stimulate more activity possibilities.

The transport systems of the first and second floors are connected through the water bridge and the roof terrace not only to enrich the landscape view, but also increase the area of public activities, providing a platform for the folk and traditional performing arts.

南投古城两街一园改造

RECONSTRUCTION OF TWO STREETS AND ONE PARK IN NANTOU ANCIENT CITY

项目地点：中国·广东 用地面积：108 445 m^2
建筑设计：深圳筑博工程设计有限公司工作室
建筑师：傅卓恒，李伟，邢果，张春亮

LOCATION: Guangdong, China SITE AREA: 108,445 m^2
DESIGN CORPORATION: Shenzhen Zhubo Architectural & amp,Engineering Design Co., Ltd.
ARCHITECTS: Fu Zhuoheng, Li Wei, Xing Guo, Zhang Chunliang

我们关注的不是建筑，而是城内的生活。

作为深圳之根的南投古城，从最初驻兵御敌之城，到管辖深港地区的新安县城，再演变为如今的城中村，已经走过了600多年。

不同时代的建筑和不同地域的生活在一起达到平衡，这种多样性是古城的最大特色。改造古城必须遵循此原则，在不破坏古城生态的前提下，植入新业态带来新生活，使其更加多样化，促进自主更新。

但是保护古城又要避免两种有害的态度，一种是将古城像真空包装一样封闭起来，另一种是将古城开发为旅游观光景点，这两种做法都将使古城与生活分离，失去生命而变为死城。

我们的策略是恢复和发展古城的空间脉络和特色，将新的生活内容注入两街一园，建构古城空间与新生活的积极关系，并以此激活整个旧城区域。

Our concern is not buildings, but city life. As the root of Shenzhen, Nantou Ancient city has experienced an evolution of 600 years, from the initial defending city to the administrative region of Xin'an county, and the followed today's urban village.

The diversity that reaches ballances between buildings of different times and lives of different regions is the most significant feature of the ancient city as well as a principle the transforming has to follow, that is, to bring new blood to the city life by introducing new industries to make it more diverse and self-updating, without distroying the ecology of the ancient city.

However, the protection of the ancient city needs to avoid two attitudes: one is to make it vacuum-packed, and the other is to develop it as a tourist site, both leading to its separation from life and liveliness and makes it a dead city.

Our strategy is to restore and develop the city's spatial structures and characteristics and implant the two Streets and one Garden into the new life to construct positive relations between city space and new life, and thus to activate the entire old city area.

深大南校区

SOUTH CAMPUS OF SHENZHEN UNIVERSITY

项目地点：中国 · 深圳　用地面积：58 564 m^2　建筑面积：168 000 m^2
建筑设计：深圳筑博工程设计有限公司工作室
建筑师：傅卓恒，张臻，龚欣，李伟，黎靖，冯茜

LOCATION: Shenzhen, China　SITE AREA: 58,564 m^2　BUILDING AREA: 168,000 m^2
DESIGN CORPORATION: Shenzhen Zhubo Architectural & amp,Engineering Design Co., Ltd.
ARCHITECTS: Fu Zhuoheng, Zhang Zhen, Gong Xin, Li Wei, LI Jing, Feng Qian

立体开放的校园街区，总建筑面积 168 000 m^2。

我们的目标是构建一个开放的校园街区，体现深圳大学这一现代化学府平等、自由的学术精神。

深大南校区被白石路从主校区分开，成为一座孤岛。基地与主校区仅通过三座人行天桥连接，很大程度削弱了与主校区的联系。我们依靠这仅有的联系资源，在地块的二层标高构建公共步行空间。通过把天桥接入建筑二层，使南北校区的步行交通自然过渡。

二层步行平台与一层步行庭院共同构筑立体步行系统，此系统包含了斜坡绿地景观、一二层的街道、庭院、架空活动空间及下沉广场、成片的绿植、几何感的硬质铺地。通过草坡、踏步自然过渡，穿行于校园之中，人的空间体验处于不断的变化中，最终展现出自然简洁、层次丰富、充满活力和想象力的校园氛围。

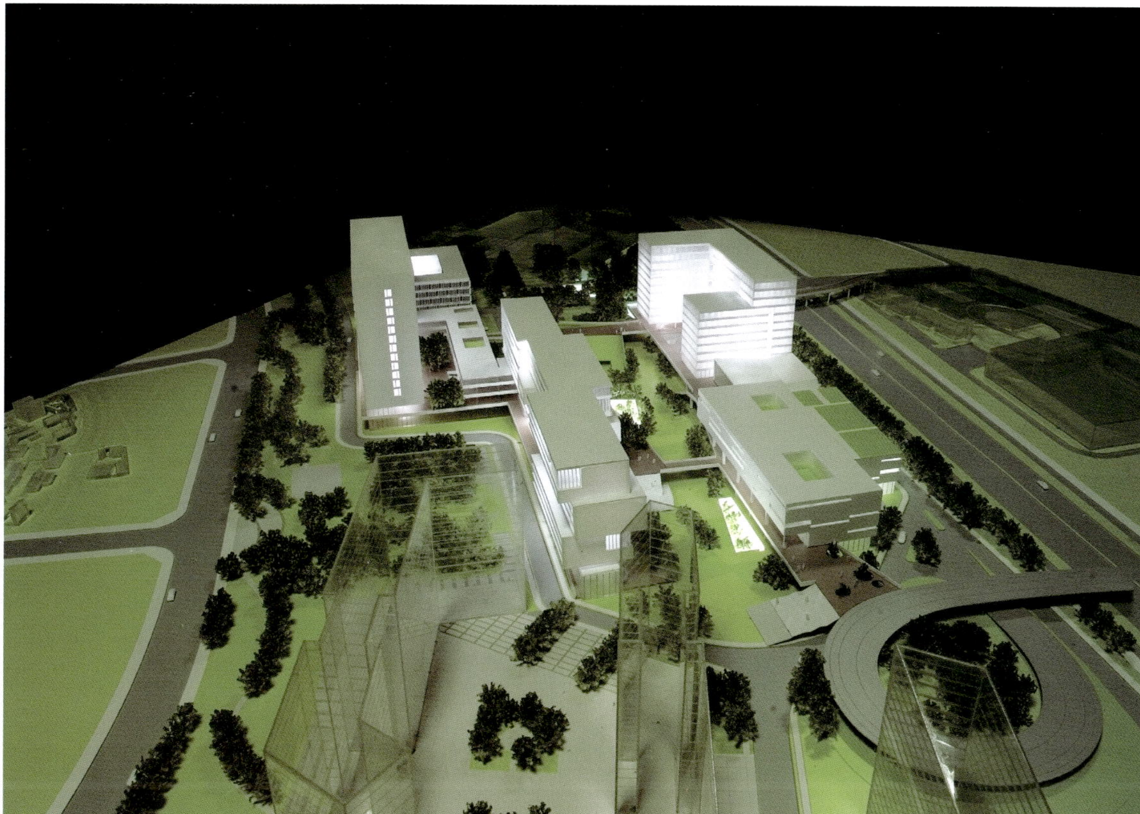

Three-dimensional open campus block, the total construction area is 168,000 m^2.

Our goal is to build an open campus neighborhood reflecting academic spirit of equality and freedom of the modern university.

The South Campus of Shenzhen University was separated from the main campus by the Whitehead Road and become an island. The base was largely weakened its connection with the main campus by connecting itself with only three pedestrian bridges. Relying on this linking resources, we built the public walking space at the second floor elevation, which accesses to the second storey by the pedestrian bridge traffic, providing natural transition between the north and south campuses.

The second-storey platform and the first-storey courtyard together constitue the three-dimensional walking system which encompasses the slope green landscape, the first and second storey avenues, the courtyard, overhead space and the sunken plaza, as well as the patches of green plants and the hard floor of geometric sense. Walking in the campus through the grass and the pedestrian, one can experience constantly changing spaces, which all finally show the campus atmosphere of natural simplicity and vitality full of imagination and hierarchy sense.

深圳华强北商业街改造

RECONSTRUCTION OF HUAQIANG NORTH SHOPPING STREET, SHENZHEN

项目地点：中国 · 深圳　用地面积：225 100 m^2　建筑面积：80 000 m^2
建筑设计：深圳筑博工程设计有限公司工作室
建筑师：冯果川，邢果，尹毓俊，张烁，龚欣，张甜甜，李俊鹏，张智勇，黎靖，詹绪勋，龚晓文

LOCATION: Shenzhen, China　SITE AREA: 225,100 m^2　BUILDING AREA: 80,000 m^2
DESIGN CORPORATION: Shenzhen Zhubo Architectural & amp, Engineering Design Co., Ltd.
ARCHITECTS: Feng Gwochuan, Xing Guo, Yin Yujun, Zhang Shuo, Gong Xin, Zhang Tiantian, Li Junpeng, Zhang Zhiyong, Li Jing, Zhan Xuxun, Gong Xiaowen

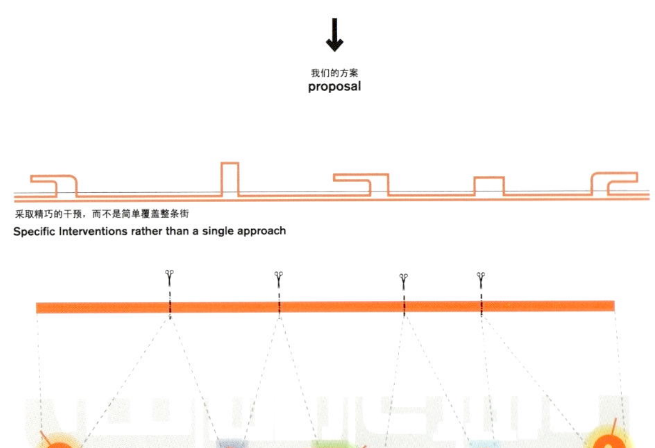

华强北是世界上独一无二的街道。无论是人行道的尺度、闪亮的楼宇，还是那些由电子与时装两大业态激活的街头生活，都是每个都市人梦寐以求的。随着城市环境的有机成长，华强北及其周边街道无疑需要改善。但我们相信任何改善方案，首先必须保护华强北的独特性，所有措施必须是提升街道生活的丰富性而不是扼杀它们。因此我们采用了一种新的城市设计方法——城市针灸：尊重华强北现有活力，以最小化的介入打通脉络，实现华强北空间品质的整体性、系统性的提升。梳理整治交通和街头活动提升华强北的交通效率和步行舒适性；完善华强北路服务配套功能，突显政府公共服务职能；创造独特而富于时代感的城市印象，为华强北点睛。我们采用系统而综合的城市设计手段，将景观设计的轻柔、交通规划的精确和建筑设计的力量融合在一起，来改善流线，强化个性，创造一个新的公共空间。

Huaqiang North is an unparalleled street in the world. Whether the scale of the sidewalk, by the shining buildings, or the street life activated by the electronic and fashion industries are dream places for every city. With organic growth of the urban environment, Huaqiang North and the surrounding streets undoubtedly need to improve. However, we believe that any improvement should remain the uniqueness of the original buildings, all actions should enhance the richness of street life instead of killing them. Therefore, we adopted a new urban design - Urban Acupuncture: to minimize the intervention with respect to the existing energy of Huaqiang North,and get through the whole structure to enhance the entirety and systematicity of the space; to regulate the traffic to upgrade the transport efficiency and walking comfort; to improve the matching facilities to highligt public service function of the government; to create a unique impression of the contemporary city to provide striking point to the city. We use systematic and comprehensive urban design techniques to integrate the tenderness of the landscape, accuracy of the traffic planning and the power of the architectural design to create a new public space with better flow lines and heavier uniqueness.

南方科技大学工学楼

ENGINEERING BUILDING OF SOUTH CHINA UNIVERSITY OF TECHNOLOGY

项目地点：中国·深圳　用地面积：21 018 m²　建筑面积：35 340 m²
建筑设计：深圳筑博工程设计有限公司工作室
建筑师：钟乔，冯茜，张碧勤

LOCATION: Shenzhen, China　SITE AREA: 21,018 m²　BUILDING AREA: 35,340 m²
DESIGN CORPORATION: Shenzhen Zhubo Architectural & amp, Engineering Design Co., Ltd.
ARCHITECTS: Zhong Qiao, Feng Qian, Zhang Biqin

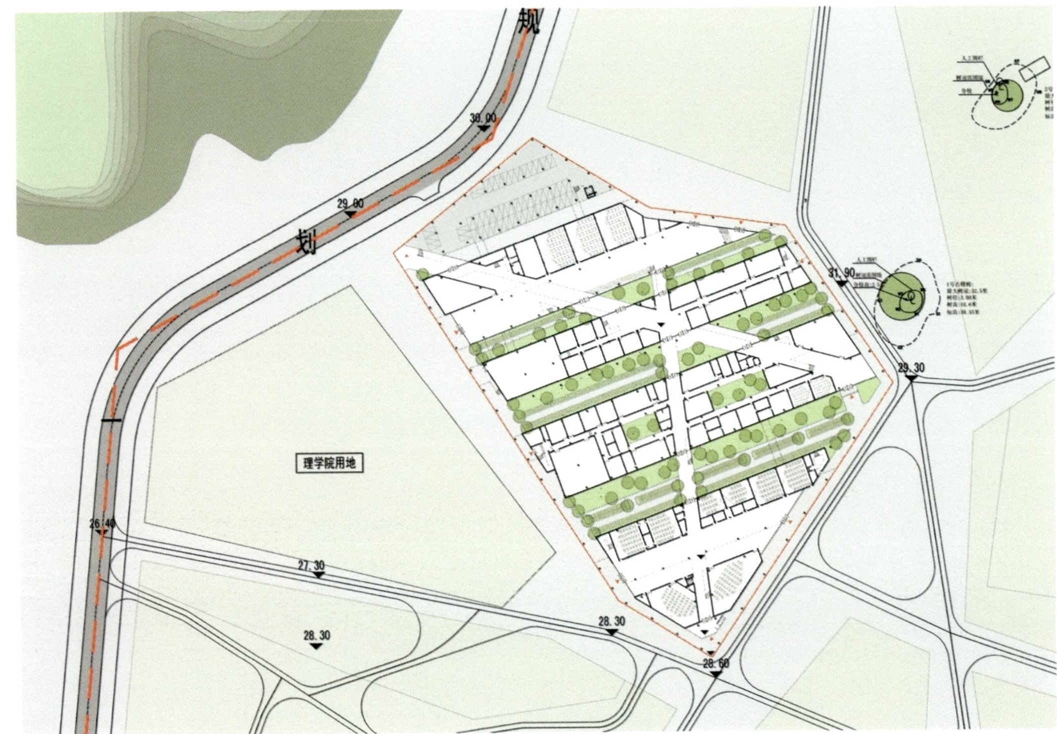

21世纪是团队的世纪，只有拥有团队精神的人才可能成功。大多企业也认为大学生综合能力最重要，他们很是看重应聘者的综合素养，而相对于专业成绩与学历，自主的学习能力和善于与团队的沟通合作和交流就变得更为重要。可大学教育里一味偏重专业知识的灌输，以封闭的管理体制鼓励学生间的盲目竞争，不注重协作和各学科间知识的融合，导致了学校教育和社会需求的严重脱节。

而南方科技大学工学楼的建筑方案设计正是在思考这些种种矛盾现象过程中挣扎所形成的产物。

1. 背景

"共享、融合、开放"是南科大校园总体规划的基本概念，打破学科分割，追求交流互动，氛围自由，促进集中和跨学科的合作是新校园规划的基本原则，以清晰的边界策略保证校园公共空间的形成和单体建筑集群设计的灵活性和可操作性。

工学楼地块位于教学区的中间位置，东临"巨环"这一学校最主要的开放性公共空间，南临理学院，北临教学区二期建筑群，西侧毗邻深圳大学新校区，东南面则是校核心区的图书馆位置，地理位置具有重要"枢纽"的地位。

地块无覆盖率要求，这是上层规划非常有建设意义的创举，使得建筑设计无需再留出本没有意义的消极开敞空间。

2. 目标

在不能改变传统教室空间形态的基础上（设计任务书有严格要求），我们希望通过创造更多的易达性"空白"空间以刺激学生之间，包括不同专业学生之间的交流和课间的非正式的对话和群聚活动。

3. 手段

切割和重组

把标准的单廊式教室和实验室切分成以层为单位的元素并依照法规和功能进行平行重组，以基本的造型元素，质朴的材料打造出生动的建筑语言，以清晰的结构秩序叠加出

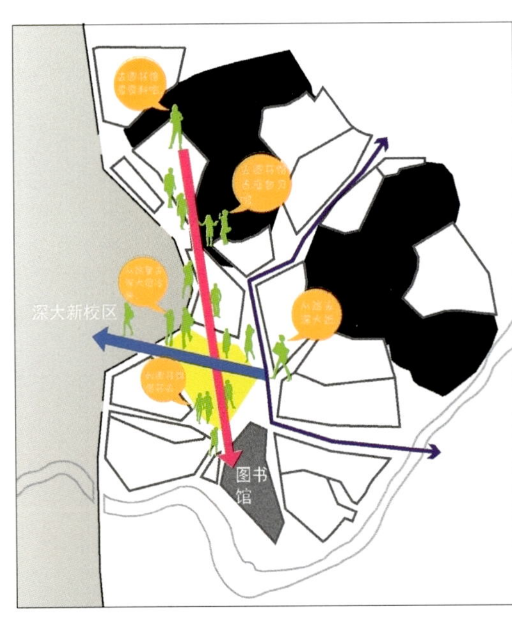

丰富的空间体验和路径感受。

"捷径"的介入

一方面在基地内人为的锲入"十"字形双向捷径：连接教学区至图书馆的直线路径和深大新校区边界至南科大中心环道的直线路径。另一方面利用平行的教室建筑之间的"间隙"形成从校园主环路至理学院的多条捷径。以开放的"枢纽"姿态引入更多的不同专业背景的人流从建筑中穿越，以刺激更广范围的交流活动。边界的策略被局限于建筑形态上，人的活动变成了一种任意的方式。

交通空间和交流空间的混合

单纯意义的交通空间被以各种方式扩大和充实，而单纯意义的开放空间因为引入了交通功能而不会担心变成没人去的消极空间。交通意义上的邂逅被刺激发展成有实质意义的非正式学术交流，专业的界限在这里被模糊了，课间的生活变得丰富和精彩。

大覆盖率的利用

在深圳这种炎热的亚热带气候条件下，空旷的所谓开放空间经常会变成无人去的、能看不能用的地带。我们的带形建筑单元几乎覆盖了整个地块，狭长的带状建筑间形成自然的风带，加速空气流动，为一层的穿越"捷径"和绿地提供阴影和舒适的小气候，让人们愿意在此停留小憩，使交流成为可能。

4. 让步

"标准盒子教室"下的蛋

我们深知道作为一名普通的建筑设计工作者并不可能仅凭借我们的专业工作和某一个学校建筑的设计而改变什么教育体制上的问题，甚至还可能背上不遵守国家规范和不符合设计任务的骂名。于是，我们在南科大工学楼的设计中仅试图通过对当代大学毕业生在现有教育体制背景下所缺乏的素质教育的系列研究，从空间手段上给予将来的大学生一些我们力所能及的帮助，而没有去撼动那些前面是黑板，下面是"排排坐"明亮却封闭的标准化的盒子群。但我们同时预留了拆除盒子变开敞式教学大空间，灵活隔断的可能性。

粗糙的"美"

对于有严格投资造价控制的校园建筑，我们宁愿把更多的资金用在建设那些阴凉的、没有具体功能定位的室内半室内的"空白"空间上，让它成为那些"有用"空间的留白。虽然它存在的理由非常模糊，但它会随着自己形态和地段条件的变化为一些未知事件的发生创造条件，让那些最具创意的使用者去发掘它的潜力。

同时我们也就放弃了在传统建筑学领域对形式和精美细节的追求，转而追求功能合理以及由此产生的建筑形式美：大量运用遮阳板，在防止太阳辐射和避免产生眩光的同时能产生丰富变化的光影立面效果；大量的屋顶平台、活动平台和共享空间的设计也能有效地组织气流，提升空间的环境品质；框架结构被大方地裸露成它本应有的形态；在建筑界面外围尽量少使用幕墙，节省造价和节约能源；用带攀爬植物的铁丝网形成对地块界面的限定；用简单的色彩作为内部功能的指示标志。总之，它呈现的是一种粗糙的"美"！

5. 结语

工学楼的设计是我们向传统的对用地覆盖率和总建筑面积控制的一次突破性尝试。我们希望通过这次在"有用空间"和"无用空间"的实质性探讨中，一方面为当代的大学教育制度提供一剂"偏方"，另一方面对国内现在土地利用控制中仅简单的采用覆盖率和总建筑面积等数据限定提出我们的质疑。

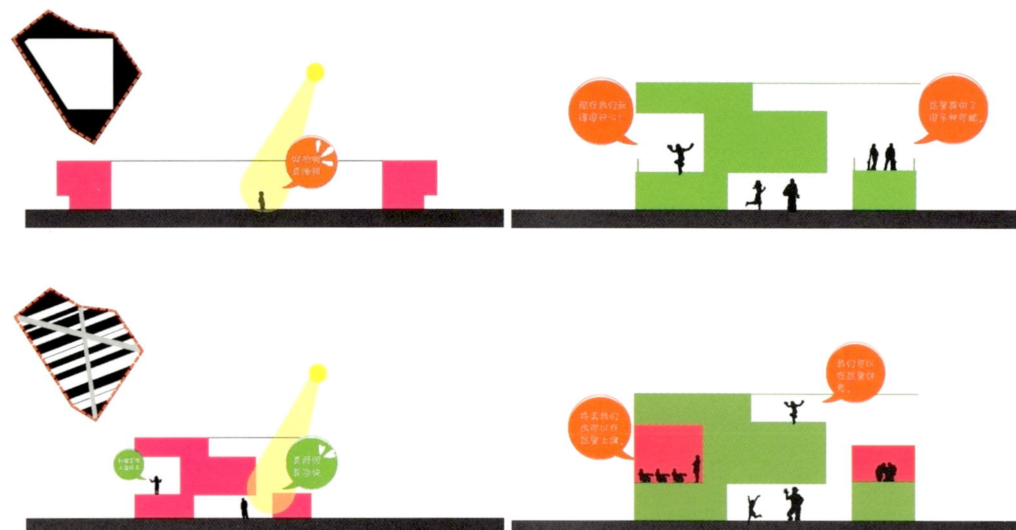

The twenty-first century belongs to a team as only team work makes perfect.Most enterprises believe students' comprehensive ability as the most important, and focus more the candidates'compehensive accomplishments, while excellence in self-learning,communication and coordination functions better than diplomas and academic achievements. However the duck-stuffing type of academic knowledge, the closed management system encouraging blind competition among students ignoring coordination and subjets integration, lead to disjunction between education and social demands. Based on the serious thinking of these contradictions and the struggle to remove them we gave birth to this engineering program of the South China University of Technology

I.Background

"Sharing, integration, open" are basic concepts of the master plan of the whole campus. The basic principles of the planning are to break the subject division and pursue interactive and free atmosphere and to promote centralized and interdisciplinary cooperation. A clear boundary is to ensure public space of the campus and flexibility and maneuverability of the design of single-building cluster.

Engineering Building Block lies in the middle of teaching area, east to the "huge loop",the most important public open space, south to the College of sience, north to the second stage buildings of the teaching area and west adjacent to the new campus of Shenzhen University,while the southeast is the library in the core area of the school.The location is at the important "hub".

Being no coverage limitation is a very initiative and constructive action of superstratum,removing the need of meaningless and passive open space.

II.Tasks

Without distroying the space form of the traditional classroom

(demanded in the design task), we hope to create more accessible "blank" space to stimulate communication and informal exchanging or group activities among students, exclusive to students of different domains.

III.Techniques

Cutting and restructuring

Cutting the standard single-gallery-style classroom and laboratory into elements of layers and parallelly restructure them in accordance with functions and regulations, and using basic modeling elements and rustic materials to create vivid architectural language. Organizing the space and path of rich sense with clear structure order.

"Shortcut" intervention

On the one hand we carve cross-type shortcuts at the base:one straight path connecting teaching area to the library and the other from the campus boundary of Shen Zhen University to the centeral ring Road of the South China University of Technology. On the other hand the "spaces" between the parallel classrooms form more than one shortcut from the main ring road of the campus to the College of Siencce. To introduce more people in various specialties across the buildings with the open the "hub" posture to stimulate a wider range of exchange activities. The boundry strategy is limited to architectural form,while the huamn activities become arbitrary.

The mix of traffic space and communication space

The pure sense of traffic space is expanded and strengthened in various ways, while the pure sense of open space is removed of the possibility of becoming a less populous negative space due to the introduction of the traffic function. The encounter of traffic sense is stimulated into a meaningful informal academic exchange,obscuring professional boundaries and enriching recess time.

Take use of the great coverage

In the tropical climate of Shenzhen, the empty so-called open space often becomes impractical unpopulated zone. Our band-shaped buildings almost covers the entire block,and it forms a natural wind area speeding up air flow between the long and narrow band type buildings, providing shade and comfortable microclimate for the passage "shortcut" and the green of the first storey so that people are willing to rest here and make exchange possible.

IV. Concessions

An egg under "Standard box classroom"

We are deeply aware that it's impossible to solve educational system problems only with professional knowledge and a school design of an ordinary architecture worker, even with the risk to be responsible for not complying with the national specification and design goals.Thus,we try to give some help to the future college students as we can in the space design on the basis of the systematic research to the lack of quality education in te existing system, without moving the front blackboards and the lines of seats in the standard bright and closed "box classrooms". However, we also reserved the possible flexibility of replacing the box with large open teaching space with partitions.

Rough beauty

Due to the strict control of school building costs, we would rather spend more money to those indoor or semi-indoor "blank" spaces which are cooler without specific functional orientation and used to be space blank for the functional spaces . Although its reason for being is very vague, but with its changing shape and location conditons allow happening of unknown events and the users to exploreit's potential. In the meantime we gave up the pursit of forms and fine details in the traditional architecture, but turned to the pursuit of rational functions and the resulting form beauty:the largely using of sun visor not only prevents solar radiation and glaring ray but also produces facades of rich light and shadow effects; the large number of roof platforms,act platform and sharing spaces can effectively improve air flow and enhance the space environment quality; the frame structure boldly exposed to show its full form and the minimum use of external curtains both save cost and energy; the mesh with climbing plants limits the block interface; the simple colors are used as the logo of interior function.In short, it presents a kind of rough beauty.

V. Conclusion

The Engineering Building design is our attempt to break the traditional control of building coverage and construction area. We hope through the practical exploration of the "functional" and "impractical space", on the one hand to provide a "recipe" for the contemporary university system, and on the other hand to put forward a doubt to the mere limitation of land coverage and construction area in land control of China.

三亚亚龙湾行政中心

SANYA YALONG BAY EXECUTIVE CENTER

项目地点：中国 · 三亚　用地面积：35 000 m^2　建筑面积：30 000 m^2
建筑设计：深圳筑博工程设计有限公司工作室
建筑师：钟乔，张甜甜，张碧勤，黎靖

LOCATION: Sanya, China　SITE AREA: 35,000 m^2　BUILDING AREA: 30,000 m^2
DESIGN CORPORATION: Shenzhen Zhubo Architectural & amp,Engineering Design Co., Ltd.
ARCHITECTS: Zhong Qiao, Zhang Tiantian, Zhang Biqin, Li Jing

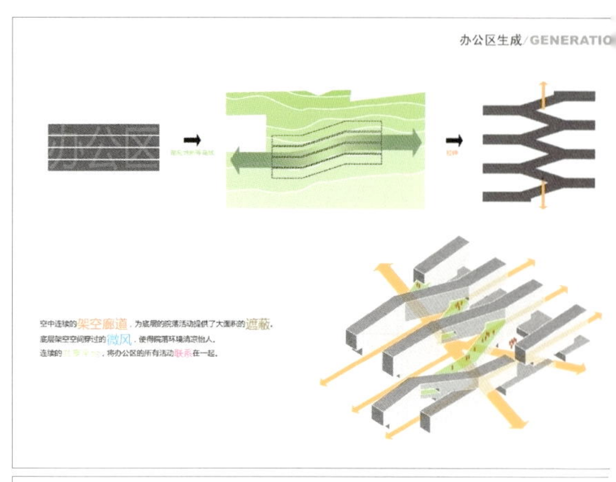

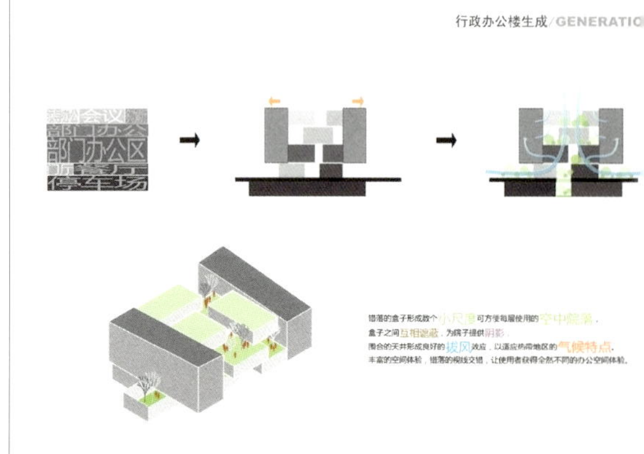

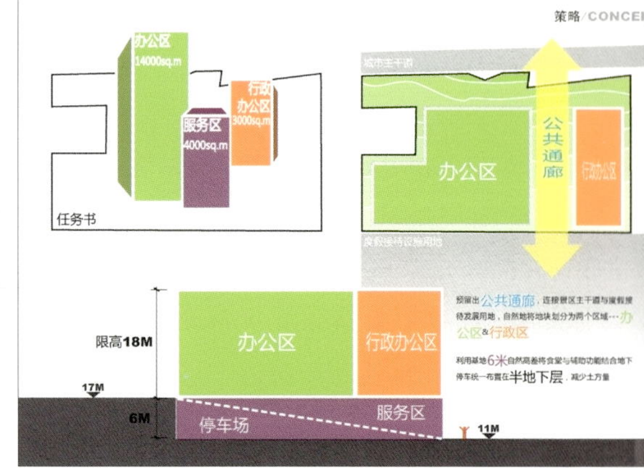

这个项目的设计基本上是基于分析权衡项目各方面的可变不可变的限制因素而推导出的结论，尝试一种纯理性的思维过程。

（1）让企业总部文化和建筑形式语言对话，充分挖掘“中粮”集团倡导的“自然之源，重塑你我”的企业核心价值观与建筑设计的内在联系。

（2）以谦逊大度的姿态在场地中留出联系北面主干道与南面休闲度假用地的公共景观通道，并以此合理划分用地功能区间。

（3）充分考虑热带气候特点，大胆突破所谓的覆盖率要求，在场地上尽量营造有阴影的地面公共交流空间，建筑底层架空，内院拔风，建筑外遮阳和利于自然通风的板式建筑形式的选择都是我们对热带气候特点的回应。

（4）充分利用地势高差，竭力减少土方开挖，推崇低碳建设。

（5）以舒展的折线形厚板建筑适应建筑将来可能的功能置换和回应建筑与周边地形等高线、周边建筑尺度的协调及远处大海的对话关系。

（6）绵延连续的流动空间增强了内部人员交流的机会，中间贯穿的建筑屋面作为第二路面来使用，并使整个建筑群联为一体，形成强烈的总部印象。

（7）灵感源自热带水果表皮肌理的遮阳体系设计最大限度地述说着热带建筑的形式语言。

The design of this project is almost resulted from the analysis of various limited factors,changing or unchanging as well as an attempt of pure rational thought.

(1) Through dialogue between headquarter culture of the enterprise and architecture forms to excavate the inner relation between the core value of "natural source, rebuild you and me" advocated by COFCO and the architectural design.

(2) To build a public landscape access linking the north main road and the south relaxing and holiday-using land with humility and magnanimity,and reasonablly divide the functional areas of the land.

(3) Fully considering the characteristics of tropical climate and boldly breaking the so-called coverage requirements to create a shadowy public ground communication space on the site. The frame structure at the bottom of the building, the wind ventilation of the inner yard, the external shading as well as the natural ventilation plate show our definite response to the characteristics of the tropical climate.

(4) Taking full use of altitude difference to reduce excavation as much as possible; to advocate low-carbon building.

(5) Using the stretching foldline-type plate construction to adapt to the possible future replacement of buildings and to respond to the coordination between the architecturre and the surrounding terrain contours and the dialogue relationship with the distant sea.

(6) The continuous flow of space increased opportunities staff to communicate, the building roof in the middle was used as a second road and integrate the building into a whole with a strong sense of headquater impression.

(7) The inspiration comes from sun proof texture of the tropical fruit skin which ultimately recounts the form language of the tropical architecture.

深圳中心区水晶岛国际竞赛

INTERNATIONAL COMPETITION ENTRY FOR SHENZHEN CBD CRYSTAL ISLAND

项目地点：中国 · 深圳　用地面积：457 100 m^2　建筑面积：297 500 m^2
建筑设计：深圳筑博工程设计有限公司工作室
建筑师：冯果川，刘慧，Ryan John Duval, Iris Jimenez, 张烁，张春亮，潘骁，詹旭勋，佘赟，尹毓俊，龚欣

LOCATION: Shenzhen, China　SITE AREA: 457,100 m^2　BUILDING AREA: 297,500 m^2
DESIGN CORPORATION: Shenzhen Zhubo Architectural & amp,Engineering Design Co., Ltd.
ARCHITECTS: Feng Guo Chuan, Liu Hui, ryan john duval, iris jimenez, Zhang Shuo, Zhang Chunliang, Pan Xiao, Zhan Xuxun, She Yun, Yin Yujun, Gong Xin

水晶岛作为深圳的核心地理中心，经历数次规划开发，仍人迹罕至，活力不足。

本案的核心是打造设计＋文化乐园，着力促成设计与市民生活的相遇。文化乐园利用周边资源，承载超大容量城市生活，突显深圳精神，成为 CBD 的活力引擎，以市民利益为支点，因此提升城市未来发展高度。

1. 压缩

水晶岛周边潜藏着优厚的文化与市民生活的契机。却各自距离遥远，我们利用高效直接的 PRT 系统，将从莲花山至会展中心的整个区域进行时空压缩，将之前无法全部享有的各项精彩生活一网打尽。

2. 乐园

保留基地原有公园，给高密度城市留出绿肺。重组公园地貌，使表面形成不同属性的场域。去除秩序和方向性，化解尺度。

（1）底层设计聚落

绿皮切开掀起，截取有趣味的城市片段作为肌理，结合各种门类的设计公司以聚落的形式塞入，制造出微缩城市的多元复杂。

（2）屋顶文化乐园

绿皮上不同的微地形诱发不同活动，运动和事件，乐园北静南动，北部为青少年和市民的生态体验基地。南部主要为活动场地，与下面设计聚落功能呼应，乐园的植物突显了深圳地域特征，与场地功能结合紧密。

呼应城市密度变化，北部置入数栋瞭望塔，根据观景视野和方向生成各自独特形体。

肌理细密，尺度宜人的聚落传递文化价值与城市风情，形态舒展的屋顶乐园显示着深圳的雄心，两者的对比形成本案的独特张力。最后一条明晰的环线将连接各个聚落，并游走于地面上下，无理中有秩序。

3. 新标志新精神

传统城市标志明确而固定，难以承载深圳活跃的城市生命力，我们不做具体的标志设计，而采用动态的，可变换的标志设计概念，制造出随时间，事件变化，时时更新的景象。

Crystal Island is located in the geographic core of Shenzhen. After numbers of redesign and redevelopment, there is still lack of positive flow of people and dynamics.

The core of our proposal is to promote creative industry and create a culture park, to create an intersection for creative industry and the urban residents' daily lives. The culture park bonded with various resources along the site perimeter to encompass the tremendously abundant city life. It is a reflection of Shenzhen spirit and becomes the dynamic engine of CBD. To focus on the majority of city residents is the leverage to recognize the project and to put the city in a forward moving position.

1. compression

Along the perimeter of the Crystal Island's site, there is abundance of rich opportunities of culture and city life. However, these positive elements were too spread out. We proposed to directly and effectively utilize the PRT system to connect all spots in the entire site from Lianhua mountain to the convention center. It is to compress the physical and psychological feeling of the space and time and extend the capacity of the site.

2. Park

We proposed to preserve the existing park, which to keep the healthy lung for this highly dense city. However, we reconstruct the topography of the park, to create a variety of characteristics, as well to remove the imposed inhuman order and break down the overwhelmingly large scale.

(1) Creative Industry Settlement on the Ground

First we cut the green "skin" and open it up. We capture the interesting segments from the city life to generate fabric, and combine them with various creative companies to form the settlements. And then we inject these settlements underneath the cut-open skin. It is to create a miniature of the city with city's complexity and diversity.

(2) Culture Park on the roof

The up and down topography on the green skin provoke various events, sports and occasions. The north part of the park was design for quite activities and the south part was for vibrant events. Accordingly, we proposed eco park for the youth and city residents on the north, and arranged space to house vibrant activities on the south. The program on the roof are corresponding to the ones underneath. The selection of vegetation for the landscape emphasized the character of Shenzhen and closely worked well with the various programs on site.

As a response to the density change of the city, we proposed a sight seeing tower on the north, in addition, the form and space of the tower changed to correspond to the different views of the city.

The fine urban fabric and human scale were embedded in the settlements, which convey the value and characteristics of the city. The park on the lofty roof reflected Shenzhen's vast ambitious. These two contrasting element generate a unique tension. At the end, a crystal clear circle connected all settlements together and the space up and down. It is natural order with diversities.

3. New Symbol and New Spirit

Conventionally, the symbol of a city is fixed, which is difficult to catch up the energetic changes of a city nowadays. Especially, a fixed symbol hardly matches Shenzhen's dynamics. Therefore, we proposed not to build a fixed symbol, but to create a changeable landmark to adept different events, occasions and ears. It is a reflection of this moving forward society and Shenzhen spirit!